Elementary Differential
Equations with Linear Algebra

# ELEMENTARY DIFFERENTIAL EQUATIONS WITH LINEAR ALGEBRA

*ALBERT L. RABENSTEIN*

*Macalester College*

ACADEMIC PRESS  New York and London

ACADEMIC PRESS, INC.
111 Fifth Avenue, New York, New York 10003

*United Kingdom Edition published by*
ACADEMIC PRESS, INC. (LONDON) LTD.
Berkeley Square House, London W1X 6BA

LIBRARY OF CONGRESS CATALOG CARD NUMBER: 76–107574
AMS 1968 SUBJECT CLASSIFICATION 1501; 3401

*Second Printing March, 1971*
*Third Printing May, 1971*

PRINTED IN THE UNITED STATES OF AMERICA

# Contents

*v*

## 3. Vector Spaces and Linear Transformations

## 4. Characteristic Values

## 5. Linear Differential Equations

## 6. Systems of Differential Equations

## 7. Series Solutions

## 8. Existence and Uniqueness of Solutions

## References

## Answers to Selected Exercises

# Preface

The purpose of this book is to provide an introduction to differential equations and linear algebra. There are advantages in treating the two subjects in a single volume. Linear and matrix algebra are useful tools in both the computational and theoretical aspects of differential equations. On the other hand, differential equations provide examples and applications of the concepts of linear algebra.

The book is intended for use in a course that follows two or three terms of calculus. Numerous physical applications of differential equations are presented, with an eye to the science or engineering student. Several sections deal with the more theoretical aspects of differential equations. Either the applications or the higher theory can be omitted without loss of continuity.

Most of the material on linear algebra is in Chapters 2, 3, and 4. In a one-term course in which differential equations are to be emphasized, parts of Chapters 3 and 4 can be omitted, depending on the methods of solution for linear systems (Chapter 6) to be covered. In any case, Sections 3.9–3.11, 4.3, and 4.4 can be omitted, and Chapter 5 can be taken up before Chapter 4 if desired.

There are many routine problems and some hard ones. Answers to about half the computational problems have been placed at the end of the book.

The author would like to thank Miss Lois Wadd for her expert typing of the manuscript. He also gratefully acknowledges the help of the editorial and production staff of Academic Press in the preparation of the book.

# Elementary Differential
# Equations with Linear Algebra

# I

## Introduction
## to Differential Equations

### 1.1 INTRODUCTION

An ordinary differential equation may be defined as an equation that involves a single unknown function of a single variable and some finite number of its derivatives. For example, a simple problem from calculus is that of finding all functions $f$ for which

$$f'(x) = 3x^2 - 4x + 5 \qquad (1.1)$$

for all $x$. If we write $y = f(x)$, thus letting $y$ be the value that the function $f$ associates with the number $x$, Eq. (1.1) may be written

$$\frac{dy}{dx} = 3x^2 - 4x + 5.$$

Clearly a function $f$ satisfies the condition (1.1) if and only if it is of the form

$$f(x) = x^3 - 2x^2 + 5x + c,$$

where $c$ is an arbitrary number. A more difficult problem is that of finding all functions $g$ for which

$$g'(x) + 2[g(x)]^2 = 3x^2 - 4x + 5. \qquad (1.2)$$

*1*

Another difficult problem is that of finding all functions $y$ for which (we use the abbreviation $y$ for $y(x)$)

$$x^2 \frac{d^2 y}{dx^2} - 3x \left(\frac{dy}{dx}\right)^2 + 4y = \sin x. \tag{1.3}$$

In each of the problems (1.1), (1.2), and (1.3) we are asked to find all functions that satisfy a certain condition, where the condition involves one or more *derivatives* of the function. We can reformulate our definition of a differential equation as follows. Let $F$ be a function of $n + 2$ variables. Then the equation

$$F[x, y, y', y'', \ldots, y^{(n)}] = 0 \tag{1.4}$$

is called an ordinary differential equation of order $n$ for the unknown function $y$. The *order* of the equation is the order of the highest order derivative that appears in the equation. Thus, Eqs. (1.1) and (1.2) are first-order equations, while Eq. (1.3) is of second order.

A *partial* differential equation (as distinguished from an *ordinary* differential equation) is an equation that involves an unknown function of more than one independent variable, together with partial derivatives of the function. An example of a partial differential equation for an unknown function $u(x, t)$ of two variables is

$$\frac{\partial^2 u}{\partial x^2} = \frac{\partial u}{\partial t} + u.$$

In this book, almost all the differential equations that we shall consider will be ordinary.

By a *solution* of an ordinary differential equation of order $n$, we mean a function that, on some interval,[1] possesses at least $n$ derivatives and satisfies the equation. For example, a solution of the equation

$$\frac{dy}{dx} - 2y = 6$$

is given by the formula

$$y = e^{2x} - 3, \qquad \text{for all } x,$$

---

[1] We shall use the notations $(a, b)$, $[a, b]$, $(a, b]$, $[a, b)$, $[a, b)$, $(a, \infty)$, $[a, \infty)$, $(-\infty, a)$, $(-\infty, a]$, $(-\infty, \infty)$ for intervals. Here $(a, b)$ is the set of all real numbers $x$ such that $a < x < b$, $[a, b]$ is the set of all real numbers $x$ such that $a \leq x \leq b$, $[a, b)$ is the set of all real numbers $x$ such that $a \leq x < b$, and so on.

because

$$\frac{d}{dx}(e^{2x} - 3) - 2(e^{2x} - 3) = 2e^{2x} - 2e^{2x} + 6 = 6$$

for all $x$. The set of all solutions of a differential equation is called the *general solution* of the equation. For instance, the general solution of the equation

$$\frac{dy}{dx} = 3x^2 - 4x$$

consists of all functions that are of the form

$$y = x^3 - 2x^2 + c, \qquad x \text{ in } \mathscr{I},$$

where $c$ is an arbitrary constant and $\mathscr{I}$ is an arbitrary interval. To *solve* a differential equation is to find its general solution.

Let us now solve the second-order equation

$$\frac{d^2y}{dx^2} = 12x + 8.$$

Integrating, we find that

$$\frac{dy}{dx} = 6x^2 + 8x + c_1,$$

where $c_1$ is an arbitrary constant. A second integration yields

$$y = 2x^3 + 4x^2 + c_1 x + c_2$$

for the general solution. Here $c_2$ is a second arbitrary constant.

The general solution of the third-order equation

$$y''' = 16e^{-2x}$$

can be found by three successive integrations. We find easily that

$$y'' = -8e^{-2x} + c_1',$$
$$y' = 4e^{-2x} + c_1' x + c_2',$$

and

$$y = -2e^{-2x} + \tfrac{1}{2}c_1'x^2 + c_2 x + c_3, \tag{1.5}$$

where $c_1'$, $c_2$, and $c_3$ are arbitrary constants. If we replace the constant $c_1'$ in the last formula by $2c_1$, Eq. (1.5) becomes

$$y = -2e^{-2x} + c_1 x^2 + c_2 x + c_3. \tag{1.6}$$

This last formula is slightly simpler in appearance. The two formulas (1.5) and (1.6) describe the same set of functions because the coefficient of $x^2$ is completely arbitrary in both cases. Since $c_1' = 2c_1$, we see that to any arbitrarily assigned value for $c_1$ there corresponds a value for $c_1'$ and vice versa.

If a formula can be found that describes the general solution of an $n$th-order equation, it usually involves $n$ arbitrary constants. We note that this principle has been borne out in the last three examples, which admittedly are rather simple. Actually it is possible to find a simple formula that describes the general solution only for relatively few types of differential equations. Several such classes of first-order equations are discussed in the following three sections. In cases where it is not possible to find explicit formulas for the solutions, it still may be possible to discover certain properties of the solutions. For instance, it may be possible to show that a solution is bounded (or unbounded), to find its limiting value as the independent variable becomes infinite, or to establish that it is a periodic function. Much advanced work in differential equations is concerned with such matters.

Perhaps some reasons should now be given as to why we want to solve differential equations. Briefly, many experimentally discovered laws of science can be formulated as relations that involve not only magnitudes of quantities but also rates of change (usually with respect to time) of these magnitudes. Thus, the laws can be formulated as differential equations. A number of examples of problems that give rise to differential equations are presented in this book. Some applications will be described in Sections 1.6 and 1.7, after we have learned how to solve several kinds of first-order equations.

We have seen that ordinary differential equations can be classified as to order. We shall also categorize them in one more way. An equation of order $n$ is said to be a *linear* equation if it is of the special form

$$a_0(x)y^{(n)} + a_1(x)y^{(n-1)} + \cdots + a_{n-1}(x)y' + a_n(x)y = f(x), \tag{1.7}$$

where $a_0, a_1, \ldots, a_n$ and $f$ are given functions that are defined on an interval $\mathscr{I}$. Thus the general $n$th-order equation (1.4) is linear if the function $F$ is a first-degree polynomial in $y, y', \ldots, y^{(n)}$. An equation that is not of the form (1.7) is said to be a *nonlinear* equation. For example, each of the equations

$$y' + (\cos x)y = e^x,$$
$$xy'' + y' = x^2,$$
$$xy''' - e^x y' + (\sin x)y = 0,$$

is linear, while each of the equations

$$y' + y^2 = 1,$$
$$y'' + (\cos x)yy' = \sin x,$$
$$y''' - x(y')^3 + y = 0,$$

is nonlinear. Because linear equations possess special properties, they will be treated in a separate chapter, Chapter 5.

In most applications that involve differential equations, the unknown function is required not only to satisfy the differential equation but also to satisfy certain other auxiliary conditions. These auxiliary conditions often specify the values of the function and some of its derivatives at one or more points. As an example, suppose we are asked to find a solution of the equation

$$\frac{dy}{dx} = 3x^2 \tag{1.8}$$

that satisfies the auxiliary condition $y = 1$ when $x = 2$, or

$$y(2) = 1. \tag{1.9}$$

Thus, we require the graph of our solution (which is called a *solution curve* or *integral curve*) to pass through the point (2, 1) in the $xy$ plane. The general solution of Eq. (1.8) is

$$y = x^3 + c, \tag{1.10}$$

where $c$ is an arbitrary constant. In order to find a specific solution that satisfies the condition (1.9), we set $x = 2$ and $y = 1$ in Eq. (1.10), finding that $1 = 8 + c$ or $c = -7$. Thus, there is only one value of $c$ for which the condition (1.9) is satisfied. The equation (1.8) possesses one, and only one solution (defined for all $x$) that satisfies the condition (1.9):

$$y = x^3 - 7.$$

For an $n$th-order equation of the form

$$y^{(n)} = G[x, y, y', y'', \ldots, y^{(n-1)}],\tag{1.11}$$

auxiliary conditions of the type

$$y(x_0) = k_0, \quad y'(x_0) = k_1, \quad y''(x_0) = k_2, \ldots, \quad y^{(n-1)}(x_0) = k_{n-1},\tag{1.12}$$

where the $k_i$ are given numbers, are common. We note that there are $n$ conditions for the $n$th-order equation. These conditions specify the values of the unknown function and its first $n - 1$ derivatives at a single point $x_0$. For a first-order equation

$$y' = H(x, y),$$

we would have only one condition

$$y(x_0) = k_0$$

specifying the value of the unknown function itself at $x_0$. In the case of a second-order equation

$$y'' = K(x, y, y'),$$

we would have two conditions

$$y(x_0) = k_0, \qquad y'(x_0) = k_1.$$

A set of auxiliary conditions of the form (1.12) is called a set of *initial conditions* for the Eq. (1.11). The equation (1.11) together with the conditions (1.12) constitute an *initial value problem*. The reason for this terminology is that in many applications the independent variable $x$ represents time and the conditions are specified at the instant $x_0$ at which some process begins.

In specifying the values of the first $n - 1$ derivatives of a solution of Eq. (1.11) at $x_0$, we have essentially specified the values of any higher derivatives that might exist. The values of these higher derivatives can be found from the differential equation itself. For example, let us consider the initial value problem

$$y'' = x^2 - y^3$$
$$y(1) = 2, \qquad y'(1) = -1.\tag{1.13}$$

From the differential equation we see that

$$y''(1) = 1 - 8 = -7.$$

By differentiating through in the differential equation, we find that

$$y''' = 2x - 3y^2 y'$$

and hence

$$y'''(1) = 2 - (3)(4)(-1) = 14.$$

The values of higher derivatives at $x = 1$ can be found by repeated differentiation.

If a function can be expanded in a power series about a point $x_0$, a knowledge of the values of the function and its derivatves at $x_0$ completely determines the function. This discussion suggests that the initial value problem (1.11) and (1.12) can have but one solution if the function $G$ is infinitely differentiable with respect to all variables. Actually it can be shown that, under rather mild restrictions on $G$, the initial value problem possesses a solution and that it has only one solution. A fuller discussion of questions of existence and uniqueness of solutions is given in Chapter 8. In most of the problems and examples of this chapter, it is possible to actually find all the solutions of the differential equation at hand. In cases where this is impossible, it is comforting to know that the problem being considered actually has a solution and that there is only one solution. An initial value problem purporting to describe some physical process would not be very valuable without these two properties.

### Exercises for Section 1.1

1. Find the order of the differential equation and determine whether it is linear or nonlinear.

   (a) $y' = e^x$
   
   (c) $y' + e^y = 0$
   
   (e) $y'' + xyy' + y = 2$
   
   (g) $y''' = 0$

   (b) $y'' + xy = \sin x$
   
   (d) $y'' + 2y' + y = \cos x$
   
   (f) $y^{(4)} + 3(\cos x)y''' + y' = 0$
   
   (h) $yy'' + y' = 0$

2. Find the general solution of the differential equation.

   (a) $y' = 2x - 3$
   
   (c) $y' = \dfrac{4}{x(x-4)}$
   
   (e) $y'' = \sec^2 x$
   
   (g) $y''' = 24x - 6$

   (b) $y' = 3x^2 \sin x^3$
   
   (d) $y'' = 12e^{-2x} + 4$
   
   (f) $y'' = 8e^{-2x} + e^x$
   
   (h) $y^{(4)} = 32 \sin 2x$

3. Find a solution of the differential equation that satisfies the specified conditions.

   (a) $y' = 0$, $y(2) = -5$

   (b) $y' = x$, $y(2) = 9$

   (c) $y' = 4x - 3$, $y(4) = 3$

   (d) $y' = 3x^2 - 6x + 1$, $y(-2) = 0$

   (e) $y'' = 0$, $y(2) = 1$, $y'(2) = -1$

   (f) $y'' = 9e^{-3x}$, $y(0) = 1$, $y'(0) = 2$

   (g) $y'' = \cos x$, $y(\pi) = 2$, $y'(\pi) = 0$

   (h) $y''' = e^{-x}$, $y(0) = -1$, $y'(0) = 1$, $y''(0) = 3$

4. Show that a function is a solution of the equation $y' + ay = 0$, where $a$ is a constant, if, and only if, it is a solution of the equation $(e^{ax}y)' = 0$. Hence show that the general solution of the equation is described by the formula $y = ce^{-ax}$, where $c$ is an arbitrary constant.

5. Use the result of Exercise 4 to find the general solution of the given differential equation.

   (a) $y' + 3y = 0$          (b) $y' - 3y = 0$

   (c) $3y' - y = 0$          (d) $3y' + 2y = 0$

6. Verify that every function of the form $y = x^2 + c/x$, where $c$ is a constant, is a solution of the equation $xy' + y = 3x^2$ on any interval that does not contain $x = 0$.

7. Verify that each of the functions $y = e^{-x}$ and $y = e^{3x}$ is a solution of the equation $y'' - 2y' - 3y = 0$ on any interval. Then show that $c_1 e^{-x} + c_2 e^{3x}$ is a solution for every choice of the constants $c_1$ and $c_2$.

8. Suppose that a function $f$ is a solution of the initial value problem $y' = x^2 + y^2$, $y(1) = 2$. Find $f'(1), f''(1)$, and $f'''(1)$.

9. If the function $g$ is a solution of the initial value problem
   $$y'' + yy' - x^3 = 0, \ y(-1) = 1, \ y'(-1) = 2 \, ,$$
   find $g''(-1)$ and $g'''(-1)$.

10. Show that the problem $y' = 2x$, $y(0) = 0$, $y(1) = 100$, has no solution. Is this an initial value problem?

## 1.2   SEPARABLE EQUATIONS

A first-order differential equation that can be written in the form

$$p(y)\frac{dy}{dx} = q(x), \qquad\qquad (1.14)$$

where $p$ and $q$ are given functions, is called a *separable* equation. Examples of such equations are

$$y^{-2}\frac{dy}{dx} = 2x, \qquad y^{-1}\frac{dy}{dx} = (x+1)^{-1}, \qquad (3y^2 + e^y)\frac{dy}{dx} = \cos x.$$

If a function $f$ is a solution of Eq. (1.14) on an interval $\mathscr{I}$, then

$$p[f(x)]f'(x) = q(x)$$

for $x$ in $\mathscr{I}$. Taking antiderivatives, we have

$$\int p[f(x)]f'(x)\,dx = \int q(x)\,dx + c$$

or

$$\int p(y)\,dy = \int q(x)\,dx + c.$$

If $P$ and $Q$ are functions such that $P'(y) = p(y)$ and $Q'(x) = q(x)$, then the solution $f$ must satisfy the equation

$$P(y) = Q(x) + c, \tag{1.15}$$

where $c$ is a constant. That is,

$$P[f(x)] = Q(x) + c$$

for $x$ in $\mathscr{I}$. Conversely, if $y$ is any differentiable function that satisfies Eq. (1.15), we see by implicit differentiation that

$$P'(y)\frac{dy}{dx} = Q'(x)$$

or

$$p(y)\frac{dy}{dx} = q(x).$$

Thus, a function is a solution of Eq. (1.14) if, and only if, it satisfies an equation of the form (1.15) for some choice of the constant $c$. It may not be possible to find an explicit formula for $y$ in terms of $x$ from Eq. (1.15). However, we say that Eq. (1.15) determines the solutions of the differential equation *implicitly*. Let us now consider some examples of separable equations.

**Example 1**

$$\frac{dy}{dx} = 2xy^2.$$    (1.16)

"Separating the variables," we have

$$y^{-2}\frac{dy}{dx} = 2x$$

or

$$y^{-2}\,dy = 2x\,dx.$$

Taking antiderivatives, we have

$$\int y^{-2}\,dy = \int 2x\,dx + c$$

or

$$-\frac{1}{y} = x^2 + c.$$

Thus, the functions defined by the formula

$$y = \frac{-1}{x^2 + c}$$    (1.17)

are solutions of Eq. (1.16). Note, however, that the identically zero function ($y = 0$) is also a solution. In arriving at formula (1.17), we started out by dividing both sides of the original equation by $y^2$, and this procedure is not valid when $y = 0$.

Suppose that it is desired to find the solution curve that passes through the point $(2, -1)$ in the $xy$ plane. Then our initial condition is

$$y(2) = -1.$$

Setting $x = 2$ and $y = -1$ in formula (1.17), we see that

$$-1 = \frac{-1}{4 + c}$$

or $c = -3$. Then the desired solution is given by the formula

$$y = \frac{1}{3 - x^2}.$$

**Example 2**

$$(x + 1)\frac{dy}{dx} = 2y. \tag{1.18}$$

Here we have

$$\frac{dy}{y} = 2\frac{dx}{x + 1}$$

or

$$\ln|y| = \ln(x + 1)^2 + c'.$$

Then

$$|y| = e^{c'}(x + 1)^2$$

and

$$y = \pm\, e^{c'}(x + 1)^2, \tag{1.19}$$

where $c'$ is an arbitrary constant. But $\pm e^{c'}$ can have any value except zero, so the set of functions described by formula (1.19) is also described by the simpler formula

$$y = c(x + 1)^2, \tag{1.20}$$

where $c$ is a constant different from zero but otherwise is arbitrary. However, since $y = 0$ is obviously a solution of the differential equation (1.18), formula (1.20) also describes a solution when $c = 0$. This formula, with $c$ completely arbitrary, gives the general solution of the equation.

**Example 3**    Consider the initial value problem

$$\frac{dy}{dx} = \frac{\cos x}{3y^2 + e^y}, \qquad y(0) = 2.$$

From the differential equation we have

$$(3y^2 + e^y)\, dy = \cos x\, dx$$

or

$$y^3 + e^y = \sin x + c.$$

Setting $x = 0$ and $y = 2$ (these values come from the initial condition) we find that

$$8 + e^2 = c.$$

Hence, the desired solution (if such a solution exists) is implicitly determined by the equation

$$y^3 + e^y = \sin x + 8 + e^2.$$

Some differential equations that are not separable as they stand become separable after a change of variable. One such class of equations consists of those that can be written in the form

$$\frac{dy}{dx} = F\left(\frac{y}{x}\right). \tag{1.21}$$

An example of such an equation is

$$(x^4 + y^4)\frac{dy}{dx} = x^3 y,$$

which may be rewritten as

$$\frac{dy}{dx} = \frac{x^3 y}{x^4 + y^4}$$

or

$$\frac{dy}{dx} = \frac{y/x}{1 + (y/x)^4}.$$

An equation of the form (1.121) can be made separable by introducing a new independent variable, $v$ where

$$v = \frac{y}{x}.$$

For then

$$y = vx, \qquad \frac{dy}{dx} = x\frac{dv}{dx} + v,$$

and Eq. (1.21) becomes

$$x\frac{dv}{dx} + v = F(v)$$

or

$$\frac{1}{F(v) - v}\frac{dv}{dx} = \frac{1}{x},$$

which is separable.

As an example we consider the equation

$$2x^2\frac{dy}{dx} = x^2 + y^2.$$

This equation can be put in the form (1.21) since

$$\frac{dy}{dx} = \frac{x^2 + y^2}{2x^2} = \frac{1}{2}\left[1 + \left(\frac{y}{x}\right)^2\right].$$

Setting $y = vx$, we have

$$x\frac{dv}{dx} + v = \frac{1}{2}(1 + v^2),$$

or

$$2x\frac{dv}{dx} = v^2 - 2v + 1,$$

or

$$\frac{dv}{(v-1)^2} = \frac{1}{2}\frac{1}{x}\,dx\,.$$

Then

$$\frac{-1}{v-1} = \frac{1}{2}\ln|x| + c'$$

or

$$v = 1 - \frac{2}{\ln|x| + 2c'}\,.$$

Replacing $v$ by $y/x$ and setting $c = 2c'$, we have

$$y = x - \frac{2x}{\ln|x| + c}\,.$$

In determining if a given first-order equation can be written in the form (1.21), the following criterion is often useful. An equation

$$M(x,\,y) + N(x,\,y)\,\frac{dy}{dx} = 0 \qquad (1.22)$$

can be put in the form (1.21) if the functions $M$ and $N$ are *homogeneous of the same degree*. A function $g(x,\,y)$ (defined for all $(x,\,y)$) is said to be *homogeneous of degree m* if

$$g(tx,\,ty) = t^m g(x,\,y)$$

for all $t$.[2] For example, if

$$g(x,\,y) = x^3 y^2 - 3x^5\,,$$

we see that

$$g(tx,\,ty) = t^5(x^3 y^2 - 3x^5) = t^5\,g(x,\,y)\,.$$

---

[2] Some restriction must be placed on the possible values of $t$ if $g$ is not defined everywhere.

Thus, $g$ is homogeneous of degree 5. However, the function $h$, where

$$h(x, y) = x^3 y - 3x^2,$$

is not homogeneous.

Let us write Eq. (1.22) as

$$\frac{dy}{dx} = -\frac{M(x, y)}{N(x, y)}.$$

If $M$ and $N$ are both homogeneous of degree $m$, then

$$M(x, y) = t^{-m} M(tx, ty), \qquad N(x, y) = t^{-m} N(tx, ty).$$

Setting $t = x^{-1}$, we have

$$M(x, y) = x^m M(1, y/x), \qquad N(x, y) = x^m N(1, y/x)$$

and

$$\frac{dy}{dx} = -\frac{M(1, y/x)}{N(1, y/x)}.$$

This equation is of the form (1.21).

In the example

$$x^2 + y^2 - 2x^2 \frac{dy}{dx} = 0,$$

where

$$M(x, y) = x^2 + y^2, \qquad N(x, y) = -2x^2,$$

it is clear that $M$ and $N$ are both homogeneous of degree 2. Hence the equation can be written in the form (1.21). Other examples of equations that can be put in the form (1.21) are

$$x^2 \frac{dy}{dx} = y^2 - x^2$$

and

$$(xy + x^2) \sin \frac{y}{x} + y^2 \frac{dy}{dx} = 0.$$

## Exercises for Section 1.2

In Exercises 1–20, find the general solution, if possible. Otherwise find a relation that defines the solutions implicitly. If an initial condition is specified, also find the particular solution that satisfies the condition.

1. $yy' = 4x, \quad y(1) = -3$

2. $xy' = 4y, \quad y(1) = -3$

3. $y' = 2xy^2, \quad y(2) = 1$

4. $y' = e^x(1 - y^2)^{1/2}, \quad y(0) = \frac{1}{2}$

5. $y' = \dfrac{1 + y^2}{1 + x^2}, \quad y(2) = 3$

6. $e^y y' = 4, \quad y(0) = 2$

7. $2(y - 1) y' = e^x, \quad y(0) = -2$

8. $2y' = y(y - 2)$

9. $3y^2 y' = (1 + y^3) \cos x$

10. $(\cos^2 x) y' = y^2(y - 1) \sin x$

11. $(\cos y) y' = 1$

12. $(\cos^2 x) y' = (1 + y^2)^{1/2}$

13. $x^2 y' = xy - y^2$

14. $x^2 y' = y^2 + 2xy$

15. $xyy' = 2y^2 - x^2$

16. $xy' = y - xe^{y/x}$

17. $e^{y/x} y' = 2(e^{y/x} - 1) + \dfrac{y}{x} e^{y/x}$

18. $y' = \dfrac{y}{x} - 3\left(\dfrac{y}{x}\right)^{4/3}$

19. $xy' = y + (x^2 + y^2)^{1/2}$

20. $3xy^2 y' = 4y^3 - x^3$

21. Show that an equation of the form $y' = F(ay + bx + c), a \neq 0$, becomes separable under the change of dependent variable $v = ay + bx + k$, where $k$ is any number.

22. Use the result of Exercise 21 to solve the differential equation.

    (a) $y' = (y + 4x - 1)^2$

    (b) $(y - x + 1) y' = y - x$

    (c) $(y - 3x) y' = 3(y - 3x + 2)$

    (d) $(y - 2x) y' = 3y - 6x + 1$

23. Consider a first-order differential equation of the form

$$(a_1 x + b_1 y + c_1) y' = a_2 x + b_2 y + c_2.$$

    (a) If $a_1 b_2 - b_1 a_2 = 0$, show that the equation is of the type considered in Exercise 21.

    (b) If $a_1 b_2 - b_1 a_2 \neq 0$ introduce new variables $u$ and $v$, where

$$u = x + p, \qquad v = y + q.$$

    Show that the constants $p$ and $q$ can be chosen in such a way that the equation takes on the form

$$(a_1 u + b_1 v)\frac{dv}{du} = a_2 u + b_2 v.$$

Hence

$$\frac{dv}{du} = F\left(\frac{v}{u}\right).$$

24. Use the results of Exercise 23 to solve the equation.
    (a) $(x + y + 1)y' = y + 2$
    (b) $(3x - y + 1)y' = -x + 3y + 5$

## 1.3 EXACT EQUATIONS

The first-order equation

$$M(x, y) + N(x, y)\frac{dy}{dx} = 0 \qquad (1.23)$$

is said to be *exact* (in some region of the $xy$-plane) if there exists a function $\phi$ with continuous first partial derivatives such that

$$\frac{\partial\phi(x, y)}{\partial x} = M(x, y), \qquad \frac{\partial\phi(x, y)}{\partial y} = N(x, y). \qquad (1.24)$$

The relationship between the function $\phi$ and the solutions of the differential equation is described in the following theorem.

**Theorem 1.1**   If the differential equation (1.23) is exact and if the function $\phi$ has the properties (1.24) then a function $f$, with $y = f(x)$, is a solution of the differential equation if and only if it satisfies an equation of the form

$$\phi(x, y) = c, \qquad (1.25)$$

where $c$ is a constant.

PROOF   Suppose that the function $f$ is a solution of Eq. (1.23). If $y = f(x)$ we have

$$\frac{\partial\phi(x, y)}{\partial x} + \frac{\partial\phi(x, y)}{\partial y}\frac{dy}{dx} = 0$$

or

$$\frac{d\phi(x, y)}{dx} = 0.$$

Hence $\phi(x, y) = c$. Conversely, suppose that a (differentiable) function $f$ satisfies the equation $\phi(x, y) = c$. Then by implicit differentiation we have

$$\frac{\partial\phi(x, y)}{\partial x} + \frac{\partial\phi(x, y)}{\partial y}\frac{dy}{dx} = 0$$

or

$$M(x, y) + N(x, y)\frac{dy}{dx} = 0.$$

Hence the function is a solution of the differential equation.

Notice that if Eq. (1.23) is exact, then the total differential of $\phi$ is

$$d\phi = \frac{\partial\phi}{\partial x}\, dx + \frac{\partial\phi}{\partial y}\, dy = M\, dx + N\, dy.$$

Along any solution curve, $d\phi = 0$, and the solutions satisfy equations of the form $\phi(x, y) = c$.

We need a criterion for determining whether or not an equation is exact. We also need a method for finding the function $\phi$ when it is exact. In what follows we assume that the functions $M$ and $N$ are continuous together with their first partial derivatives in some region.

Suppose that Eq. (1.23) is exact. Then there exists a function $\phi$ such that $M = \partial\phi/\partial x$ and $N = \partial\phi/\partial y$. Hence

$$\frac{\partial M}{\partial y} = \frac{\partial^2\phi}{\partial y\, \partial x}, \qquad \frac{\partial N}{\partial y} = \frac{\partial^2\phi}{\partial x\, \partial y},$$

and because the mixed second partial derivatives of $\phi$ are equal (since $\phi$, $\phi_x$, $\phi_y$, $\phi_{xy}$, and $\phi_{yx}$ are continuous), we have

$$\frac{\partial M}{\partial y} = \frac{\partial N}{\partial x}. \tag{1.26}$$

Thus if the equation is exact, the condition (1.26) is satisfied.

It can be shown that if $M$ and $N$ satisfy the condition (1.26) in a *simply connected region* then the differential equation is exact. A simply connected region is such that every simple closed curve[3] in the region contains only points of the region inside it. The interior of an ellipse or a rectangle is a simply connected region but the region bounded by two concentric circles is not simply connected. We shall prove that the condition (1.26) is sufficient for exactness in the special case of a rectangle. Let $D$ be the rectangle

$$a < x < b, \quad c < y < d, \tag{1.27}$$

where any or all of $a$, $b$, $c$, and $d$ may be infinite. Such a region is simply connected.

**Theorem 1.2** Let $M$ and $N$ satisfy the condition (1.26) in the rectangle (1.27). Then the equation $M + Ny' = 0$ is exact.

PROOF   Let $(x_0, y_0)$ be any fixed point in $D$. We define a function $\phi$ of two variables by means of the formula

$$\phi(x, y) = \int_{x_0}^{x} M(s, y_0)\, ds + \int_{y_0}^{y} N(x, t)\, dt \tag{1.28}$$

for $(x, y)$ in $D$. See Fig. 1.1. We need to verify that $\partial\phi/\partial x = M$ and $\partial\phi/\partial y = N$. Differentiating with respect to $x$, we have[4]

$$\frac{\partial\phi(x, y)}{\partial x} = M(x, y_0) + \int_{y_0}^{y} \frac{\partial N(x, t)}{\partial x}\, dt.$$

Since the condition (1.26) is satisfied, $\partial N(x, t)/\partial x = \partial M(x, t)/\partial t$. Hence

$$\begin{aligned}
\frac{\partial\phi(x, y)}{\partial x} &= M(x, y_0) + \int_{y_0}^{y} \frac{\partial M(x, t)}{\partial t}\, dt \\
&= M(x, y_0) + M(x, y) - M(x, y_0) \\
&= M(x, y).
\end{aligned}$$

In similar fashion it can be shown (Exercise 1) that $\partial\phi/\partial y = N$. Since the function $\phi$ has the property (1.24) the differential equation is exact.

---

[3] A *simple* closed curve does not cross itself. A circle is a simple closed curve. A figure eight is closed but not simple.

[4] We have differentiated with respect to $x$ under the integral sign, a procedure that requires justification. This situation is covered by *Leibnitz's rule*, which is discussed in most advanced calculus books.

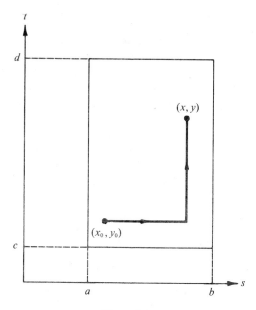

**Figure 1.1**

Formula (1.28) can be used to find a function $\phi$ when the equation (1.23) is exact. However, the function can usually be found by means of a simpler procedure which does not necessitate memorizing the formula. We shall illustrate the method with an example.

The equation

$$3x^2 - 2y^2 + (1 - 4xy)\frac{dy}{dx} = 0 \qquad (1.29)$$

is exact, since

$$M(x, y) = 3x^2 - 2y^2, \qquad N(x, y) = 1 - 4xy,$$

and

$$\frac{\partial M(x, y)}{\partial y} = -4y = \frac{\partial N(x, y)}{\partial x}$$

for all $(x, y)$. Hence, by Theorem 1.2, there exists a function $\phi$ such that

$$\frac{\partial \phi(x, y)}{\partial x} = 3x^2 - 2y^2, \qquad \frac{\partial \phi(x, y)}{\partial y} = 1 - 4xy. \qquad (1.30)$$

Integrating with respect to $x$ in the first of these relations, we see that $\phi$ is of the form

$$\phi(x, y) = x^3 - 2xy^2 + f(y), \tag{1.31}$$

where $f$ can be any function of $y$ only, since $\partial f(y)/\partial x = 0$. We must choose $f$ so that the second of the conditions (1.30) is satisfied. We require that

$$\frac{\partial \phi(x, y)}{\partial y} = -4xy + f'(y) = 1 - 4xy$$

or

$$f'(y) = 1 .$$

One possible choice for $f$ is $f(y) = y$. Then from Eq. (1.31) we have

$$\phi(x, y) = x^3 - 2xy^2 + y .$$

The solutions of Eq. (1.29) are those differentiable functions that satisfy equations of the form

$$x^3 - 2xy^2 + y = c .$$

If the equation $M + Ny' = 0$ is not exact, it may be possible to make it exact by multiplying through by some function. That is, we may be able to find a function $\rho$ such that

$$\rho(x, y)M(x, y) + \rho(x, y)N(x, y) y' = 0$$

is an exact equation. If such a function $\rho$ exists, it is called an *integrating factor* for the original equation. As an example, let us consider the equation

$$1 + x^2y^2 + y + xy' = 0 . \tag{1.32}$$

This equation is not exact, since

$$\frac{\partial M(x, y)}{\partial y} = 2x^2y + 1, \qquad \frac{\partial N(x, y)}{\partial x} = 1 .$$

However, if we multiply through in Eq. (1.32) by

$$\rho(x, y) = \frac{1}{1 + x^2y^2},$$

it becomes

$$\left(1 + \frac{y}{1 + x^2 y^2}\right) + \frac{x}{1 + x^2 y^2} \, y' = 0. \tag{1.33}$$

This equation is exact, since

$$\frac{\partial}{\partial y}\left(1 + \frac{y}{1 + x^2 y^2}\right) = \frac{\partial}{\partial x}\left(\frac{x}{1 + x^2 y^2}\right) = \frac{1 - x^2 y^2}{(1 + x^2 y^2)^2}.$$

Since

$$d(\tan^{-1} xy) = \frac{y}{1 + x^2 y^2} \, dx + \frac{x}{1 + x^2 y^2} \, dy,$$

we see from Eq. (1.33) that

$$d(x + \tan^{-1} xy) = 0$$

along a solution curve. Hence the solutions satisfy the relation

$$x + \tan^{-1} xy = c$$

and the solutions themselves are given by the formula

$$y = \frac{1}{x} \tan(c - x).$$

There is no general rule for finding integrating factors. If the solver has in mind the formula for $d(\tan^{-1} xy)$, he might be led to an integrating factor in the above example. Usually the procedure is one of trial and error, with educated guessing.

## Exercises for Section 1.3

**1.**  Show that $\partial\phi(x, y)/\partial y = N(x, y)$, where $\phi$ is defined as in Eq. (1.28).

In Exercises 2–13, first determine if the equation is exact. If it is exact, find the general solution, or at least a relation that defines the solutions implicitly.

**2.**  $3x^3 y^2 y' + 3x^2 y^3 - 5x^4 = 0$

**3.**  $(3x^2 y^2 - 4xy) \, y' + 2xy^3 - 2y^2 = 0$

4. $xe^{xy}y' + ye^{xy} - 4x^3 = 0$

5. $(x + y^2)y' + 2x^2 - y = 0$

6. $[\cos(x^2 + y) - 3xy^2]y' + 2x\cos(x^2 + y) - y^3 = 0$

7. $(x^2 - y)y' + 2x^3 + 2xy = 0$

8. $(x + y\sin x)y' + y + x\sin y = 0$

9. $(y^3 - x^2y)y' - xy^2 = 0$

10. $(y^{-1/3} - y^{-2/3}e^x)y' - 3(e^xy^{1/3} + e^{2x}) = 0$

11. $(e^{2y} - xe^y)y' - e^y - x = 0$

12. $y^2(x^6 + y^3)^{1/3}y' + 2x^5[(x^6 + y^3)^{1/3} - x^2] = 0$

13. $(y^{-3} - y^{-2}\sin x)y' + y^{-1}\cos x = 0$

14. Show that the separable equation $p(y)y' - q(x) = 0$ is exact.

15. Show that the function $\rho$ is an integrating factor for the equation $M + Ny' = 0$ if it satisfies the partial differential equation

$$N\frac{\partial\rho}{\partial x} - M\frac{\partial\rho}{\partial y} = \rho\left(\frac{\partial M}{\partial y} - \frac{\partial N}{\partial x}\right).$$

16. Show that an integrating factor for the equation $y' - F(y/x) = 0$ is

$$\rho(x, y) = \frac{1}{xF(y/x) - y}.$$

In Exercises 17–20, determine if the equation has an integrating factor of the form $\rho(x, y) = x^m y^n$. If it does, solve the equation.

17. $(1 - xy)y' + y^2 + 3xy^3 = 0$

18. $(3x^2 + 5xy^2)y' + 3xy + 2y^3 = 0$

19. $(x^2 + xy^2)y' - 3xy + 2y^3 = 0$

20. $3xy' + xy^3 + y^2 = 0$

In Exercises 21–24 find an integrating factor by inspection and solve the equation. The following formulas may be helpful:

$$\frac{x\,dx + y\,dy}{x^2 + y^2} = \frac{1}{2}d\ln(x^2 + y^2)$$

$$\frac{-y\, dx + x\, dy}{x^2 + y^2} = d\left(\tan^{-1}\frac{y}{x}\right)$$

$$y\, dx + x\, dy = d(xy)$$

$$\frac{-y\, dx + x\, dy}{x^2} = d\left(\frac{y}{x}\right)$$

$$\frac{y\, dx - x\, dy}{y^2} = d\left(\frac{x}{y}\right)$$

**21.** $(x - 4x^2y^3)y' + 3x^4 - y = 0$

**22.** $(x - 2x^2y - 2y^3)y' - y = 0$

**23.** $(2y^3 - x)y' + 3x^2y^2 + y = 0$

**24.** $yy' + x - x^2 - y^2 = 0$

**25.** Let $P(x) = \int p(x)\, dx$. Show that $e^{P(x)}$ is an integrating factor for the linear equation $y' + p(x)y - q(x) = 0$.

## 1.4  FIRST-ORDER LINEAR EQUATIONS

As defined in Section 1.1, a *linear* differential equation of order $n$ has the form

$$a_0(x)y^{(n)} + a_1(x)y^{(n-1)} + \cdots + a_{n-1}(x)y' + a_n(x)y = f(x),$$

where the functions $a_i$ and $f$ are specified on some interval. We assume that $a_0(x) \neq 0$ for all $x$ in this interval. (A solution may not exist throughout an interval on which $a_0$ vanishes. The discussion in Sections 5.11 and 8.3 explains this point more fully.)

A first-order linear equation is of the form

$$a_0(x)\, y' + a_1(x)y = f(x).$$

Since $a_0(x)$ is never zero, we can divide through by $a_0$ and write this equation in the form

$$y' + p(x)\, y = q(x), \tag{1.34}$$

where $p = a_1/a_0$ and $q = f/a_0$. A formula for the solutions of equation (1.34) is given in the following theorem.

**Theorem 1.3** Let $P$ be any function such that $P'(x) = p(x)$; that is,

$$P(x) = \int p(x)\, dx,$$

and let $\rho$ be specified by the relation

$$\rho(x) = \pm e^{P(x)}, \tag{1.35}$$

where either the plus or the minus sign may be chosen. Then the solutions of Eq. (1.34) are given by the formula

$$\rho(x)\, y = \int \rho(x)\, q(x)\, dx + c, \tag{1.36}$$

where $c$ is an arbitrary constant.

PROOF   We treat the case where the plus sign is chosen. The other case is similar. If Eq. (1.34) is multiplied through by $e^{P(x)}$, it becomes

$$[y' + p(x)y]e^{P(x)} + e^{P(x)}\, q(x)$$

or, since $P'(x) = p(x)$,

$$\frac{d[ye^{P(x)}]}{dx} = e^{P(x)}q(x).$$

Taking antiderivatives, we have the relation

$$e^{P(x)}y = \int e^{P(x)}\, q(x)\, dx + c, \tag{1.37}$$

which is the same as formula (1.36). Thus if a solution of the differential equation (1.34) exists, it must be of the form (1.37). Conversely, any function defined by Eq. (1.37) is a solution of the Eq. (1.34), as can be verified by starting with formula (1.37) and retracing steps.

The trick in solving Eq. (1.34) was to multiply through first by $\pm e^{P(x)}$. Some motivation for doing this is provided by the following reasoning. Suppose we attempt to find an integrating factor for equation (1.34) that depends on $x$ only. Multiplying through in the equation by $\rho(x)$ and collecting all terms on one side of the equals sign, we have

$$[p(x)y - q(x)]\, \rho(x) + \rho(x)y' = 0.$$

For this equation to be exact we must have

$$p(x)\,\rho(x) = \rho'(x)\,.$$

This is a separable equation for $\rho$. We find that

$$\frac{\rho'(x)}{\rho(x)} = p(x)\,, \qquad \ln|\rho(x)| = \int p(x)\,dx\,,$$

and

$$\rho(x) = \pm\exp\!\left(\int p(x)\,dx\right) = \pm e^{P(x)}.$$

An example of a first-order linear equation is

$$(x + 1)y' - y = x\,. \tag{1.38}$$

Dividing through by $x + 1$ to put it in the form (1.34), we have

$$y' + \frac{-1}{x + 1}\,y = \frac{x}{x + 1}\,.$$

Here

$$p(x) = \frac{-1}{x + 1}\,, \qquad q(x) = \frac{x}{x + 1}\,,$$

and

$$P(x) = \int p(x)\,dx = -\ln|x + 1| + c\,.$$

Our integrating factors are of the form

$$\rho(x) = \pm e^{P(x)} = \pm\frac{e^c}{|x + 1|}\,.$$

For simplicity we may as well choose $c = 0$. Then

$$\rho(x) = \pm\frac{1}{|x + 1|}\,. \tag{1.39}$$

This function is not defined when $x = -1$. (Notice that the coefficient $a_0(x) = x + 1$ in Eq. (1.38) vanishes when $x = -1$.) We may consider the two intervals $(-\infty, -1)$ and $(-1, \infty)$ separately. If we choose the minus sign for the first interval in Eq. (1.39) and the plus sign for the second interval, we obtain the simple formula

$$\rho(x) = \frac{1}{x+1}, \qquad x \neq -1$$

for our integrating factor. From formula (1.36) we have

$$\frac{y}{x+1} = \int \frac{x}{(x+1)^2} \, dx + c.$$

Partial fractions can be used to evaluate the interval. We find that

$$\frac{y}{x+1} = \int \left[ \frac{1}{x+1} - \frac{1}{(x+1)^2} \right] dx + c,$$

$$\frac{y}{x+1} = \ln|x+1| + \frac{1}{x+1} + c,$$

or

$$y = c(x+1) + 1 + (x+1) \ln|x+1|.$$

Sometimes a nonlinear equation can be put in the form (1.34) by means of a change of variable. One set of equations for which this can always be accomplished is the class of *Bernoulli equations*. These are of the form

$$y' + p(x)y = q(x)y^n, \tag{1.40}$$

where $n$ is any number other than 0 or 1. Division by $y^n$ yields the equation

$$y^{-n} y' + p(x) y^{1-n} = q(x). \tag{1.41}$$

If we let $u = y^{1-n}$, then $u' = (1-n)y^{-n}y'$ and Eq. (1.41) becomes

$$\frac{1}{1-n} u' + p(x) u = q(x).$$

This is a linear equation that can be solved by the method described earlier in this section.

An example of a Bernoulli equation is

$$y' + \frac{3}{x} y = x^2 y^2. \tag{1.42}$$

Dividing through by $y^2$, we have

$$y^{-2} y' + \frac{3}{x} y^{-1} = x^2. \tag{1.43}$$

If we set $u = y^{-1}$, then $u' = -y^{-2} y'$ and Eq. (1.43) becomes

$$u' - \frac{3}{x} u = -x^2.$$

An integrating factor is

$$\pm \exp\left(-3\int x^{-1} \, dx\right) = \pm |x|^{-3}.$$

If we choose the plus sign for $x > 0$ and the minus sign for $x < 0$, then

$$\rho(x) = x^{-3}.$$

Using formula (1.36), we find that

$$ux^{-3} = -\int x^{-1} \, dx = -\ln|x| + c$$

and

$$u = x^3(c - \ln|x|).$$

Since $u = y^{-1}$ we have

$$y = x^{-3}(c - \ln|x|)^{-1}.$$

It should be noted that $y = 0$ is also a solution of the original equation (1.42). In dividing through by $y^2$ we tacitly assumed that $y$ was never zero.

## Exercises for Section 1.4

In Exercises 1–12, find the general solution of the equation. If an initial condition is given, also find the solution that satisfies the condition.

1. $xy' + 2y = 4x^2$, $y(1) = 4$    2. $xy' - 3y = x^3$, $y(1) = 0$
3. $xy' + (x - 2)y = 3x^3 e^{-x}$    4. $y' - 2y = 4x$, $y(0) = 1$
5. $y' - 2xy = 1$, $y(a) = b$    6. $y' + (\cos x)y = \cos x$, $y(\pi) = 0$
7. $x(\ln x)y' + y = 2 \ln x$    8. $(x^2 + 1)y' - 2xy = x^2 + 1$, $y(1) = \pi$
9. $y' + 2xy = 2x$    10. $y' + (\cot x)y = 3 \sin x \cos x$
11. $x(x + 1)y' - y = 2x^2(x + 1)$    12. $xy' - y = x \sin x$
13. Show that the solution of the initial value problem

$$y' + p(x)y = q(x), \qquad y(a) = b$$

is given by the formula

$$y = be^{-P(x)} + \int_a^x e^{-[P(x) - P(t)]}q(t) \, dt,$$

where

$$P(x) = \int_a^x p(t) \, dt.$$

Suggestion: integrate from $a$ to $x$ in the equation preceding (1.37).

14. A function $f$ is said to be *bounded* on an interval $I$ if there exists a number $M$ such that $|f(x)| \le M$ for $x$ in $I$. Let the function $q$ be continuous and bounded on the interval $[0, \infty)$. Let $k$ be a positive constant.
    (a) Show that every solution of the equation $y' + ky = q(x)$ is bounded on the interval $[0, \infty)$.
    (b) Show that the equation $y' - ky = q(x)$ possesses solutions that are not bounded on $[0, \infty)$.

15. Let $q$ be continuous on $[0, \infty)$ and let $\lim_{x \to \infty} q(x) = L$. If $k$ is a positive number, show that every solution of the equation $y' + ky = q(x)$ tends to the limit $L/k$ as $x$ becomes infinite. Suggestion: given $\varepsilon > 0$ there is a positive number $x_0$ such that $|q(x) - L| < \varepsilon$ if $x \ge x_0$. Let $h(x) = q(x) - L$ and use the result of Exercise 13, with $a = x_0$.

In Exercises 16–21, solve the differential equation.

16. $xy' + y + x^2 y^2 e^x = 0$
17. $xy' - (3x + 6)y = -9xe^{-x}y^{4/3}$
18. $3xy^2 y' - 3y^3 = x^4 \cos x$

**19.** $xyy' = y^2 - x^2$

**20.** $y' - 2(\sin x) y = -2y^{3/2} \sin x$

**21.** $2y' + \dfrac{1}{x+1} y + 2(x^2 - 1) y^3 = 0$

In Exercises 22–24 find a new dependent variable such that the equation becomes linear in that variable. Then solve the equation.

**22.** $xe^y y' - e^y = 3x^2$   (Suggestion: let $u = e^y$)

**23.** $\dfrac{1}{y^2 + 1} y' + \dfrac{2}{x} \tan^{-1} y = \dfrac{2}{x}$

**24.** $y' - \dfrac{1}{x+1} y \ln y = (x+1)y$

## 1.5   ORTHOGONAL TRAJECTORIES

If $c$ is an arbitrary constant, the equation

$$y = cx^2 \tag{1.44}$$

describes a family of parabolas. Some of these are shown by the solid curves in Fig. 1.2. Through every point $(x_0, y_0)$ in the plane, except those points on the $y$-axis, there passes exactly one curve of the family. For if we specify $(x_0, y_0)$ with $x_0 \neq 0$, then $c$ is determined by the condition $y_0 = cx_0^2$ or $c = y_0/x_0^2$.

The slope of the curve through the point $(x, y)$ is

$$y' = 2cx.$$

But since $c = y/x^2$, we have the formula

$$y' = 2\frac{y}{x} \tag{1.45}$$

for the slope of the curve of the family (1.44) that passes through the point $(x, y)$ at the point $(x, y)$.

Suppose now that we wish to find a second family of curves, with exactly one curve of the family passing through each point $(x, y)$ and such that at each point the curve of this second family is orthogonal or perpendicular to the curve of the original family (1.44) that passes through the point. The slope

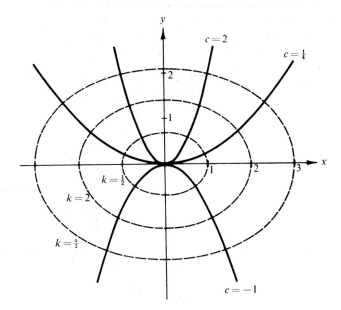

**Figure 1.2**

of the curve of the second family must be the negative reciprocal of the slope of the curve of the first family. In view of formula (1.45), we must find a family of curves for which

$$y' = -\frac{x}{2y}.$$

This is a separable equation. We have

$$2y\,dy = -x\,dx$$

and hence

$$y^2 = -\tfrac{1}{2}x^2 + k$$

or

$$\frac{x^2}{2k} + \frac{y^2}{k} = 1.$$

(Here $k$ must be a positive constant, otherwise there is no curve.) This is a family of ellipses, a few of which are shown by the dotted curves in Fig. 1.2.

To consider a slightly different problem, suppose we wish to find a third family of curves, with one curve of the family through each point, such that at $(x, y)$ the curve of the third family makes an angle of $\pi/4$ with the curve of the first family (1.44). The angle is to be measured counterclockwise from the curve of the third family to the curve of the first family, as shown in Fig. 1.3. Using the notation shown in the figure, we must have

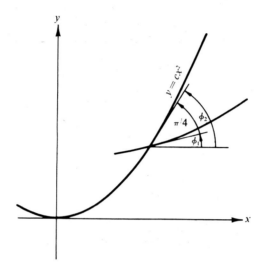

**Figure 1.3**

$$\frac{\tan \phi_2 - \tan \phi_1}{1 + \tan \phi_2 \tan \phi_1} = \tan \frac{\pi}{4}.$$

But $\tan \phi_2 = 2y/x$, and if we set $\tan \phi_1 = y'$, we require that

$$\frac{(2y/x) - y'}{1 + (2y/x)y'} = 1.$$

Simplification yields the differential equation

$$(2y + x)y' = 2y - x.$$

This equation can be written in the form $y' = F(y/x)$. The change of variable $y = vx$ leads to the separable equation

$$\frac{4v + 2}{2v^2 - v + 1} v' = -\frac{2}{x}.$$

Integration yields the relation

$$\ln(2v^2 - v + 1) + \frac{6}{\sqrt{7}} \tan^{-1}\left(\frac{4v - 1}{\sqrt{7}}\right) = -\ln x^2 + k.$$

In terms of $x$ and $y$ this relation is

$$\ln(2y^2 - xy + x^2) + \frac{6}{\sqrt{7}} \tan^{-1}\left(\frac{4y - x}{\sqrt{7}\,x}\right) = k.$$

## Exercises for Section 1.5

In Exercises 1–10 find the orthogonal trajectories of the family of curves. In Exercises 1–6, also sketch several curves of each family.

**1.** $y = e^x + c$          **2.** $y = ce^x$

**3.** $y = cx$          **4.** $y = \tan^{-1} x + c$

**5.** $x^2 + \dfrac{y^2}{4} = c^2$          **6.** $x^2 + (y - c)^2 = c^2$

**7.** $x^2 - y^2 = 2cx$          **8.** $y = x \ln cx$

**9.** $y = \dfrac{x}{cx + 1}$          **10.** $y = \dfrac{1 + cx}{1 - cx}$

In Exercises 11–14, find a family of curves making an angle of $\pi/4$ with the curves of the given family. The angle is to be measured counterclockwise *to* the curve of the given family.

**11.** $y = c\,x^{-1}$          **12.** $y = \dfrac{1}{x + c}$

**13.** $y = \dfrac{x^3}{3} + c$          **14.** $y = \sin x + c$

**15.** Find a family of curves that cuts the family $y = x^2 + c$ at an angle of $\pi/6$. The angle is to be measured counterclockwise toward the curve of the given family.

**16.** Find a family of curves that cuts the family $y = x^3 + c$ at an angle of $\pi/3$. The angle is to be measured counterclockwise toward the curve of the given family.

**17.** Let $(r, \theta)$ be polar coordinates. If $\psi$ is the positive angle measured counterclockwise from the radius vector to the tangent line to a curve

$r = f(\theta)$, then it is shown in calculus that

$$\tan \psi = \frac{r}{dr/d\theta}.$$

Show that two curves $r = f_1(\theta)$ and $r = f_2(\theta)$ are orthogonal at a point if the corresponding values of $r/(dr/d\theta)$ are negative reciprocals, provided that the same coordinates $(r, \theta)$ are used to describe the same point on both curves.

Use the results of Exercise 17 to find the orthogonal trajectories of the family of curves in Exercises 18–21. Sketch a few curves of each family.

**18.** $r = c(1 + \cos \theta), \quad c > 0$     **19.** $r = c \sin \theta$

**20.** $r = c \sin 2\theta$     **21.** $r^2 = c \cos 2\theta$

## 1.6   DECAY AND MIXING PROBLEMS

In Section 1.1, we stated that many natural laws of science could be formulated as differential equations. In this and the next section we consider some phenomena whose mathematical description leads to first-order equations.

Our first example concerns the rate of decay of a radioactive substance. It is to be expected that the more of the undecayed substance present, the greater will be the rate of decay. Experiment indicates that a radioactive substance decays at a rate directly proportional to the amount of undecayed matter remaining. Thus if $x(t)$ is the amount (mass) of undecayed substance present at time $t$, we have

$$\frac{dx}{dt} = -kx, \tag{1.46}$$

where $k$ is a positive constant of proportionality. The minus sign occurs because $x$ is a decreasing function, and $dx/dt$ is negative.

Suppose that at time $t_0$ the amount of undecayed substance is $x_0$ and that at some later time $t_1$ it is $x_1$. Thus we have the auxiliary conditions

$$x(t_0) = x_0, \qquad x(t_1) = x_1. \tag{1.47}$$

Separating the variables in Eq. (1.46) we have

$$\frac{dx}{x} = -k \, dt. \tag{1.48}$$

Integrating, we have

$$\ln x = -kt + \ln c$$

(here $x > 0$ and $\ln c$ is an arbitrary constant for $c > 0$) or

$$x = ce^{-kt}. \tag{1.49}$$

We could now use the two conditions (1.47) to determine the numbers $c$ and $k$ in formula (1.49). However, let us proceed in another way. Starting with Eq. (1.48) and using definite integrals, we may write

$$\int_{x_0}^{x_1} \frac{dx}{x} = -k \int_{t_0}^{t_1} dt.$$

Then

$$\ln \frac{x_1}{x_0} = -k(t_1 - t_0)$$

and so

$$k = \frac{1}{t_1 - t_0} \ln \frac{x_0}{x_1}. \tag{1.50}$$

Having determined $k$, we go back to Eq. (1.48) and write

$$\int_{x_0}^{x} \frac{dx}{x} = -k \int_{t_0}^{t} dt.$$

(We could have used $t_1$ and $x_1$ in place of $t_0$ and $x_0$, respectively.) Carrying out the integration, we find that

$$\ln \frac{x}{x_0} = -k(t - t_0)$$

or

$$x = x_0 e^{-k(t - t_0)}.$$

The use of the value (1.50) for $k$ yields the result

$$x = x_0 \exp\left[ -\frac{t - t_0}{t_1 - t_0} \ln \frac{x_0}{x_1} \right].$$

As a second example of a situation that leads to a first-order equation, we consider a mixing problem. Suppose that we start with a tank that contains 20 gal of a solution of a certain chemical and that 10 lb of the chemical are in the solution. Starting at a certain instant, a solution of the same chemical, with a concentration of 2 lb/gal, is allowed to flow into the tank at the rate of 3 gal/min. The mixture is drained off at the same rate so that the volume of solution in the tank remains constant. (We make the simplifying assumption that the concentration of the solution in the tank is kept uniform, perhaps by stirring.) The problem we wish to solve is this. How many gallons of the solution should be pumped into the tank to raise the amount of dissolved chemical to 15 lb?

It is true that each gallon of the solution coming in brings with it 2 lb of the chemical, but the mixture leaving takes some of the chemical with it. We can solve our problem if we can obtain a formula for the amount of chemical in the tank at time $t$. If $x(t)$ is the amount in the tank at time $t$, then the rate of change of $x$, $dx/dt$, is given by the following rule: $dx/dt$ is equal to the rate at which the chemical enters the tank minus the rate at which the chemical leaves the tank. The rate at which the chemical enters is $2 \times 3 = 6$ lb/min, since 3 gal of solution flow in per minute and each gallon contains 2 lb of the chemical. At time $t$ the concentration of the solution in the tank is $x(t)/20$, since the volume of solution in the tank is always 20 gal. Hence the rate at which the chemical is leaving the tank is $3 \times x/20$ gal/min. Thus we arrive at the differential equation

$$\frac{dx}{dt} = 6 - \frac{3}{20}x.$$

This is a linear equation. It is also separable. The general solution is

$$x = ce^{-3t/20} + 40.$$

Using the fact that $x = 10$ when $t = 0$, we have $10 = c + 40$ or $c = -30$. Then

$$x = 40 - 30e^{-3t/20}.$$

We want to find the value of $t$ when $x$ is 15. Setting $x = 15$ in the above equation, we have

$$15 = 40 - 30e^{-3t/20},$$

$$e^{-3t/20} = \frac{25}{30} = \frac{5}{6},$$

or

$$t = \frac{20}{3} \ln \frac{6}{5}.$$

Using a table of natural logarithms we find that $\ln 6/5 = 1.8232$ and that $t = 12.15$ min. Since 3 gal of solution enter the tank per minute, we must pump in 36.45 gal to bring the concentration up to the desired value.

### Exercises for Section 1.6

1. Let $x_0$ be the amount of radioactive substance present at $t = 0$ and let $T$ be the time required for one-half of the substance to decay. Show that $T$ is independent of $x_0$. The time $T$ is called the *half-life* of the radioactive substance.

2. Let $x(t)$ be the amount of a radioactive substance present at time $t$ and let $x(0) = x_0$. If $T$ is the half-life (see Exercise 1) show that

$$x(t) = x_0 2^{-t/T}.$$

3. At a certain instant 100 gm of a radioactive substance are present. After 4 yr, 20 gm remain. How much of the substance remains after 8 years?

4. After 6 hr, 60 gm of a radioactive substance are present. After 8 hr (2 hr later) 50 gm are present. How much of the substance was present initially?

5. If the half-life (see Exercise 1) of a radioactive substance is 10 yr, when does 25 percent of the substance remain? Think!

6. Let $N(t)$ denote the number of bacteria in a culture at time $t$. Assuming that $N$ increases at a rate proportional to the number of bacteria present, find a formula for $N$ in terms of $t$.

7. A conical tank of height 12 ft and radius 4 ft is initially filled with fluid. After 6 hr the height of the fluid is 10.5 ft. If the fluid evaporates at a rate proportional to the surface area exposed to the air, find a formula for the volume of fluid in the tank as a function of time.

8. A tank initially contains 100 gal of a solution that holds 30 lb of a chemical. Water runs into the tank at the rate of 2 gal/min and the solution runs out at the same rate. How much of the chemical remains in the tank after 20 min?

9. A tank initially contains 50 gal of a solution that holds 30 lb of a chemical. Water runs into the tank at the rate of 3 gal/min and the mixture

runs out at the rate of 2 gal/min. After how long will there be 25 lb of the chemical in the tank?

**10.** A tank initially contains 100 gal of a solution that holds 40 lb of a chemical. A solution containing 2 lb/gal of the chemical runs into the tank at the rate of 2 gal/min and the mixture runs out at the rate of 3 gal/min. How much chemical is in the tank after 50 min?

**11.** A tank initially holds 100 liters of a solution in which is dissolved 200 gm of a radioactive substance. Assume that the substance decays at a rate equal to $k$ times the amount present. If water flows in at the rate of 2 liters/min and the solution flows out at the same rate, find a formula for the amount of radioactive substance in the tank after $t$ min.

## 1.7  COOLING; THE RATE OF A CHEMICAL REACTION

The first problem that we consider in this section has to do with the change in temperature in a cooling body. If a body cools in a surrounding medium (such as air or water) it might be expected that the rate of change of the temperature of the body would depend on the difference between the temperature of the body and that of the surrounding medium. *Newton's law of cooling* asserts that the rate of change is directly proportional to the difference of the temperatures. Thus if $u(t)$ is the temperature of the body at time $t$ and if $u_0$ is the (constant) temperature of the surrounding medium, we have

$$\frac{du}{dt} = -k(u - u_0), \tag{1.51}$$

where $k$ is a positive constant. The minus sign occurs because $du/dt$ will be negative when $u > u_0$.

As an example, suppose that an object is heated to 300°F and allowed to cool in a room whose air temperature is 80°F. If after 10 minutes the temperature of the body is 250°, what will be its temperature after 20 minutes? The differential equation (1.51) becomes

$$\frac{du}{dt} = -k(u - 80) \tag{1.52}$$

and the initial condition is $u(0) = 300$. We also know that $u(10) = 250$.

Equation (1.52) is linear; it is also separable. Treating it as a separable equation, we write

$$\frac{du}{u - 80} = -k \, dt. \tag{1.53}$$

Using the specified conditions to determine $k$, we have

$$\int_{300}^{250} \frac{du}{u - 80} = -k \int_0^{10} dt.$$

From this equation we find that

$$k = \frac{1}{10} \ln \frac{220}{170} = 0.0258.$$

To determine $u$ when $t = 20$, we go back to Eq. (1.53) and write

$$\int_{300}^{u} \frac{du}{u - 80} = -k \int_0^{20} dt.$$

We find that

$$\ln \frac{u - 80}{220} = -20k$$

or

$$u = 220e^{-20k} + 80.$$

Since $e^{-20k} = e^{-0.516} = 0.60$, we have

$$u(20) = 212°.$$

Our next application of differential equations concerns the rate of a chemical reaction. Suppose that A and B are two chemicals that react in solution. Let $x(t)$ and $y(t)$ denote the concentrations (in moles[5] per liter) at time $t$ of A and B, respectively.

In our first example, we shall assume that one molecule of A combines with one molecule of B to form a new product or products. Then $x$ and $y$ decrease at the same rate, so that $dx/dt = dy/dt$. Let $z(t)$ be the amount by which $x$ and $y$ have decreased in time $t$. If $a$ and $b$ are the initial concentrations of A and B, respectively, then

$$x(t) = a - z(t), \qquad y(t) = b - z(t) \qquad (1.54)$$

[5] If the molecular weight of a chemical is $w$, then one mole of that chemical consists of $w$ grams.

and

$$\frac{dz}{dt} = -\frac{dx}{dt} = -\frac{dy}{dt}.$$    (1.55)

The quantity $dz/dt$ is called the *rate of reaction*. For many reactions of this type (one molecule of A combining with one molecule of B), it is found that[6] under conditions of constant temperature

$$\frac{dz}{dt} = kxy$$    (1.56)

or

$$\frac{dz}{dt} = k(a - z)(b - z),$$    (1.57)

where $k$ is a positive constant of proportionality. Thus the rate of reaction is directly proportional to the concentration of each reactant.

The differential equation (1.57) is nonlinear, but separable. Separating the variables and using the fact that $z(0) = 0$, we have

$$\int_0^z \frac{dz}{(a - z)(b - z)} = k \int_0^t dt.$$

The integral on the left can be evaluated by the use of partial fractions. We find that

$$\frac{1}{a - b} \int_0^z \left( \frac{-1}{a - z} + \frac{1}{b - z} \right) dz = kt$$

or

$$\frac{1}{a - b} \ln \frac{b(a - z)}{a(b - z)} = kt.$$

The value of $k$ can be determined by experiment, in which $z$ is measured for

---

[6] The formula (1.57) does not always apply. In some cases nonreacting substances (catalysts) influence the rate of reaction. In any case the rate of reaction must ultimately be determined by experiment.

various values of $t$. From the above relation we obtain the formula

$$z = ab \frac{e^{k(a-b)t} + 1}{ae^{k(a-b)t} - b}$$

for $z$. The concentrations of A and B can be found from Eq. (1.54).

Let us next consider a reaction in which two molecules of chemical B combine with one molecule of chemical A to form new products. Then two moles of B are used up for every mole of A, so that

$$\frac{dy}{dt} = 2 \frac{dx}{dt}.$$

If $z(t)$ is the decrease in chemical A in time $t$, then

$$x = a - z, \qquad y = b - 2z. \tag{1.58}$$

In this case (two molecules of B combining with one of A), it is found that[7]

$$\frac{dz}{dt} = kxy^2 \tag{1.59}$$

or

$$\frac{dz}{dt} = k(a - z)(b - 2z)^2. \tag{1.60}$$

Here $dz/dt$ is directly proportional to $x$ and to the square of $y$. The integration of Eq. (1.60) is left to the exercises.

In the general case where $m$ molecules of A combine with $n$ molecules of B, we have

$$n \frac{dx}{dt} = m \frac{dy}{dt}.$$

If we set

$$z = \frac{1}{m}(a - x) = \frac{1}{n}(b - y),$$

[7] See footnote 6, p. 40.

then $z(0) = 0$ and

$$\frac{dz}{dt} = -\frac{1}{m}\frac{dx}{dt} = -\frac{1}{n}\frac{dy}{dt}.$$

The equation for $z$ is found to be

$$\frac{dz}{dt} = kx^m y^n$$

or

$$\frac{dz}{dt} = k(a - mz)^m (b - nz)^n.$$

The exponents $m$ and $n$ are often called the *orders* of the reaction with respect to the concentrations of A and B.

## Exercises for Section 1.7

1. An object whose initial temperature is 150°F is allowed to cool in a room where the temperature of the air is 75°F. After 10 min the temperature of the object is 125°. When will its temperature be 100°?

2. A heated object is allowed to cool in air whose temperature is 20°C. After 5 min its temperature is 200°C. After 10 min (five minutes later) its temperature is 160°. What was the temperature of the object initially?

3. An object whose temperature is 220°F is placed in a room where the temperature is 60°F. After 10 minutes the temperature of the object is 200°. At this point refrigeration equipment, which lowers the temperature of the room at the rate of 1°F/min, is turned on. What is the temperature of the object $t$ min after the equipment is turned on?

4. An object with a temperature of 10°F is placed in a room where the temperature is 80°F. After 10 min the temperature of the object is 30°. What will be the temperature of the object after it has been in the room for 30 min?

5. Suppose that in a chemical reaction where one molecule of A combines with one molecule of B, the rule (1.56) applies. Assume that A and B have the same initial concentration $a$.

   (a) Find a formula for the concentrations of A and B at time $t$.

   (b) Find a formula for the half-life of the reaction, which is the time required for the concentrations of the reactants to be halved.

6.  Suppose that in a chemical reaction where 2 molecules of B combine with one of A, the rule (1.59) applies. Find a formula that expresses $k$ in terms of $z$ and $t$ in the case where

    (a)  $b \neq 2a$          (b)  $b = 2a$.

7.  A chemical A breaks down when heated, with $n$ molecules of A reacting to form new products. The law of reaction is

$$\frac{dx}{dt} = -kx^n,$$

where $x(t)$ is the amount of the chemical remaining at time $t$. If half the chemical decomposes after $T$ min, find a formula for $x$ in terms of $t$. Let $a$ denote the initial amount of the chemical.

## 1.8   TWO SPECIAL TYPES OF SECOND-ORDER EQUATIONS

A second-order differential equation is of the form

$$F\left(t, x, \frac{dx}{dt}, \frac{d^2x}{dt^2}\right) = 0.$$

In this section we shall consider two classes of second-order equations that can be solved by successively solving two first-order equations. Thus the methods of solution for first-order equations that were presented earlier in this chapter may be used.

We consider first the class of second-order equations in which the independent variable $x$ is absent. Such an equation is of the form

$$G\left(t, \frac{dx}{dt}, \frac{d^2x}{dt^2}\right) = 0. \tag{1.61}$$

Suppose that $x(t)$ is a solution of Eq. (1.61). If we set $v = dx/dt$, then $v$ must be a solution of the first-order equation

$$G\left(t, v, \frac{dv}{dt}\right) = 0. \tag{1.62}$$

If we can solve this equation for $v$, then the solutions of Eq. (1.61) can be found from the relation

$$\frac{dx}{dt} = v(t) \tag{1.63}$$

by integration.

As an example, let us consider the equation

$$t\frac{d^2x}{dt^2} = 2\left[\left(\frac{dx}{dt}\right)^2 - \frac{dx}{dt}\right]. \tag{1.64}$$

Note that $x$ itself is absent. Setting $v = dx/dt$, we obtain the first-order equation

$$t\frac{dv}{dt} = 2(v^2 - v) \tag{1.65}$$

for $v$. This equation is separable and we have

$$\frac{dv}{v^2 - v} = 2\frac{dt}{t}$$

or

$$\left(\frac{1}{v-1} - \frac{1}{v}\right) dv = 2\frac{dt}{t}.$$

Integrating, we find that

$$\ln\left|\frac{v-1}{v}\right| = 2\ln|t| + c_1'$$

or

$$\frac{v-1}{v} = c_1 t^2.$$

Then

$$v = \frac{dx}{dt} = \frac{1}{1 - c_1 t^2}.$$

If $c_1$ is positive, say $c_1 = a^2$, we have

$$\frac{dx}{dt} = \frac{1}{1 - a^2 t^2},$$

so that

$$x = \frac{1}{2a} \ln \left| \frac{1 + at}{1 - at} \right| + c_2.$$

If $c_1$ is negative, say $c_1 = -b^2$, then

$$\frac{dx}{dt} = \frac{1}{1 + b^2 t^2}$$

and

$$x = \frac{1}{b} \tan^{-1} bt + c_2.$$

Finally, we observe that since the constant functions $v = 0$ and $v = 1$ are solutions of Eq. (1.65), the functions

$$x = c, \qquad x = t + c$$

are solutions of Eq. (1.64).

The second class of second-order equations that we shall consider are those in which the independent variable $t$ is missing. Such equations are of the form

$$H\left( x, \frac{dx}{dt}, \frac{d^2 x}{dt^2} \right) = 0. \tag{1.66}$$

Suppose that $x(t)$ is a solution of Eq. (1.66) and let $v = dx/dt$. On an interval where $x$ is a strictly increasing, or decreasing, function, $t$ can be regarded as a function of $x$ and we can write

$$\frac{d^2 x}{dt^2} = \frac{dv}{dt} = \frac{dv}{dx}\frac{dx}{dt} = v \frac{dv}{dx}.$$

Then Eq. (1.66) becomes

$$H\left( x, v, v \frac{dv}{dx} \right) = 0 \tag{1.67}$$

and this is a first-order equation for $v$. If we can solve it, finding a solution $v(x)$, then a solution of the original equation (1.66) can be found by solving the first-order equation

$$\frac{dx}{dt} = v(x). \tag{1.68}$$

As an illustration, we consider the equation

$$x \frac{d^2x}{dt^2} = \left(\frac{dx}{dt}\right)^2 + 2\frac{dx}{dt}, \tag{1.69}$$

in which $t$ is missing. Setting

$$\frac{dx}{dt} = v, \qquad \frac{d^2x}{dt^2} = v\frac{dv}{dx},$$

we have

$$xv \frac{dv}{dx} = v^2 + 2v. \tag{1.70}$$

By inspection we see that $v = 0$ and $v = -2$ are solutions of this equation, so that

$$x = c, \qquad x = -2t + c$$

are solutions of Eq. (1.69). To find the remaining solutions, we divide through by $v$ in Eq. (1.70), obtaining the separable equation

$$x \frac{dv}{dx} = v + 2.$$

We easily find that

$$v = c_1 x - 2.$$

Now we must solve the equation

$$\frac{dx}{dt} = c_1 x - 2.$$

For $c_1 \neq 0$ we find that

$$x = \frac{1}{c_1}(c_2 e^{c_1 t} + 2).$$

(When $c_1 = 0$ we have $v = -2$, which was considered previously.)

Some applications that give rise to the types of differential equations discussed here are presented in the next section.

### Exercises for Section 1.8

1. If $v_0$ is a zero of the function $f$, show that the differential equation

$$\frac{d^2x}{dt^2} = f\left(\frac{dx}{dt}\right)g(t).$$

   possesses the solutions $x = v_0 t + c$.

In Exercises 2–20, find the general solution.

2. $\dfrac{d^2x}{dt^2} = \dfrac{dx}{dt} + 2t$

3. $2t\dfrac{dx}{dt}\dfrac{d^2x}{dt^2} = \left(\dfrac{dx}{dt}\right)^2 + 1$

4. $\dfrac{d^2x}{dt^2} = -2t\left(\dfrac{dx}{dt}\right)^2$

5. $2t\dfrac{d^2x}{dt^2} = \left(\dfrac{dx}{dt}\right)^2 - 1$

6. $t^2\dfrac{d^2x}{dt^2} + \left(\dfrac{dx}{dt}\right)^2 = 2t\dfrac{dx}{dt}$

7. $\left(\dfrac{dx}{dt} - t\right)\dfrac{d^2x}{dt^2} - \dfrac{dx}{dt} = 0$

8. $t\exp\left(\dfrac{dx}{dt}\right)\dfrac{d^2x}{dt^2} = \exp\left(\dfrac{dx}{dt}\right) - 1$

9. $t\dfrac{d^2x}{dt^2} = \dfrac{dx}{dt} + 2\sqrt{t^2 + \left(\dfrac{dx}{dt}\right)^2}$

10. $\dfrac{d^2x}{dt^2} = \dfrac{1}{t}\dfrac{dx}{dt} + \tanh\left(\dfrac{dx}{dt}\Big/t\right)$

11. $\dfrac{d^2x}{dt^2} + x^{-3} = 0$

12. $x\dfrac{d^2x}{dt^2} = \left(\dfrac{dx}{dt}\right)^2$

13. $\dfrac{d^2x}{dt^2} + \left(\dfrac{dx}{dt}\right)^3 = 0$

14. $3x\dfrac{dx}{dt}\dfrac{d^2x}{dt^2} = \left(\dfrac{dx}{dt}\right)^3 - 1$

15. $\dfrac{d^2x}{dt^2} + e^{-x}\dfrac{dx}{dt} = 0$

16. $(x^2 + 1)\dfrac{d^2x}{dt^2} = 2x\left(\dfrac{dx}{dt}\right)^2$

17. $x^3\dfrac{d^2x}{dt^2} = 2\left(\dfrac{dx}{dt}\right)^3$

**18.**  $\dfrac{d^2x}{dt^2} = \left(\dfrac{dx}{dt}\right)^2 \tanh x$        **19.**  $x\dfrac{d^2x}{dt^2} = \dfrac{dx}{dt}\left(\dfrac{dx}{dt} + 2\right)$

**20.**  $\dfrac{d^2x}{dt^2} + 2\left(\dfrac{dx}{dt}\right)^2 \tan x = 0$

## 1.9  FALLING BODIES

The applications in this section involve the motion of a solid body whose center of mass moves in a straight line. Let us denote by $x$ the directed distance of the center of mass from some fixed point on the line of motion. Then $x$ depends on time $t$. The velocity and acceleration of the center of mass are $dx/dt$ and $d^2x/dt^2$, respectively. The notations

$$\dot{x} = \frac{dx}{dt}, \qquad \ddot{x} = \frac{d^2x}{dt^2}$$

are commonly used.

According to *Newton's second law of motion*, the mass of the body times the acceleration of the center of mass is proportional to the force acting on the body. Actually the commonly used systems of units for measuring mass, distance, time, and force are arranged so that the constant of proportionality may be taken as unity. Thus

$$m\frac{d^2x}{dt^2} = F, \tag{1.71}$$

where $m$ is the mass of the body and $F$ is the force. When $F$ depends on $t$, $x$, and $dx/dt$, Eq. (1.71) is a second-order differential equation for $x$. The units in Eq. (1.71) must be chosen appropriately. Equation (1.71) is sometimes called the *equation of motion* of the body.

In the *centimeter-gram-second* system of units (c.g.s.) distance is in *centimeters*, time is in *seconds*, mass in in *grams*, and force is in *dynes*. In the *British* system of units, distance is in *feet*, time is in *seconds*, mass is in *slugs*, and force is in *pounds*.

In the examples of this section we shall examine the motion of falling (and rising) bodies. One of the forces present is that of gravity, The *weight* of a body is very nearly[8] the force exerted on the body by the earth's gravitational field. Near the surface of the earth the force due to gravity is $mg$, where $g$ is

---

[8] The rotation of the earth complicates an exact definition of weight, but for most practical considerations the weight of a body may be taken to be the force due to gravity.

approximately 980 cm/sec² or 32 ft/sec². Actually the value of $g$ (the acceleration due to gravity) varies slightly over the surface of the earth, being slightly larger at the poles than at the equator. If the weight of a body is $w$ lb, then the mass $m$ of the body in slugs is given by the formula

$$m = \frac{w}{g}.$$

In our first example let us suppose that an object is thrown directly upward from the surface of the earth, with an initial velocity $v_0$. Let $x$ denote the directed distance upward of the object from the surface of the earth. If we assume that the only force acting on the object is that due to gravity, then Eq. (1.71) becomes

$$m\ddot{x} = -mg. \tag{1.72}$$

The minus sign occurs because the force acts in the direction of decreasing $x$. The initial conditions are

$$x(0) = 0, \qquad \dot{x}(0) = v_0, \tag{1.73}$$

assuming that $t = 0$ is the time at which the object is thrown.

A first integration of Eq. (1.72) yields the relation

$$\dot{x} = -gt + c_1,$$

where $c_1$ is a constant. The condition $x(0) = v_0$ tells us that $c_1 = v_0$, so we have

$$\dot{x} = -gt + v_0. \tag{1.74}$$

Integrating again, we have

$$x = -\tfrac{1}{2}gt^2 + v_0 t + c_2.$$

Since $x(0) = 0$ we must have $c_2 = 0$ and

$$x = -\tfrac{1}{2}gt^2 + v_0 t. \tag{1.75}$$

Formulas (1.74) and (1.75) describe the velocity and position of the object at time $t$, $t \geq 0$. From formula (1.74) we see that the velocity is positive until $t = v_0/g$, after which time it becomes negative, with the object descending. Thus the time required for the object to reach its maximum height is $v_0/g$

and the maximum height, as found from formula (1.75), is

$$h = \frac{v_0^2}{2g}.$$

The time when the object returns to earth can be found by setting $x = 0$ in Eq. (1.75) and solving for $t$. We find that the time is $2v_0/g$ so that the time that it takes the object to fall back to earth is the same as the time going up. The velocity with which the object strikes the earth is found by setting $t = 2v_0/g$ in formula (1.74). This velocity is found to be $-v_0$. The magnitude of the final velocity is therefore the same as that of the initial velocity. The symmetry that occurs in this problem (time going up equals time coming down, and final velocity equals initial velocity) does not always arise when forces other than gravity are considered. Examples are presented in the exercises.

If an object is dropped from a specified height $h$ above the earth, it is probably more convenient to let $x$ denote the directed distance *downward* from the point of release. Then the equation of motion becomes

$$m\ddot{x} = mg. \tag{1.76}$$

Here the force acts in the direction of increasing $x$, and there is no minus sign. The initial conditions are

$$x(0) = 0, \qquad \dot{x}(0) = 0. \tag{1.77}$$

Two integrations of Eq. (1.76) yield the formula

$$x = \tfrac{1}{2}gt^2 \tag{1.78}$$

for the distance through which the object falls in time $t$. If a stone is dropped from a bridge and 3 sec elapse before it hits the water below, we can estimate the height of the bridge above the water from formula (1.78). Using the value $g = 32$ ft/sec$^2$, we have

$$h = \tfrac{1}{2}(32)(3^2) = 144 \quad \text{ft.}$$

Actually, when a body moves through the air (or other surrounding medium) the air exerts a damping force $F_d$ on the body. This force depends on the velocity of the body (and on the shape of the body and the nature of the surrounding medium). In some situations the damping force is proportional to the velocity. In others it is more nearly proportional to the square or cube of the velocity.

Let us now reexamine the problem of a falling object that is released from a height $h$ above the earth. Let $x$ be the directed distance downward from the point of release. If the damping force $F_d$ is proportional to the velocity, then

$$F_d = -c\dot{x},\qquad(1.79)$$

where $c$ is a positive constant constant of proportionality called the *damping constant*. The minus sign indicates that the force acts in a direction opposite to that of the velocity vector. The resultant force $F$ is the sum of the forces acting, and in this case is

$$F = -c\dot{x} + mg.$$

The equation of motion becomes

$$m\ddot{x} = -c\dot{x} + mg \qquad(1.80)$$

and the initial conditions are

$$x(0) = 0, \qquad \dot{x}(0) = 0. \qquad(1.81)$$

In the second-order equation (1.80) both $x$ and $t$ are absent. Setting $x = v$ and $\ddot{x} = \dot{v}$, we obtain the first-order linear equation

$$m\dot{v} + cv = mg \qquad(1.82)$$

for $v$. Solving, and using the initial condition $v(0) = 0$, we find that

$$v = \dot{x} = \frac{mg}{c}(1 - e^{-ct/m}). \qquad(1.83)$$

An integration yields the formula

$$x = \frac{mg}{c}\left(t + \frac{m}{c}e^{-ct/m}\right) - g\left(\frac{m}{c}\right)^2 \qquad(1.84)$$

for $x$.

We notice from formula (1.83) that as $t$ becomes infinite the velocity tends to the limiting value

$$v_\infty = \frac{mg}{c}. \qquad(1.85)$$

Given that there is a limiting velocity, its value could have been found from the equation of motion (1.82). As the velocity of the falling object increases, the damping force increases in magnitude until it balances the force due to gravity. The acceleration thus tends to zero. Setting $\dot{v} = 0$ in Eq. (1.82), we find that $cv = mg$ or $v = mg/c$, which is the value given in formula (1.85).

When the damping constant $c$ in Eq. (1.80) is small, we might expect the formula (1.84) to agree closely with the formula (1.78), at least when the time interval is short. The latter formula was derived in the absence of a damping force ($c = 0$). To make a comparison, we expand the exponential function that appears in formula (1.84) in a Maclaurin series,

$$e^{-ct/m} = 1 - \frac{ct/m}{1!} + \frac{(ct/m)^2}{2!} - \frac{(ct/m)^3}{3!} + \cdots.$$

Formula (1.84) becomes

$$x = g\left(\frac{m}{c}\right)^2 \left[\frac{1}{2}\left(\frac{ct}{m}\right)^2 - \frac{1}{6}\left(\frac{ct}{m}\right)^3 + \cdots\right]$$

or

$$x = \frac{1}{2}gt^2 - \frac{1}{6}\frac{cg}{m}t^3 + \cdots.$$

When $ct/m$ is small compared with unity, we see that formula (1.78) is a good approximation to the more complicated formula (1.84).

In the case of a body falling from a great height we can no longer assume that the force due to gravity is constant. Instead we must use the more accurate "inverse square law" of Newton. This implies that the force of attraction between any two spherically symmetric bodies is directly proportional to the product of their masses and inversely proportional to the square of the distance between their centers of mass. Consider the case of a body of mass $m$ falling toward the earth. We assume that the earth is a sphere of mass $M$ and radius $R$. If $r$ is the distance from the center of the earth to the center of mass of the falling body, then the force is

$$F = k\frac{mM}{r^2}, \tag{1.86}$$

where $k$ is a constant of proportionality. To determine $k$ we use the fact that, when $r = R$, $F$ has the value $mg$. Setting $r = R$ in formula (1.86) we have

$$mg = k\frac{mM}{R^2}$$

or

$$k = \frac{gR^2}{M}.$$

Substituting this value for $k$ into Eq. (1.86), we obtain the formula

$$F = \frac{mgR^2}{r^2}. \tag{1.87}$$

As a check we observe that $F = mg$ when $r = R$.

Let us denote by $x$ the directed distance from the center of the earth to the center of mass of the falling body. Then the equation of motion of the body is

$$m\ddot{x} = -\frac{mgR^2}{x^2}, \tag{1.88}$$

neglecting air resistance and other forces.

As an example, suppose that a projectile is fired directly upward from the surface of the earth with a velocity $v_0$. Then the initial conditions are

$$x(0) = R, \qquad \dot{x}(0) = v_0. \tag{1.89}$$

In the second-order equation (1.88), the independent variable $t$ is missing. Setting $\dot{x} = v$ we have

$$\ddot{x} = \frac{dv}{dt} = \frac{dv}{dx}\frac{dx}{dt} = v\frac{dv}{dx}$$

and the equation becomes

$$v\frac{dv}{dx} = -\frac{gR^2}{x^2}.$$

Then

$$\int_{v_0}^{v} v\, dv = -gR^2 \int_{R}^{x} \frac{dx}{x^2},$$

$$\frac{1}{2}(v^2 - v_0^2) = gR^2\left(\frac{1}{x} - \frac{1}{R}\right),$$

or

$$v^2 = v_0^2 + 2gR^2\left(\frac{1}{x} - \frac{1}{R}\right).$$

Initially $v$ is positive and must be the positive square root of the right-hand side of this equation. Thus

$$v = \frac{dx}{dt} = \left[ v_0^2 - 2gR + \frac{2gR^2}{x} \right]^{1/2} \tag{1.90}$$

As $x$ increases we see that the velocity decreases. If the quantity in brackets in Eq. (1.90) vanishes, the velocity becomes negative and the body starts to fall back to earth. However, if $v_0^2 - 2gR \geq 0$, $v$ can never become negative no matter how large $x$ becomes. The critical value $v_0 = \sqrt{2gR}$ is called the *escape velocity* of the earth. Unless $v_0$ is greater than, or equal to, this value, the projectile will ultimately fall back to earth. The value of the escape velocity is approximately 7 m/sec. It should be pointed out that we have ignored a number of forces acting on the body, such as air resistance and the attractive forces of other planets. We have also ignored the effect of the rotation of the earth.

## Exercises for Section 1.9

1.  Suppose that a particle of mass $m$ moves along a straight line. Let $x(t)$ denote the directed distance of the particle from a fixed-point on the line of motion at time $t$. Suppose that the force acting on the particle depends only on $x$ (and not on $t$ or $\dot{x}$), so that the equation of motion has the form $m\ddot{x} = F(x)$. If

$$T = \tfrac{1}{2}m\dot{x}^2, \qquad V(x) = -\int_0^x F(s)\, ds,$$

    show that $T + V$ is constant during the motion. (Here $T$ and $V$ are the kinetic and potential energies, respectively, of the particle.)

2.  An object of mass $m$ is thrown directly upward from the surface of the earth with velocity $v_0$. Assume that the acceleration due to gravity has the constant value $g$ and that the air resists the motion of the object with a force equal to a constant $c$ times the velocity.

    (a) Find the equation of motion.

    (b) Find the velocity and position of the object as functions of time.

    (c) Find the maximum height $h$ attained by the object and the time $t_1$ required to reach this height.

    (d) Show that the velocity of impact (the velocity when the object returns to earth) is less in magnitude than the initial velocity $v_0$. Suggestion: show that

$$\int_{t'}^{t''} (\dot{x}\ddot{x} + g\dot{x})\, dt < 0$$

if $t' < t''$. Note that $2\dot{x}\ddot{x} = (d/dt)\dot{x}^2$.

(e) Show that the time $t_1$ required for the object to attain its maximum height is less than the time required for the object to fall back to earth from this height. Suggestion: show that $x(2t_1) > 0$.

3. An object of mass $m$ is thrown directly upward from the surface of the earth with velocity $v_0$. Assume that the acceleration due to gravity has the constant value $g$ and that the air resists the motion with a force equal to $c$ times the square of the velocity.

    (a) Find the equation of motion. Suggestion: treat the time intervals when the object is going up and when it is coming down separately.

    (b) Find the maximum height $h$ attained by the object and the time $t_1$ required to reach this height.

    (c) Find the time $T$ required for the object to fall back to earth from its maximum height and find the velocity of impact.

4. An object of mass $m$ is thrown directly upward from the surface of the earth with velocity $v_0$, where $v_0 < (2gR)^{1/2}$, $R$ being the radius of the earth. Assume that Newton's inverse square law applies and neglect air resistance.

    (a) Find the maximum height $h$ attained by the object.

    (b) Find the time $t_1$ required for the object to attain its maximum height.

    (c) Show that the time required for the object to fall back to earth from its maximum height is the same as the time required for it to reach this height. Also show that the velocity of impact is the same in magnitude as the initial velocity.

5. A body of mass $m$ falls from a height $h$ above the earth. Assuming that the acceleration due to gravity has the constant value $g$ and that the force due to air resistance is equal to $c$ times the $n$th power of the velocity, find:

    (a) the equation of motion

    (b) the limiting value of the velocity of the body.

6. A ship of mass $m$ traveling in a straight line with speed $v_0$ shuts off its engines and coasts. Assume that the water resistance is equal to $c$ times the $\alpha$ power of the velocity.

    (a) Show that the ship travels a finite distance before coming to rest if $0 < \alpha < 2$ and find this distance.

    (b) What is the time required for the ship to come to rest?

7.  An object of mass $m$ is thrown directly upward from the surface of the earth with velocity $v_0$. Assume that the acceleration due to gravity has the constant value $g$ and that the force due to air resistance is equal to $c$ times the fourth power of the velocity.

    (a) Find the maximum height attained by the object.

    (b) From what height must an object of mass $m$ be dropped in order to achieve an impact velocity of $v_0$ ?

8.  A particle of mass $m$ moves along a straight line, with $x(t)$ its directed distance from a fixed point on the line of motion at time $t$. The particle is repelled from the point $x = 0$ by a force of magnitude $m/x^2$. It starts at $x = 1$ with a speed $v_0$ toward $x = 0$.

    (a) How close does the particle come to the origin?

    (b) What is the time required for the particle to reach the position where it is closest to the origin?

9.  A simple pendulum consists of an object of mass $m$ attached to the end of a massless rod of length $L$. The other end of the rod is connected to a frictionless pivot, as in Fig. 1.4. Assume that the acceleration due to gravity has the constant value $g$ and that air resistance is negligible.

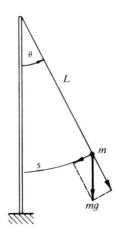

**Figure 1.4**

    (a) By equating $m\ddot{s}$ to the tangential component of force, derive the equation of motion

$$\ddot{\theta} + \frac{g}{L} \sin \theta = 0.$$

(b) To what physical situations do the constant solutions $\theta = n\pi$, $n$ an integer, of the equation in part (a) correspond?

(c) Suppose that the pendulum is released from rest at time $t = 0$ from the position $\theta = -\alpha$, where $0 < \alpha < \pi$. Show that on the first half-swing of the pendulum ($\theta$ increasing)

$$t = \left(\frac{L}{2g}\right)^{1/2} \int_{-\alpha}^{\theta} (\cos\theta - \cos\alpha)^{-1/2} \, d\theta.$$

(d) Show that the formula in part (c) can be written as

$$t = \left(\frac{L}{g}\right)^{1/2} \int_{-\pi/2}^{\psi} (1 - k^2 \sin^2 z)^{-1/2} \, dz,$$

where $k = \sin \alpha/2$ and $\psi = \sin^{-1}(k^{-1} \sin \theta/2)$. Suggestion: let $k \sin z = \sin \theta/2$.

(e) The function $F$, where

$$F(\phi, k) = \int_0^{\phi} (1 - k^2 \sin^2 z)^{-1/2} \, dz,$$

is called an elliptic integral of the first kind. It has been tabulated for various values of $\phi$ and $k$. (See, for example, Pierce [1929], or Jahnke and Emde [1945].) By using the result of part (d), show that the period of the pendulum is

$$4\left(\frac{L}{g}\right)^{1/2} F\left(\frac{\pi}{2}, k\right).$$

# II

## Matrices
## and Determinants

### 2.1 SYSTEMS OF LINEAR EQUATIONS

Let us begin by considering a linear system of two equations with two unknowns. Such a system is of the form

$$a_{11}x_1 + a_{12}x_2 = b_1,$$
$$a_{21}x_1 + a_{22}x_2 = b_2, \tag{2.1}$$

where $a_{11}$, $a_{12}$, $a_{21}$, $a_{22}$, $b_1$, and $b_2$ are given numbers.[1] A *solution* of the system (2.1) is an ordered pair of numbers $(x_1, x_2)$ that satisfies the system. For example, a solution of the system

$$2x_1 - 3x_2 = 5,$$
$$-x_1 + 2x_2 = -2, \tag{2.2}$$

is the ordered pair $(4, 1)$, because

$$2(4) - 3(1) = 5 \quad \text{and} \quad -(4) + 2(1) = -2.$$

Note, however, that the ordered pair $(1, 4)$ is *not* a solution.

The system (2.1) possesses a simple geometrical interpretation. Each equation of the system can be regarded as that of a straight line in a rectan-

---

[1] We assume for the moment that all numbers are real. This restriction will be removed shortly.

gular $x_1 x_2$ coordinate system. A solution of the system represents the coordinates of a point of intersection of the two lines. In general, two lines intersect in one point and there will be one, and only one, solution of the system (2.1). However, other possibilities are apparent. The two lines may be parallel and noncoincident. In this case, no solution of the system will exist. If the two lines are coincident, the system (2.1) will possess infinitely many solutions. In fact, the coordinates of every point on the common line will constitute a solution.

The system (2.1) involves the same number of equations as unknowns. But this need not always be the case. Let us consider a system of the form

$$\begin{aligned}
a_{11}x_1 + a_{12} x_2 &= b_1 , \\
a_{21}x_1 + a_{22} x_2 &= b_2 , \\
a_{31}x_1 + a_{32} x_2 &= b_3 , \\
a_{41}x_1 + a_{42} x_2 &= b_4 .
\end{aligned} \tag{2.3}$$

It is likely that the system (2.3) will have no solution, because four arbitrary lines are unlikely to pass through a common point. But if the four lines do have a common point a solution will exist. If all four lines are coincident, the system (2.3) will possess infinitely many solutions.

Real systems of equations that involve three unknowns, such as

$$\begin{aligned}
a_{11}x_1 + a_{12} x_2 + a_{13} x_3 &= b_1 , \\
a_{21}x_1 + a_{22} x_2 + a_{23} x_3 &= b_2 , \\
&\cdots\cdots\cdots\cdots\cdots\cdots \\
a_{m1}x_1 + a_{m2} x_2 + a_{m3} x_3 &= b_m ,
\end{aligned} \tag{2.4}$$

also permit a geometric interpretation. Each equation can be regarded as that of a plane in a three dimensional $x_1 x_2 x_3$ rectangular coordinate system. A solution of the system (2.4) is an ordered triple $(x_1, x_2, x_3)$ of numbers that satisfies the system. If the $m$ planes pass through a common point $P$, then the coordinates of $P$ constitute a solution of the system. A little reflection should convince the reader that the system (2.4) may have exactly one solution, no solution, or infinitely many solutions (the latter when all the planes pass through a line, or are coincident).

The most general linear system is of the form

$$\begin{aligned}
a_{11}x_1 + a_{12} x_2 + a_{13} x_3 + \cdots + a_{1n} x_n &= b_1 , \\
a_{21}x_1 + a_{22} x_2 + a_{23} x_3 + \cdots + a_{2n} x_n &= b_2 , \\
a_{31}x_1 + a_{32} x_2 + a_{33} x_3 + \cdots + a_{3n} x_n &= b_3 , \\
&\cdots\cdots\cdots\cdots\cdots\cdots\cdots\cdots \\
a_{m1}x_1 + a_{m2} x_2 + a_{m3} x_3 + \cdots + a_{mn} x_n &= b_m .
\end{aligned} \tag{2.5}$$

The system (2.5) involves $m$ equations and $n$ unknowns. The $mn$ numbers[2] $a_{ij}$, $1 \leq i \leq m$, $1 \leq j \leq n$, are called the *coefficients* of the system. The pattern of the subscripts should be noted carefully. The first subscript indicates the number of the equation to which the coefficient belongs and the second subscript indicates the number of the unknown by which the coefficient is multiplied. Thus all the coefficients $a_{2j}$, $1 \leq j \leq n$, belong to the second equation (or row). Each coefficient of the form $a_{i3}$, $1 \leq i \leq m$, is multiplied by $x_3$ (and belongs to the third column, so to speak, of the system (2.5)). The $m$ numbers $b_1, b_2, \ldots, b_m$ are called the *constants* of the system. When these numbers are all zero, the system is said to be a *homogeneous* system. Otherwise the system is said to be *nonhomogeneous*.

A solution of the system (2.5) is simply an ordered $n$-tuple of numbers[3] $(x_1, x_2, \ldots, x_n)$ that satisfies the system. If the system possesses at least one solution it is said to be *consistent*. If no solution exists, the system is said to be *inconsistent*. The set of all solutions of a system of equations is called the *general solution* or *complete solution* of the system. In regard to the number of solutions of the system (2.5), it turns out that there are three possibilities:

1. No solution exists.

2. Exactly one solution exists.

3. Infinitely many solutions exist.

Thus it can never happen that the system possesses exactly two or exactly 50 solutions. Either there is none, one, or infinitely many. In the cases $n = 2$ and $n = 3$, when all numbers are restricted to be real, these facts are fairly apparent from the geometric interpretation of the systems. We shall elaborate on the general case below. When infinitely many solutions exist, it is still possible to describe the "number" of solutions, in a sense to be explained in Chapter 3.

One important property of the systems of equations that we have been considering should not escape notice. Each equation of the system (2.5) is *linear*. That is, each equation is a *polynomial equation of first degree* in the unknowns $x_1, x_2, \ldots, x_n$. Systems of linear equations are called *linear systems*. It is possible to consider nonlinear systems, such as

$$x_1^2 x_2 - 2x_1^2 = 3,$$
$$2x_1 + 4x_2 = 5 \tag{2.6}$$

---

[2] We now allow all numbers to be complex. Although we shall present examples that involve only real numbers, all the theory in this chapter is valid when complex numbers are permitted, except where noted. We have made the restriction to real numbers until now in order to obtain geometric interpretations in the cases $n = 2$ and $n = 3$.

[3] Usually when the coefficients and constants of the system are real we are interested only in real solutions.

or

$$e^{x_1} + x_2 \sin x_1 = 0,$$
$$\cos(x_1 x_2) - x_2 = 3 ; \tag{2.7}$$

however, we shall not be concerned with such systems in this chapter.

Two systems of equations are said to be *equivalent* if they have exactly the same solutions. For example, the two systems

$$2x_1 - 3x_2 = 1, \qquad x_1 - 7x_2 = -5,$$
$$x_1 + 4x_2 = 6, \qquad 2x_1 + 8x_2 = 12, \tag{2.8}$$

are equivalent because each possesses the single solution (2, 1). Many of the standard procedures for solving a linear system involve the finding of an equivalent but simpler system that can be solved easily. Let us write our system as

$$f_1(x_1, x_2, \ldots, x_n) = b_1,$$
$$f_2(x_1, x_2, \ldots, x_n) = b_2,$$
$$\cdots\cdots\cdots\cdots\cdots\cdots$$
$$f_m(x_1, x_2, \ldots, x_n) = b_m, \tag{2.9}$$

where

$$f_i(x_1, x_2, \ldots, x_n) = a_{i1}x_1 + a_{i2}x_2 + \cdots + a_{in}x_n, \quad 1 \le i \le m.$$

In finding an equivalent system, three types of operations are used. These are

1.  Interchanging two equations.
2.  Multiplying through in an equation by a number that is not zero.
3.  Adding to one equation an equation that is formed by multiplying through in another equation by a number.

It is apparent that the first type of operation leads to an equivalent system. To justify the second type of operation, we merely observe that an ordered $n$-tuple $(x_1, x_2, \ldots, x_n)$ that satisfies either of the equations

$$f_i(x_1, x_2, \ldots, x_n) = b_i, \tag{2.10}$$

$$cf_i(x_1, x_2, \ldots, x_n) = cb_i, \tag{2.11}$$

must also satisfy the other if $c$ is a number other than zero. The restriction $c \ne 0$ is necessary; see Exercise 17 at the end of this section.

We now consider the third type of operation. Taking the $j$th and $k$th equations

$$f_j(x_1, x_2, \ldots, x_n) = b_j,$$
$$f_k(x_1, x_2, \ldots, x_n) = b_k, \tag{2.12}$$

of the system (2.9), let us add $c$ times the $j$th equation to the $k$th. The resulting pair of equations is

$$f_j(x_1, x_2, \ldots, x_n) = b_j,$$
$$f_k(x_1, x_2, \ldots, x_n) + cf_j(x_1, x_2, \ldots, x_n) = b_k + cb_j. \tag{2.13}$$

The reader can easily verify that any ordered $n$-tuple that satisfies the pair of equations (2.12) also satisfies the pair (2.13) and conversely. Hence the third type of operation always yields an equivalent system.

The method of solving linear systems that we shall now describe is known as the *Gauss reduction* or *Gaussian elimination* method. We consider some examples before examining the general case.

**Example 1**

$$x_1 - 2x_2 = 1,$$
$$-2x_1 + x_2 = 4, \tag{2.14}$$
$$x_1 + 3x_2 = -9.$$

We first eliminate $x_1$ from the second and third equations by adding appropriate multiples of the first equation to the others. Thus we add 2 times the first equation to the second and $(-1)$ times the first equation to the third. The result is the equivalent system

$$x_1 - 2x_2 = 1,$$
$$-3x_2 = 6, \tag{2.15}$$
$$5x_2 = -10.$$

We eliminate $x_2$ from the third equation by adding $\frac{5}{3}$ times the second equation to the third. The resulting system is

$$x_1 - 2x_2 = 1,$$
$$-3x_2 = 6, \tag{2.16}$$
$$0 = 0.$$

The second equation is satisfied if and only if $x_2 = -2$. With this choice for $x_2$, the first equation is satisfied if and only if $x_1 = -3$. Thus the ordered pair $(-3, -2)$ is a solution, and the only solution, of the system (2.16) and hence of the system (2.14).

**Example 2**

$$
\begin{aligned}
2x_1 - 4x_2 &= -4, \\
3x_1 - 4x_2 &= -2, \\
-4x_1 + 3x_2 &= 6.
\end{aligned}
\tag{2.17}
$$

We multiply through in the first equation by $\frac{1}{2}$. Then we eliminate $x_1$ from the remaining equations. The result is

$$
\begin{aligned}
x_1 - 2x_2 &= -2, \\
2x_2 &= 4, \\
-5x_2 &= -2.
\end{aligned}
\tag{2.18}
$$

Adding $\frac{5}{2}$ times the second equation to the third to eliminate $x_2$ from the second equation, we have

$$
\begin{aligned}
x_1 - 2x_2 &= -2, \\
2x_2 &= 4, \\
0 &= 8.
\end{aligned}
\tag{2.19}
$$

Evidently this system has no solution, since the last equation cannot be satisfied for any choice of $x_1$ and $x_2$.

**Example 3**

$$
\begin{aligned}
x_1 - x_2 + x_3 - 2x_4 &= -1, \\
2x_1 - 2x_2 + x_3 - 2x_4 &= -3, \\
-x_1 + x_2 - 2x_3 + 4x_4 &= 0, \\
-3x_1 + 3x_2 - x_3 + 2x_4 &= 5.
\end{aligned}
\tag{2.20}
$$

We first eliminate $x_1$ from all equations below the first equation, obtaining the system

$$
\begin{aligned}
x_1 - x_2 + x_3 - 2x_4 &= -1, \\
-x_3 + 2x_4 &= -1, \\
-x_3 + 2x_4 &= -1, \\
2x_3 - 4x_4 &= 2.
\end{aligned}
\tag{2.21}
$$

Every coefficient of $x_2$ is zero in the equations below the first, but there is a nonzero coefficient of $x_3$. We place the unknown $x_3$ in the second position, so that we have

$$
\begin{aligned}
x_1 + x_3 - 2x_4 - x_2 &= -1, \\
-x_3 + 2x_4 &= -1, \\
-x_3 + 2x_4 &= -1, \\
2x_3 - 4x_4 &= 2.
\end{aligned}
\tag{2.22}
$$

We use the second equation to eliminate $x_3$ in all equations below the second. The resulting system is

$$
\begin{aligned}
x_1 + x_3 - 2x_4 - x_2 &= -1, \\
-x_3 + 2x_4 &= -1, \\
0 &= 0, \\
0 &= 0.
\end{aligned}
\tag{2.23}
$$

If we assign any values whatsoever to the unknowns $x_2$ and $x_4$, the remaining unknowns $x_1$ and $x_3$ are completely determined by the Eqs. (2.23). Setting $x_2 = c_1$ and $x_4 = c_2$, where $c_1$ and $c_2$ are arbitrary constants, we see that $x_3 = 2c_2 + 1$ and $x_1 = c_1 + 2c_2 - 1 - (2c_2 + 1) = c_1 - 2$. Hence the general solution is described by the formulas

$$
\begin{aligned}
x_1 &= c_1 - 2, \\
x_2 &= c_1, \\
x_3 &= 2c_2 + 1, \\
x_4 &= c_2,
\end{aligned}
\tag{2.24}
$$

or by the ordered 4-tuple

$$
(c_1 - 2, c_1, 2c_2 + 1, c_2).
$$

We now describe the general case, without going into the details of the derivation. It can be shown that the system (2.5) is equivalent to a system of the form

$$
\begin{aligned}
\tilde{a}_{11}x_{i_1} + \tilde{a}_{12}x_{i_2} + \cdots + \tilde{a}_{1r}x_{i_r} + \cdots + \tilde{a}_{1n}x_{i_n} &= \tilde{b}_1, \\
\tilde{a}_{22}x_{i_2} + \cdots + \tilde{a}_{2r}x_{i_r} + \cdots + \tilde{a}_{2n}x_{i_n} &= \tilde{b}_2, \\
\cdots\cdots\cdots\cdots\cdots\cdots\cdots\cdots\cdots\cdots\cdots\cdots\cdots\cdots\cdots\cdots \\
\tilde{a}_{rr}x_{i_r} + \cdots + \tilde{a}_{rn}x_{i_n} &= \tilde{b}_r, \\
0 &= \tilde{b}_{r+1}, \\
\cdots \\
0 &= \tilde{b}_m,
\end{aligned}
\tag{2.25}
$$

where none of $\tilde{a}_{11}, \tilde{a}_{22}, \ldots, \tilde{a}_{rr}$ is zero. The order of the unknowns may be changed. In Example 3, for instance, we see from Eqs. (2.23) that

$$x_{i_1} = x_1, \qquad x_{i_2} = x_3, \qquad x_{i_3} = x_4, \qquad x_{i_4} = x_2.$$

If $m > r$, the system (2.25) possesses no solution unless $\tilde{b}_{r+1}, \tilde{b}_{r+2}, \ldots, \tilde{b}_m$ are all zero. The system is consistent if $m = r$, or if $m > r$ and $\tilde{b}_{r+1}, \tilde{b}_{r+2}, \ldots, \tilde{b}_m$ are all zero. If the system is consistent and $r = n$, there is exactly one solution. If the system is consistent and $r < n$, the $n - r$ unknowns $x_{i_{r+1}}, x_{i_{r+2}}, \ldots, x_{i_n}$ can be assigned arbitrary values and the unknowns $x_{i_1}, x_{i_2}, \ldots, x_{i_r}$ are then completely determined. In this case we say that the system has an $n - r$ parameter family of solutions.

The case of a homogeneous system ($b_1 = b_2 = \cdots = b_m = 0$) is of particular interest. Such a system always has the solution

$$x_1 = x_2 = \cdots = x_n = 0.$$

This solution is called the *trivial solution* of the system. A homogeneous system can be reduced to a system of the form (2.25) in which

$$\tilde{b}_1 = \tilde{b}_2 = \cdots = \tilde{b}_m = 0.$$

If $r = n$, the only solution is the trivial solution. However, if $r < n$, there are infinitely many nontrivial solutions because we can assign arbitrary values to $n - r$ of the unknowns. In particular, *when there are fewer equations than unknowns* ($m < n$), *a homogeneous system always possesses nontrivial solutions* (because $r \leq m < n$). We now present some examples of homogeneous systems.

**Example 4**

$$
\begin{aligned}
x_1 - x_2 &= 0, \\
2x_1 - 3x_2 + x_3 &= 0, \\
x_1 + x_2 - 5x_3 &= 0.
\end{aligned}
$$

We subtract the first equation from the third and two times the first equation from the second (eliminating $x_1$) to obtain the system

$$
\begin{aligned}
x_1 - x_2 &= 0, \\
-x_2 + x_3 &= 0, \\
2x_2 - 5x_3 &= 0.
\end{aligned}
$$

In this system we add twice the second equation to the third. The result is

$$
\begin{aligned}
x_1 - x_2 \quad\;\; &= 0, \\
-x_2 + x_3 &= 0, \\
-3x_3 &= 0.
\end{aligned}
$$

Thus $m = n = r = 3$ for our system. From the last set of equations, we see from the third equation that $x_3 = 0$, from the second equation that $x_2 = 0$, and from the first equation that $x_1 = 0$. Thus the only solution is the trivial solution.

**Example 5**

$$
\begin{aligned}
x_1 - \quad x_2 - 2x_3 + x_4 &= 0, \\
-3x_1 + 3x_2 + \quad x_3 - x_4 &= 0, \\
2x_1 - 2x_2 + \quad x_3 \quad\quad &= 0.
\end{aligned}
$$

Here $m = 3$ and $n = 4$. Since there are more unknowns than equations, the existence of nontrivial solutions is guaranteed. Using the first equation to eliminate $x_1$ from the remaining two equations, we have

$$
\begin{aligned}
x_1 - x_2 - 2x_3 + \quad x_4 &= 0, \\
-5x_3 + 2x_4 &= 0, \\
5x_3 - 2x_4 &= 0.
\end{aligned}
$$

We add the second equation to the third, and then write the resulting system as

$$
\begin{aligned}
x_1 - 2x_3 - x_2 + \quad x_4 &= 0, \\
-5x_3 \quad\quad + 2x_4 &= 0, \\
0 &= 0.
\end{aligned}
$$

The unknowns $x_2$ and $x_4$ may be assigned arbitrary values. Then $x_1$ and $x_3$ may be expressed in terms of these. Setting

$$
x_2 = c_1, \qquad x_4 = c_2,
$$

we find that

$$
x_3 = \tfrac{2}{5}c_2, \qquad x_1 = c_1 - \tfrac{1}{5}c_2.
$$

If we replace $c_2$ by $5c_2$ we obtain the simpler form

$$x_1 = c_1 - c_2,$$
$$x_2 = c_1,$$
$$x_3 = 2c_2,$$
$$x_4 = 5c_2.$$

## Exercises for Section 2.1

In Exercises 1–16, find all real solutions of the system, using the Gauss reduction method.

**1.** $x_1 + 2x_2 = 3$
$3x_1 - x_2 = -5$

**2.** $-2x_1 + 6x_2 = -8$
$x_1 - 3x_2 = 4$

**3.** $-2x_1 + x_2 = 5$
$4x_1 - 2x_2 = -1$

**4.** $2x_1 + 8x_2 = 14$
$x_1 - 3x_2 = 0$
$4x_1 + 2x_2 = 14$

**5.** $3x_1 - x_2 = 4$
$6x_1 - 2x_2 = 8$
$-9x_1 + 3x_2 = -12$

**6.** $x_1 - 2x_2 + x_3 = 5$
$-x_1 + x_2 - 4x_3 = -7$
$3x_1 + 3x_2 + x_3 = 4$

**7.** $2x_1 - x_2 + x_3 = 5$
$x_1 - x_2 - x_3 = 4$
$-2x_1 + 2x_2 + x_3 = -6$

**8.** $3x_1 - x_2 + x_3 = 0$
$x_1 - x_2 - x_3 = 0$
$x_1 + x_2 + x_3 = 0$

**9.** $x_1 + 2x_2 - x_3 = -3$
$3x_1 - x_2 - 2x_3 = 13$
$x_1 - 5x_2 = 19$

**10.** $-2x_1 + 2x_2 - 2x_3 = -8$
$x_1 - x_2 + x_3 = 4$
$2x_1 - 2x_2 + 2x_3 = 8$

**11.** $x_1 - x_2 - 3x_3 = 0$
$x_1 + x_2 + x_3 = 0$
$2x_1 + 2x_2 + x_3 = 0$

**12.** $x_1 + x_2 - x_3 = 5$
$2x_1 + x_3 = -2$
$x_1 - x_2 + 2x_3 = 0$

**13.** $2x_1 - x_2 + x_3 = -1$
$x_1 + x_2 + x_3 = 3$

**14.** $x_1 + x_2 - x_3 + x_4 = 4$
$2x_1 + x_2 - x_3 + x_4 = 5$
$x_2 + x_3 = -1$
$x_2 + 2x_4 = 4$

**15.**
$$2x_1 + x_2 - 2x_3 \qquad = 0$$
$$x_1 + 2x_2 - x_3 + x_4 = 0$$
$$x_2 \qquad + x_4 = 0$$
$$x_1 \qquad - x_3 \qquad = 0$$

**16.**
$$x_1 + 2x_2 - x_3 \qquad = 1$$
$$2x_1 \qquad + 2x_3 + x_4 = 4$$
$$x_1 - 2x_2 + 3x_3 + x_4 = 3$$
$$3x_1 - 2x_2 + 5x_3 + 2x_4 = 7$$

**17.** Suppose that the first equation of the system $2x_1 - x_2 = -1$, $x_1 + x_2 = 7$, is multiplied through by zero, yielding the system $0x_1 + 0x_2 = 0$, $x_1 + x_2 = 7$. Show that the two systems are not equivalent.

**18.** Show that the system of equations $x_1 - x_2 = -7$, $x_1^2 + x_2 = 9$, possesses the two solutions $(1, 8)$ and $(-2, 5)$, and no others. Does this fact contradict the assertions made about systems of the form $(2.5)$? Explain.

## 2.2  MATRICES AND VECTORS

The coefficients of a linear system

$$a_{11}x_1 + a_{12}x_2 + \cdots + a_{1n}x_n = b_1 ,$$
$$a_{21}x_1 + a_{22}x_2 + \cdots + a_{2n}x_n = b_2 ,$$
$$\cdots\cdots\cdots\cdots\cdots\cdots\cdots\cdots\cdots\cdots\cdots \qquad (2.26)$$
$$a_{m1}x_1 + a_{m2}x_2 + \cdots + a_{mn}x_n = b_m ,$$

can be displayed in a rectangular array of the form

$$\begin{bmatrix} a_{11} & a_{12} & \cdots & a_{1n} \\ a_{21} & a_{22} & \cdots & a_{2n} \\ \vdots & \vdots & & \vdots \\ a_{m1} & a_{m2} & \cdots & a_{mn} \end{bmatrix} \qquad (2.27)$$

with $m$ rows and $n$ columns. Such an array is called a *matrix of size* $m \times n$. To be more precise, a matrix of size $m \times n$ is a function of two variables whose domain is the set of ordered pairs of integers $(i, j)$, where $1 \le i \le m$, $1 \le j \le n$. For example, in the above arrangement $a_{23}$ is the value of the function that corresponds to the ordered pair $(2, 3)$. The numbers $a_{ij}$ are called the *elements* of the matrix. A matrix with the same number of rows as columns is called a *square matrix*. A square $n \times n$ matrix is said to be a matrix of *order* $n$. With the exceptions noted later in this section, we shall denote matrices by capital letters $A$, $B$, $C$, and so on. Usually the elements of a matrix will be denoted by corresponding lower case letters with subscripts. Thus $b_{ij}$ will be the element in the $i$th row and $j$th column of a matrix $B$.

The matrix (2.27) is called the *coefficient matrix* of the linear system (2.26).

Two matrices $A$ and $B$ are said to be *equal*, written $A = B$, if they are of the same size and have equal corresponding elements. Thus two matrices $A$ and $B$ of size $m \times n$ are equal if $a_{ij} = b_{ij}$ for $1 \leq i \leq m$, $1 \leq j \leq n$.

If $A$ is a matrix and $c$ a number, we define the *product* $cA = Ac$ to be the matrix obtained from $A$ by multiplying each element of $A$ by $c$. Thus if

$$A = \begin{bmatrix} 3 & -1 & 1 \\ 4 & 2 & -3 \end{bmatrix},$$

then

$$2A = \begin{bmatrix} 6 & -2 & 2 \\ 8 & 4 & -6 \end{bmatrix}, \qquad -A = (-1)A = \begin{bmatrix} -3 & 1 & -1 \\ -4 & -2 & 3 \end{bmatrix}.$$

If $A$ and $B$ are matrices of the same size, we define the *sum* of $A$ and $B$, written $A + B$, to be the matrix of the same size as $A$ and $B$ whose elements are the sums of the corresponding elements of $A$ and $B$. Thus if $A$ and $B$ are $m \times n$ matrices and $C = A + B$, then $C$ is an $m \times n$ matrix with elements $c_{ij} = a_{ij} + b_{ij}$. For example, if

$$A = \begin{bmatrix} 1 & 2 & 3 \\ 4 & 5 & 6 \end{bmatrix}, \qquad B = \begin{bmatrix} 3 & -7 & 1 \\ 0 & 1 & -8 \end{bmatrix},$$

then

$$A + B = \begin{bmatrix} 4 & -5 & 4 \\ 4 & 6 & -2 \end{bmatrix}.$$

Note that the sum of matrices of different sizes is not defined. We define *subtraction* of matrices of the same size by means of the formula

$$A - B = A + (-1)B.$$

We state a number of easily verifiable properties as a theorem. The proofs are left as an exercise at the end of this section.

**Theorem 2.1** If $A$, $B$, and $C$ are matrices of the same size and if $a$, $b$, and $c$ are numbers, then

$$B + A = A + B,$$
$$(A + B) + C = A + (B + C),$$
$$a(bA) = (ab)A = (ba)A = b(aA),$$
$$(a + b)A = aA + bA,$$
$$c(A + B) = cA + cB.$$

A matrix with one column and $n$ rows, such as

$$\begin{bmatrix} x_1 \\ x_2 \\ \vdots \\ x_n \end{bmatrix} \tag{2.28}$$

is also called an $n$-dimensional *column vector*. A matrix with one row and $n$ columns, such as

$$[y_1 \quad y_2 \quad \cdots \quad y_n] \tag{2.29}$$

is called an $n$-dimensional *row vector*. The elements of a row or column vector are sometimes called the *components* of the vector. We shall designate row and column vectors by lower case letters in bold face type. Thus we might write

$$\mathbf{u} = \begin{bmatrix} u_1 \\ u_2 \\ u_3 \end{bmatrix}, \qquad \mathbf{v} = [v_1 \quad v_2 \quad v_3]. \tag{2.30}$$

Associated with the $m \times n$ matrix (2.27) are the $n$ column vectors

$$\begin{bmatrix} a_{11} \\ a_{21} \\ \vdots \\ a_{m1} \end{bmatrix}, \quad \begin{bmatrix} a_{12} \\ a_{22} \\ \vdots \\ a_{m2} \end{bmatrix}, \dots, \quad \begin{bmatrix} a_{1n} \\ a_{2n} \\ \vdots \\ a_{mn} \end{bmatrix},$$

called the *column vectors of the matrix*. The $m$ row vectors

$$[a_{11} \quad a_{12} \cdots a_{1n}], [a_{21} \quad a_{22} \cdots a_{2n}], \dots, [a_{m1} \quad a_{m2} \cdots a_{mn}]$$

are called the *row vectors of the matrix*.

If $\mathbf{u}_1, \mathbf{u}_2, \dots, \mathbf{u}_n$ are each $m$-dimensional column vectors, we write

$$[\mathbf{u}_1, \mathbf{u}_2, \dots, \mathbf{u}_n]$$

to denote the $m \times n$ matrix whose column vectors are $\mathbf{u}_1, \mathbf{u}_2, \dots, \mathbf{u}_n$. If $\mathbf{v}_1, \mathbf{v}_2, \dots, \mathbf{v}_m$ are $n$-dimensional row vectors, we write

$$\begin{bmatrix} \mathbf{v}_1 \\ \mathbf{v}_2 \\ \vdots \\ \mathbf{v}_m \end{bmatrix}$$

for the $m \times n$ matrix whose row vectors are $v_1, v_2, \ldots, v_m$. For example, if

$$u_1 = \begin{bmatrix} 1 \\ 0 \end{bmatrix}, \qquad u_2 = \begin{bmatrix} -2 \\ 3 \end{bmatrix}, \qquad u_3 = \begin{bmatrix} 3 \\ 1 \end{bmatrix},$$

then

$$[u_1, u_2, u_3] = \begin{bmatrix} 1 & -2 & 3 \\ 0 & 3 & 1 \end{bmatrix}.$$

We denote by $R^n$ the set of all ordered $n$-tuples of real numbers. Thus $R^1$ is the set of all real numbers, and $R^2$ is the set of all ordered pairs $(a, b)$ of real numbers $a$ and $b$. A particular ordered $n$-tuple $(x_1, x_2, \ldots, x_n)$ is called a *vector* in $R^n$, or an *n-dimensional vector*. We also denote $n$-dimensional vectors by bold face symbols. If

$$x = (x_1, x_2, \ldots, x_n), \qquad y = (y_1, y_2, \ldots, y_n)$$

are $n$-dimensional vectors, we define the *product* of a real number $c$ and the vector x to be

$$cx = xc = (cx_1, cx_2, \ldots, cx_n). \tag{2.31}$$

We also define the *sum*

$$x + y = (x_1 + y_1, x_2 + y_2, \ldots, x_n + y_n). \tag{2.32}$$

The particular vector $(0, 0, \ldots, 0)$ is called the $n$-dimensional *zero vector* and is denoted by 0.

Associated with a vector in $R^n$ are a column vector and a row vector. We write

$$\begin{bmatrix} x_1 \\ x_2 \\ \vdots \\ x_n \end{bmatrix} \leftrightarrow (x_1, x_2, \ldots, x_n) \leftrightarrow [x_1 \quad x_2 \cdots x_n]$$

to indicate the correspondence. This correspondence is perserved under addition and multiplication by a number. To see this for column vectors, we observe that

$$c\begin{bmatrix} x_1 \\ x_2 \\ \vdots \\ x_n \end{bmatrix} = \begin{bmatrix} cx_1 \\ cx_2 \\ \vdots \\ cx_n \end{bmatrix} \leftrightarrow (cx_1, cx_2, \ldots cx_n) = c(x_1, x_2, \ldots, x_n)$$

and

$$\begin{bmatrix} x_1 \\ x_2 \\ \vdots \\ x_n \end{bmatrix} + \begin{bmatrix} y_1 \\ y_2 \\ \vdots \\ y_n \end{bmatrix} = \begin{bmatrix} x_1 + y_1 \\ x_2 + y_2 \\ \vdots \\ x_n + y_n \end{bmatrix} \leftrightarrow (x_1 + y_1, x_2 + y_2, \dots, x_n + y_n)$$

$$= (x_1, x_2, \dots, x_n) + (y_1, y_2, \dots, y_n).$$

For convenience, we shall sometimes refer to the column (or row) vector $\mathbf{u} = (u_1, u_2, \dots, u_n)$, when actually we mean the column (or row) vector associated with the *n*-tuple $(u_1, u_2, \dots, u_n)$.

Associated with a vector $(x_1, x_2, x_3)$ in $R^3$ is the geometric vector

$$x_1 \mathbf{i} + x_2 \mathbf{j} + x_3 \mathbf{k},$$

where $x_1, x_2$, and $x_3$ are the components with respect to some rectangular coordinate system. It is natural to define the *inner product* $\mathbf{x} \cdot \mathbf{y}$ of two vectors $\mathbf{x} = (x_1, x_2, x_3)$ and $\mathbf{y} = (y_1, y_2, y_3)$ in $R^3$ as

$$\mathbf{x} \cdot \mathbf{y} = x_1 y_1 + x_2 y_2 + x_3 y_3 .$$

The *length* $\|\mathbf{x}\|$ of $\mathbf{x}$ is defined as

$$\|\mathbf{x}\| = (x_1^2 + x_2^2 + x_3^2)^{1/2} = (\mathbf{x} \cdot \mathbf{x})^{1/2}.$$

For *n*-dimensional vectors $\mathbf{x} = (x_1, x_2, \dots, x_n)$ and $\mathbf{y} = (y_1, y_2, \dots, y_n)$ we define the *inner product* $\mathbf{x} \cdot \mathbf{y}$ of $\mathbf{x}$ and $\mathbf{y}$ as

$$\mathbf{x} \cdot \mathbf{y} = x_1 y_1 + x_2 y_2 + \cdots + x_n y_n.$$

The *length* $\|\mathbf{x}\|$ of $\mathbf{x}$ is defined as

$$\|\mathbf{x}\| = (\mathbf{x} \cdot \mathbf{x})^{1/2} = (x_1^2 + x_2^2 + \cdots + x_n^2)^{1/2}.$$

When $n = 2$ or $n = 3$, the cosine of the angle $\theta$ between the geometric vectors associated with $\mathbf{x}$ and $\mathbf{y}$ is given by the formula

$$\cos \theta = \frac{\mathbf{x} \cdot \mathbf{y}}{\|\mathbf{x}\| \, \|\mathbf{y}\|}.$$

For $n > 3$, there is no longer a geometrical interpretation of inner product and length. Nevertheless, the above definitions are useful. We leave it to the reader (Exercise 14) to verify the following properties of the inner product for $R^n$:

$$\mathbf{x} \cdot \mathbf{x} \geq 0 \qquad (2.33a)$$

and $\mathbf{x} \cdot \mathbf{x} = 0$ if and only if $\mathbf{x} = \mathbf{0}$;

$$\mathbf{x} \cdot \mathbf{y} = \mathbf{y} \cdot \mathbf{x}, \qquad (2.33b)$$

$$(c\mathbf{x}) \cdot \mathbf{y} = \mathbf{x} \cdot (c\mathbf{y}), \qquad (2.33c)$$

and

$$(c\mathbf{x}) \cdot \mathbf{y} = c(\mathbf{x} \cdot \mathbf{y}) \qquad (2.33d)$$

for every real number $c$; and

$$(\mathbf{x} + \mathbf{y}) \cdot \mathbf{z} = \mathbf{x} \cdot \mathbf{z} + \mathbf{y} \cdot \mathbf{z}. \qquad (2.33e)$$

Let us denote the set of all ordered $n$-tuples of complex numbers by $C^n$. If

$$\mathbf{u} = (u_1, u_2, \ldots, u_n), \qquad \mathbf{v} = (v_1, v_2, \ldots, v_n)$$

are elements of $C^n$, we define

$$c\mathbf{u} = (cu_1, cu_2, \ldots, cu_n)$$

for every complex number $c$ and

$$\mathbf{u} + \mathbf{v} = (u_1 + v_1, u_2 + v_2, \ldots, u_n + v_n).$$

We define the *inner product* $\mathbf{u} \cdot \mathbf{v}$ of $\mathbf{u}$ and $\mathbf{v}$ as

$$\mathbf{u} \cdot \mathbf{v} = u_1 \bar{v}_1 + u_2 \bar{v}_2 + \cdots + u_n \bar{v}_n,$$

where the bar indicates the complex conjugate. The *length* $\|\mathbf{u}\|$ of $\mathbf{u}$ is defined to be

$$\|\mathbf{u}\| = (\mathbf{u} \cdot \mathbf{u})^{1/2} = (|u_1|^2 + |u_2|^2 + \cdots + |u_n|^2)^{1/2}.$$

Note that $\|\mathbf{u}\|$ is a nonnegative real number. The reader may verify the following properties:

$$\mathbf{u} \cdot \mathbf{u} \geq 0 \qquad (2.34a)$$

and $\mathbf{u} \cdot \mathbf{u} = 0$ if and only if $\mathbf{u} = \mathbf{0}$;

$$\mathbf{u} \cdot \mathbf{v} = \overline{\mathbf{v} \cdot \mathbf{u}}, \tag{2.34b}$$

$$(c\mathbf{u}) \cdot \mathbf{v} = \mathbf{u} \cdot (\bar{c}\mathbf{v}), \tag{2.34c}$$

and

$$(c\mathbf{u}) \cdot \mathbf{v} = c(\mathbf{u} \cdot \mathbf{v}) \tag{2.34d}$$

for every complex number $c$; and

$$(\mathbf{u} + \mathbf{v}) \cdot \mathbf{w} = \mathbf{u} \cdot \mathbf{w} + \mathbf{v} \cdot \mathbf{w}. \tag{2.34e}$$

Two vectors $\mathbf{x}$ and $\mathbf{y}$ (in $R^n$ or in $C^n$) are said to be *orthogonal* if

$$\mathbf{x} \cdot \mathbf{y} = 0.$$

A set of nonzero vectors $\mathbf{x}_1, \mathbf{x}_2, \ldots, \mathbf{x}_m$ is called an *orthogonal set* of vectors if $\mathbf{x}_i \cdot \mathbf{x}_j = 0$, whenever $i \neq j$. For example, the vectors

$$\mathbf{x}_1 = (1, -1, 2, 0), \qquad \mathbf{x}_2 = (3, 1, -1, 0),$$
$$\mathbf{x}_3 = (0, 0, 0, 1), \qquad \mathbf{x}_4 = (-1, 7, 4, 0)$$

constitute an orthogonal set in $R^4$.

A vector $\mathbf{x}$ such that $\|\mathbf{x}\| = 1$ is called a *unit vector*. An orthogonal set of unit vectors is called an *orthonormal set* of vectors. If $\mathbf{u}$ is any nonzero vector, then $\mathbf{v} = \mathbf{u}/\|\mathbf{u}\|$ is a unit vector. To see this, we observe that

$$\|\mathbf{v}\|^2 = \mathbf{v} \cdot \mathbf{v} = \left(\frac{1}{\|\mathbf{u}\|}\mathbf{u}\right) \cdot \left(\frac{1}{\|\mathbf{u}\|}\mathbf{u}\right) = 1.$$

Thus $\mathbf{v}$ has unit length. If

$$\mathbf{y}_1 = \frac{1}{\sqrt{6}}\mathbf{x}_1, \qquad \mathbf{y}_2 = \frac{1}{\sqrt{11}}\mathbf{x}_2, \qquad \mathbf{y}_3 = \mathbf{x}_3, \qquad \mathbf{y}_4 = \frac{1}{\sqrt{66}}\mathbf{x}_4,$$

where the $\mathbf{x}_i$ are as in the preceding example, then $\mathbf{y}_1, \mathbf{y}_2, \mathbf{y}_3, \mathbf{y}_4$ form an orthonormal set.

## Exercises for Section 2.2

1. Find the coefficient matrix of the given linear system.

   (a) $\begin{aligned} 2x_1 - x_2 &= 4 \\ -x_1 + 3x_2 &= 0 \end{aligned}$

   (b) $\begin{aligned} x_1 + 2x_2 - x_3 &= 0 \\ 3x_1 - 3x_2 &= 1 \end{aligned}$

   (c) $\begin{aligned} x_2 + 2x_3 &= 4 \\ x_1 + x_2 + x_3 &= 0 \\ 2x_1 - x_3 &= -1 \end{aligned}$

   (d) $\begin{aligned} x_1 - x_2 &= 0 \\ 3x_1 + 2x_2 &= 5 \\ -x_1 + x_2 &= -1 \\ -2x_1 + 2x_2 &= -2 \end{aligned}$

2. Find the linear *homogeneous* system of equations that has the given matrix as its coefficient matrix.

   (a) $\begin{bmatrix} 3 & -5 \\ -1 & 2 \end{bmatrix}$

   (b) $\begin{bmatrix} 2 & 0 & 1 \\ 0 & 1 & -1 \end{bmatrix}$

   (c) $\begin{bmatrix} 1 & 1 & 0 \\ 0 & 5 & -1 \\ 0 & 4 & 2 \end{bmatrix}$

   (d) $\begin{bmatrix} 4 & -2 \\ -2 & 1 \\ -6 & 3 \\ 2 & -1 \end{bmatrix}$

3. If the given matrix is denoted by $A$, find $3A$, $-A$, $-2A$, and $0A$.

   (a) $\begin{bmatrix} 2 & -5 \\ 1 & 0 \end{bmatrix}$

   (b) $\begin{bmatrix} -1 & 2 & 0 \\ 3 & 0 & 1 \end{bmatrix}$

4. Given the matrices $A$ and $B$ below, find (a) $A + B$, (b) $A - B$, (c) $2A - 3B$.

$$A = \begin{bmatrix} 1 & 0 & 2 \\ -3 & 1 & 4 \end{bmatrix}, \qquad B = \begin{bmatrix} -2 & 2 & 4 \\ 1 & 1 & 3 \end{bmatrix}$$

5. Find the sum of any two of the matrices below for which the sum is defined.

$$A = \begin{bmatrix} 2 & 0 \\ 1 & 4 \\ 1 & 1 \end{bmatrix}, \qquad B = \begin{bmatrix} -2 & 1 & 0 \\ 3 & 0 & 0 \\ 0 & 0 & 0 \end{bmatrix}$$

$$C = \begin{bmatrix} 3 & 4 \\ 5 & 6 \end{bmatrix}, \qquad D = \begin{bmatrix} 1 & 2 \\ 3 & 4 \end{bmatrix}$$

6. Are any two of the following matrices equal?

$$A = \begin{bmatrix} 0 & 0 \\ 0 & 0 \end{bmatrix}, \qquad B = \begin{bmatrix} 0 & 0 \\ 0 & 0 \\ 0 & 0 \end{bmatrix}$$

$$C = \begin{bmatrix} 2 & 2 \\ 2 & 2 \end{bmatrix}, \qquad D = \begin{bmatrix} 2 & 2 & 2 \\ 2 & 2 & 2 \end{bmatrix}$$

7.  Prove Theorem 2.1.

8.  Find the column vectors of the given matrix.

    (a) $\begin{bmatrix} 2 & 1 & -3 & 0 \\ 5 & 0 & 0 & 2 \\ 0 & 1 & 4 & 0 \end{bmatrix}$,    (b) $\begin{bmatrix} 6 & -1 \\ 3 & 0 \\ 2 & 2 \end{bmatrix}$

9.  Find the row vectors for each of the matrices of Exercise 8.

10. Display the elements of the matrix $A = [\mathbf{u}, \mathbf{v}, \mathbf{w}]$, where $\mathbf{u}$, $\mathbf{v}$, and $\mathbf{w}$ are the column vectors $\mathbf{u} = (2, -1, 3, 0)$, $\mathbf{v} = (0, 0, 0, 2)$, $\mathbf{w} = (1, 4, 0, 0)$.

11. Display the elements of the matrix whose row vectors (in order) are $[0\,2\,-1\,0]$, $[1\,1\,0\,4]$ and $[0\,0\,1\,0]$.

12. If $\mathbf{u}$ and $\mathbf{v}$ are as given, find $\mathbf{u} \cdot \mathbf{v}$, $\|\mathbf{u}\|$, and $\|\mathbf{v}\|$.

    (a)  $\mathbf{u} = (1, -2, 0)$,        $\mathbf{v} = (2, 5, -4)$

    (b)  $\mathbf{u} = (1, 1, -2, 0)$,     $\mathbf{v} = (0, 2, -2, 1)$

    (c)  $\mathbf{u} = (2, 0, 1, -3)$,     $\mathbf{v} = (0, 0, -1, 5)$

13. For $\mathbf{u}$ and $\mathbf{v}$ as given, find $\mathbf{u} \cdot \mathbf{v}$, $\mathbf{v} \cdot \mathbf{u}$, $\|\mathbf{u}\|$, and $\|\mathbf{v}\|$.

    (a)  $\mathbf{u} = (2 - i, 1)$        $\mathbf{v} = (-i, -2i)$

    (b)  $\mathbf{u} = (1, i, 2 + i)$,     $\mathbf{v} = (1 - 3i, 0, 1 + i)$

14. Verify the properties (2.33) of the inner product on $R^n$.

15. Verify the properties (2.34) of the inner product on $C^n$.

16. (a)  If $\mathbf{x}$ and $\mathbf{y}$ are in $R^n$, verify that
    $$\|\mathbf{x} + \mathbf{y}\|^2 = \|\mathbf{x}\|^2 + 2\mathbf{x} \cdot \mathbf{y} + \|\mathbf{y}\|^2.$$

    (b)  If $\mathbf{u}$ and $\mathbf{v}$ are in $C^n$, verify that
    $$\|\mathbf{u} + \mathbf{v}\|^2 = \|\mathbf{u}\|^2 + 2\,\mathrm{Re}(\mathbf{u} \cdot \mathbf{v}) + \|\mathbf{u}\|^2.$$

    (Here Re denotes the real part.)

## 2.3  MATRIX MULTIPLICATION

Let us again consider the system of equations

$$a_{11}x_1 + a_{12}x_2 + \cdots + a_{1n}x_n = b_1,$$
$$\cdots\cdots\cdots\cdots\cdots\cdots\cdots\cdots$$
$$a_{i1}x_1 + a_{i2}x_2 + \cdots + a_{in}x_n = b_i, \qquad (2.35)$$
$$\cdots\cdots\cdots\cdots\cdots\cdots\cdots\cdots$$
$$a_{m1}x_1 + a_{m2}x_2 + \cdots + a_{mn}x_n = b_m.$$

We form the matrices

$$A = \begin{bmatrix} a_{11} & a_{12} & \cdots & a_{1n} \\ \cdots\cdots\cdots\cdots\cdots \\ a_{i1} & a_{i2} & \cdots & a_{in} \\ \cdots\cdots\cdots\cdots\cdots \\ a_{m1} & a_{m2} & \cdots & a_{mn} \end{bmatrix}, \qquad \mathbf{x} = \begin{bmatrix} x_1 \\ x_2 \\ \vdots \\ x_n \end{bmatrix}, \qquad \mathbf{b} = \begin{bmatrix} b_1 \\ b_2 \\ \vdots \\ b_m \end{bmatrix}. \qquad (2.36)$$

Notice that the left-hand member of the *i*th equation of the system (2.35) is formed by taking the products of the *n* elements in the *i*th row of *A* with the corresponding components of the column vector **x** and summing. This sum can be written as

$$\sum_{k=1}^{n} a_{ik} x_k .$$

The number of elements in each row of *A* is the same as the number of com-ponents of **x**. We now define the *product* $A\mathbf{x}$ of an $m \times n$ matrix *A* and an *n*-dimensional column vector **x** to be the *m*-dimensional column vector with components

$$\sum_{k=1}^{n} a_{ik} x_k, \qquad 1 \le i \le m. \qquad (2.37)$$

Then the system (2.35) can be written as

$$A\mathbf{x} = \mathbf{b}. \qquad (2.38)$$

Equation (2.38) asserts the equality of two *m*-dimensional column vectors. It is satisfied if and only if each equation of the system (2.35) is satisfied.

As an example, let

$$A = \begin{bmatrix} 2 & -1 & 3 \\ 1 & 4 & -2 \end{bmatrix}, \qquad \mathbf{x} = \begin{bmatrix} x_1 \\ x_2 \\ x_3 \end{bmatrix}, \qquad \mathbf{b} = \begin{bmatrix} 0 \\ 8 \end{bmatrix}.$$

Since the number of elements in each row of *A* is the same as the number of components of **x**, the product $A\mathbf{x}$ is defined and

$$A\mathbf{x} = \begin{bmatrix} 2 & -1 & 3 \\ 1 & 4 & -2 \end{bmatrix} \begin{bmatrix} x_1 \\ x_2 \\ x_3 \end{bmatrix} = \begin{bmatrix} 2x_1 - x_2 + 3x_3 \\ x_1 + 4x_2 - 2x_3 \end{bmatrix}.$$

The vector equation $A\mathbf{x} = \mathbf{b}$ corresponds to the system

$$2x_1 - x_2 + 3x_3 = 0,$$
$$x_1 + 4x_2 - 2x_3 = 8.$$

The particular column vector $\mathbf{c} = (2, 1, -1)$ is a *solution* of the equation $A\mathbf{x} = \mathbf{b}$ because

$$A\mathbf{c} = \begin{bmatrix} 2 & -1 & 3 \\ 1 & 4 & -2 \end{bmatrix} \begin{bmatrix} 2 \\ 1 \\ -1 \end{bmatrix} = \begin{bmatrix} 4-1-3 \\ 2+4+2 \end{bmatrix} = \begin{bmatrix} 0 \\ 8 \end{bmatrix} = \mathbf{b}.$$

Notice that if

$$B = \begin{bmatrix} 3 & -1 & 1 \\ 0 & 2 & 4 \end{bmatrix}, \qquad \mathbf{u} = \begin{bmatrix} 2 \\ 3 \end{bmatrix},$$

the product $B\mathbf{u}$ is *not* defined because the number of elements in a row of $B$ is not the same as the number of components of $\mathbf{u}$.

Having defined the product of a matrix and a column vector, we now attempt a more general definition of the product of two matrices. Let $A$ be an $m \times n$ matrix and let $B$ be an $n \times p$ matrix. Let us write

$$B = [\mathbf{b}_1, \mathbf{b}_2, \ldots, \mathbf{b}_p], \tag{2.39}$$

where each column vector $\mathbf{b}_j$ of $B$ has $n$ components. Since each row of $A$ has $n$ elements ($A$ has $n$ columns) each of the products

$$A\mathbf{b}_1, A\mathbf{b}_2, \ldots, A\mathbf{b}_p \tag{2.40}$$

is defined and is an $m$-dimensional column vector. We now define the *product $AB$* to be the $m \times p$ matrix with the column vectors (2.40); that is,

$$AB = [A\mathbf{b}_1, A\mathbf{b}_2, \ldots, A\mathbf{b}_p]. \tag{2.41}$$

Let $C = AB$. Then $C$ is an $m \times p$ matrix. The element $c_{ij}$ in the $i$th row and $j$th column of $C$ is the $i$th component of the vector $A\mathbf{b}_j$. This $i$th component is formed by multiplying the elements in the $i$th row of $A$ by the corresponding elements in the $j$th column of $B$. Thus

$$c_{ij} = \sum_{k=1}^{n} a_{ik} b_{kj}, \qquad 1 \le i \le m, \quad 1 \le j \le p. \tag{2.42}$$

Thus if $A$ is an $m \times n$ matrix and $B$ is an $n \times p$ matrix (the number of columns of $A$ must be the same as the number of rows of $B$) the product $AB$ is an $m \times p$ matrix $C$ whose elements are given by formula (2.42). To illustrate, let

$$A = \begin{bmatrix} 1 & -1 \\ 3 & 0 \\ 2 & 4 \end{bmatrix}, \quad B = \begin{bmatrix} 2 & 1 \\ -3 & -2 \end{bmatrix}.$$

Since $A$ is of size $3 \times 2$ and $B$ is of size $2 \times 2$, the product $AB$ is defined and is a $3 \times 2$ matrix. We have

$$AB = \begin{bmatrix} (1)(2)+(-1)(-3) & (1)(1)+(-1)(-2) \\ (3)(2)+(0)(-3) & (3)(1)+(0)(-2) \\ (2)(2)+(4)(-3) & (2)(1)+(4)(-2) \end{bmatrix} = \begin{bmatrix} 5 & 3 \\ 6 & 3 \\ -8 & -6 \end{bmatrix}.$$

Notice that the product $BA$ is not defined in this example, since the number of columns of $B$ is not the same as the number of rows of $A$. Even when $AB$ and $BA$ are both defined, the products are not necessarily the same. For instance, if

$$A = \begin{bmatrix} 1 & 2 \\ 3 & 1 \end{bmatrix}, \quad B = \begin{bmatrix} 1 & -1 \\ 2 & 0 \end{bmatrix},$$

the reader can verify that

$$AB = \begin{bmatrix} 5 & -1 \\ 5 & -3 \end{bmatrix}, \quad BA = \begin{bmatrix} -2 & 1 \\ 2 & 4 \end{bmatrix},$$

so that $AB \neq BA$. Thus matrix multiplication is not a commutative operation. When $AB = BA$, we say that $A$ and $B$ *commute*.

Some properties of matrices that involve multiplication are given in the following theorem.

**Theorem 2.2**   If $c$ is a number and if $A$, $B$, $C$, and $D$ are matrices such that the indicated sums and products are defined, then

$$(cA)B = A(cB) = c(AB) \tag{2.43a}$$
$$A(B + C) = AB + AC \tag{2.43b}$$
$$(B + C)D = BD + CD \tag{2.43c}$$
$$A(BD) = (AB)D. \tag{2.43d}$$

We shall prove properties (2.43b) and (2.43d) only, leaving the proofs of the remaining properties as exercises. First, consider property (2.43b). Let the size of $A$ be $m \times n$ and let that of $B$ and $C$ be $n \times p$. Then $AB$, $AC$, and $A(B + C)$ are defined and have size $m \times p$. Let $E = A(B + C)$, $F = AB$, and $G = AC$. We want to show that $E = F + G$ or that $e_{ij} = f_{ij} + g_{ij}$ for all $i$ and $j$. We have

$$e_{ij} = \sum_{k=1}^{n} a_{ik}(b_{kj} + c_{kj})$$

$$= \sum_{k=1}^{n} a_{ik} b_{kj} + \sum_{k=1}^{n} a_{ik} c_{kj}$$

$$= f_{ij} + g_{ij},$$

as we wished to prove.

In order to prove property (2.43d), we need a preliminary result. Let $S$ be an $m \times n$ matrix,

$$S = \begin{bmatrix} s_{11} & s_{12} & \cdots & s_{1n} \\ s_{21} & s_{22} & \cdots & s_{2n} \\ \vdots & \vdots & & \vdots \\ s_{m1} & s_{m2} & \cdots & s_{mn} \end{bmatrix},$$

and let $s$ denote the sum of all the elements of $S$. The sum of the elements in the $i$th row of $S$ is

$$\sum_{j=1}^{n} s_{ij}, \qquad 1 \leq i \leq m.$$

Consequently

$$s = \sum_{i=1}^{m} \sum_{j=1}^{n} s_{ij}. \tag{2.44}$$

But the sum of the elements in the $j$th column of $S$ is

$$\sum_{i=1}^{m} s_{ij}, \qquad 1 \leq j \leq n.$$

Hence we also have

$$s = \sum_{j=1}^{n} \sum_{i=1}^{m} s_{ij}. \tag{2.45}$$

Comparing the formulas (2.44) and (2.45), we see that

$$\sum_{i=1}^{m} \sum_{j=1}^{n} s_{ij} = \sum_{j=1}^{n} \sum_{i=1}^{m} s_{ij}. \tag{2.46}$$

This relation says that the order of summation in a repeated sum can be reversed without changing its value.

We now turn to a proof of property (2.43d). Let the size of $A$ be $m \times n$, that of $B$ be $n \times p$, and that of $D$ be $p \times q$. Then $BD$ is of size $n \times q$ and $AB$ is of size $m \times p$. Both $A(BD)$ and $(AB)D$ are of size $m \times q$. Let $E = BD$ $F = AB$, $G = AE$, and $H = FD$. Now

$$g_{ij} = \sum_{k=1}^{n} a_{ik} e_{kj} = \sum_{k=1}^{n} a_{ik} \sum_{r=1}^{p} b_{kr} d_{rj}.$$

Since $a_{ik}$ does not depend on $r$, we may write

$$g_{ij} = \sum_{k=1}^{n} \sum_{r=1}^{p} a_{ik} b_{kr} d_{rj}.$$

Next,

$$h_{ij} = \sum_{r=1}^{p} f_{ir} d_{rj} = \sum_{r=1}^{p} \left( \sum_{k=1}^{n} a_{ik} b_{kr} \right) d_{rj} = \sum_{r=1}^{p} \sum_{k=1}^{n} a_{ik} b_{kr} d_{rj}.$$

But, according to the relation (2.46), this is the same as

$$\sum_{k=1}^{n} \sum_{r=1}^{p} a_{ik} b_{kr} d_{rj},$$

which is $g_{ij}$. Thus $G = H$ or $A(BD) = (AB)D$.

One additional feature of matrix multiplication should be noted. Let

$$A = [\mathbf{a}_1, \mathbf{a}_2, \ldots, \mathbf{a}_n]$$

be an $m \times n$ matrix and let

$$\mathbf{c} = \begin{bmatrix} c_1 \\ c_2 \\ \vdots \\ c_n \end{bmatrix}$$

be an $n$-dimensional column vector. Then the matrix product $A\mathbf{c}$ of the $m \times n$ matrix $A$ and the $n \times 1$ matrix $\mathbf{c}$ is defined and

$$A\mathbf{c} = c_1\mathbf{a}_1 + c_2\mathbf{a}_2 + \cdots c_n\mathbf{a}_n. \tag{2.47}$$

We leave the verification to the reader.

## Exercises for Section 2.3

In Exercises 1–10, find all the products $AB$ and $BA$ that are defined.

1. $A = \begin{bmatrix} 4 & 2 \\ 1 & 3 \end{bmatrix}$, $B = \begin{bmatrix} 1 & -1 \\ -2 & 2 \end{bmatrix}$

2. $A = \begin{bmatrix} -2 & 5 \\ 1 & 4 \end{bmatrix}$, $B = \begin{bmatrix} 3 & 2 \\ -3 & 4 \end{bmatrix}$

3. $A = \begin{bmatrix} 1 & 2 & 0 \\ 2 & -1 & 3 \\ 0 & 1 & 2 \end{bmatrix}$, $B = \begin{bmatrix} 3 & -3 & 2 \\ 2 & 1 & -1 \\ -3 & 0 & 0 \end{bmatrix}$

4. $A = \begin{bmatrix} 1 & 4 & -3 \\ 2 & 0 & 1 \\ -3 & -2 & 0 \end{bmatrix}$, $B = \begin{bmatrix} 2 & 3 & 4 \\ -3 & -2 & 3 \\ 0 & 1 & 0 \end{bmatrix}$

5. $A = \begin{bmatrix} 1 & 2 \\ -1 & 3 \end{bmatrix}$, $B = \begin{bmatrix} 3 & 2 & 1 \\ 2 & 1 & 4 \end{bmatrix}$

6. $A = \begin{bmatrix} 2 & 1 & 0 \\ -1 & -2 & 2 \end{bmatrix}$, $B = \begin{bmatrix} 2 & 4 \\ 1 & -1 \\ 3 & 1 \end{bmatrix}$

7. $A = \begin{bmatrix} -4 & 3 \\ 2 & 0 \end{bmatrix}$, $B = \begin{bmatrix} 1 & 2 \\ -1 & 1 \\ 0 & 0 \end{bmatrix}$

8. $A = [2 \quad 3 \quad -1]$, $B = \begin{bmatrix} 2 & 0 \\ 1 & 4 \\ -2 & 1 \end{bmatrix}$

9. $A = [4 \quad 2 \quad 3]$, $B = [2 \quad 1 \quad -5]$

10. $A = [6 \quad -2 \quad 1]$, $B = \begin{bmatrix} 1 \\ 3 \\ 2 \end{bmatrix}$

11. Find a matrix $A$ and column vectors $\mathbf{x}$ and $\mathbf{b}$ such that the vector equation $A\mathbf{x} = \mathbf{b}$ corresponds to the given system.

(a)  $3x_1 + x_2 = 7$
     $4x_1 - 2x_2 = -3$

(b)  $2x_1 + 5x_2 = -1$
     $3x_1 - 2x_2 = 0$
     $x_1 + x_2 = 3$

(c)  $2x_1 - x_2 + x_3 = 4$
     $-x_1 + x_2 + 5x_3 = -2$
     $2x_1 + x_2 \quad = 3$

(d)  $3x_1 + x_2 - x_3 = 2$
     $2x_1 \quad + x_3 = -4$

12. Find the system of equations that corresponds to the vector equation $Ax = b$, where $A$ and $b$ are as given.

(a)  $A = \begin{bmatrix} 2 & -3 \\ 1 & 4 \end{bmatrix}$,  $b = \begin{bmatrix} 2 \\ -5 \end{bmatrix}$

(b)  $A = \begin{bmatrix} -2 & 1 & 3 \\ 3 & 0 & 1 \end{bmatrix}$,  $b = \begin{bmatrix} 0 \\ 0 \end{bmatrix}$

(c)  $A = \begin{bmatrix} 2 & -1 & 3 \\ 0 & 3 & -2 \\ 1 & 0 & 4 \end{bmatrix}$,  $b = \begin{bmatrix} 1 \\ -1 \\ 0 \end{bmatrix}$

13. Let $A$ be a matrix of size $4 \times 6$, with elements $a_{ij}$. Express the sum of the elements of $A$ in two ways, using summation notation.

14. If $A$ is the matrix in Exercise 13, are the following sums equal?

$$\sum_{i=1}^{4} \sum_{j=1}^{6} a_{ij}, \quad \sum_{i=1}^{4} \sum_{k=1}^{6} a_{ik}, \quad \sum_{n=1}^{4} \sum_{m=1}^{6} a_{nm}$$

15. Prove properties (2.43a) and (2.43c) in Theorem 2.2.

16. (a)  If $u$ and $v$ are solutions of the equation $Ax = 0$, show that $cu$ (where $c$ is any number) and $u + v$ are also solutions.

    (b)  If $y$ is a solution of the equation $Ax = b$ and $u$ is a solution of the equation $Ax = 0$, verify that $cu + y$ is a solution of the equation $Ax = b$.

17. Let $A$ be an $m \times n$ matrix. Let $x_1, x_2, \ldots, x_k$ be $n$-dimensional column vectors, and let $X = [x_1, x_2, \ldots, x_k]$. Show that $AX = 0$ if and only if each of the vectors $x_1, x_2, \ldots, x_k$ is a solution of the equation $Ax = 0$.

18. Verify formula (2.47).

19. If $C$ is a square matrix, we write $C^2 = C \cdot C$. If $A$ and $B$ are square matrices of the same size, show that in general $(A+B)^2 \neq A^2 + 2AB + B^2$.

## 2.4  SOME SPECIAL MATRICES

The $m \times n$ matrix whose elements are all zero is called the *zero matrix* of size $m \times n$. We denote it by $0_{m \times n}$, or simply by 0 when the size is clear. If

$A$ is any $m \times n$ matrix, then

$$A + 0_{m \times n} = A$$

and

$$A + (-1)A = 0_{m \times n}.$$

We also see that

$$A0_{n \times p} = 0_{m \times p}$$

and

$$0_{q \times m} A = 0_{q \times n}.$$

Let us define the useful symbol $\delta_{ij}$, where $i$ and $j$ are positive integers, by means of the formula

$$\delta_{ij} = \begin{cases} 0 & \text{if } i \neq j, \\ 1 & \text{if } i = j. \end{cases} \tag{2.48}$$

The symbol $\delta_{ij}$ is called the *Kronecker delta*. We have, for example, $\delta_{11} = \delta_{22} = \delta_{33} = 1$ and $\delta_{12} = \delta_{13} = \delta_{23} = 0$. The matrix of size $n \times n$ (same number of rows as columns) whose element in the $i$th row and $j$th column is $\delta_{ij}$, $1 \leq i \leq n$, $1 \leq j \leq n$, is called the $n \times n$ *identity matrix*. We denote it by $I_n$, or simply by $I$ when the size is evident. Thus

$$I_3 = \begin{bmatrix} 1 & 0 & 0 \\ 0 & 1 & 0 \\ 0 & 0 & 1 \end{bmatrix}.$$

The identity matrix has the following important properties.

**Theorem 2.3**   If $A$ is any $m \times n$ matrix, then

$$AI_n = A \tag{2.49a}$$

and

$$I_m A = A. \tag{2.49b}$$

PROOF   We shall prove the first of these properties. The establishment of the second is left as an exercise. Let $B = AI_n$. Then $B$ is an $m \times n$ matrix with elements

$$b_{ij} = \sum_{k=1}^{n} a_{ik} \delta_{kj}.$$

Each term in the sum is zero except the one where $k$ is equal to $j$. Hence

$$b_{ij} = a_{ij} \delta_{jj} = a_{ij}$$

so $B = A$ or $AI_n = A$.

If $A$ is a square $n \times n$ matrix, the elements $a_{11}, a_{22}, \ldots, a_{nn}$ are called the *diagonal* elements of $A$. If all the elements of a square matrix $D$ that are not diagonal elements are zero, then $D$ is called a *diagonal matrix*. We write $D = \text{diag}(d_1, d_2, \ldots, d_n)$ to indicate the diagonal matrix $D$ with diagonal elements $d_1, d_2, \ldots, d_n$. For example, if $D = \text{diag}(2, -1, 0)$ then

$$D = \begin{bmatrix} 2 & 0 & 0 \\ 0 & -1 & 0 \\ 0 & 0 & 0 \end{bmatrix}.$$

If $D = \text{diag}(d_1, d_2, \ldots, d_n)$ then the element in the $i$th row and $j$th column of $D$ is $d_{ij} = d_i \delta_{ij} = d_j \delta_{ij}$. A diagonal matrix with identical diagonal elements is called a *scalar matrix*. Thus a $3 \times 3$ scalar matrix has the form

$$\begin{bmatrix} d & 0 & 0 \\ 0 & d & 0 \\ 0 & 0 & d \end{bmatrix}$$

The $n \times n$ identity matrix and the $n \times n$ zero matrix are examples of scalar matrices. Note that each $n \times n$ scalar matrix is of the form $cI$, where $I$ is the $n \times n$ identity matrix.

**Theorem 2.4**   Let $A$ be an $m \times n$ matrix, let $C = \text{diag}(c_1, c_2, \ldots, c_m)$ and let $D = \text{diag}(d_1, d_2, \ldots, d_n)$. Then $CA$ is the matrix obtained from $A$ by multiplying every element in the $i$th row of $A$ by $c_i$, and $AD$ is the matrix obtained from $A$ by multiplying every element in the $j$th column of $A$ by $d_j$.

PROOF   We shall prove the part for $CA$, leaving the proof of the remaining part as an exercise. Let $E = CA$. Then $E$ is an $m \times n$ matrix, and

$$e_{ij} = \sum_{k=1}^{n} c_{ik} a_{kj} = \sum_{k=1}^{n} c_i \delta_{ik} a_{kj} = c_i a_{ij}.$$

Thus every element $e_{ij}$ in the $i$th row of $E$, $1 \le j \le n$, is formed by multiplying $a_{ij}$ by $c_i$. The proof of the following corollary is left as an exercise.

**Corollary**    Let $A$ be an $m \times n$ matrix. Let $C$ be an $m \times m$ scalar matrix with diagonal elements equal to $c$, and let $D$ be an $n \times n$ scalar matrix with diagonal elements equal to $d$. Then

$$CA = cA$$
$$AD = dA .$$

If $A$ is a square matrix, it is said to be *upper triangular* if $a_{ij} = 0$ whenever $i > j$. It is said to be *lower triangular* if $a_{ij} = 0$ whenever $i < j$. For example, the first of the matrices

$$\begin{bmatrix} 2 & 1 & -6 \\ 0 & 0 & 2 \\ 0 & 0 & 5 \end{bmatrix}, \qquad \begin{bmatrix} 4 & 0 & 0 \\ 0 & 1 & 0 \\ 2 & 3 & 2 \end{bmatrix}$$

is upper triangular, and the second is lower triangular.

If $A$ is an $m \times n$ matrix, we define the *transpose* of $A$, written $A^T$, to be the $n \times m$ matrix whose elements $a_{ij}^T$ are given by the relation

$$a_{ij}^T = a_{ji}, \qquad 1 \le i \le n, \qquad 1 \le j \le m .$$

The elements in the $i$th row of $A^T$ are the corresponding elements in the $i$th column of $A$. The elements in the $j$th column of $A^T$ are the corresponding elements in the $j$th row of $A$. For example, if

$$B = \begin{bmatrix} 1 & 2 & 3 \\ 4 & 5 & 6 \end{bmatrix},$$

then

$$B^T = \begin{bmatrix} 1 & 4 \\ 2 & 5 \\ 3 & 6 \end{bmatrix}.$$

The derivations of the properties

$$(A^T)^T = A \qquad\qquad\qquad (2.50a)$$

$$(A + B)^T = A^T + B^T \qquad\qquad (2.50b)$$

$$(AB)^T = B^T A^T \qquad\qquad\qquad (2.50c)$$

are left as exercises.

A matrix $A$ with the property that $A^T = A$ is said to be *symmetric*. If $A^T = -A$, then $A$ is said to be *skew-symmetric*.

The transpose of a row vector is a column vector and the transpose of a column vector is a row vector. If $\mathbf{u}$ and $\mathbf{v}$ are both $n$-dimensional column vectors, then $\mathbf{u}^T$ is a row vector and we have

$$\mathbf{u}^T\mathbf{v} = [u_1 \quad u_2 \cdots u_n] \begin{bmatrix} v_1 \\ v_2 \\ \vdots \\ v_n \end{bmatrix} = [u_1 v_1 + u_2 v_2 + \cdots + u_n v_n].$$

If $\mathbf{u}$ and $\mathbf{v}$ are both real then

$$\mathbf{u}^T\mathbf{v} = [\mathbf{u} \cdot \mathbf{v}].$$

## Exercises for Section 2.4

In Exercises 1–6, determine whether or not the given matrix is (a) diagonal, (b) scalar, (c) upper triangular, or (d) lower triangular.

**1.** $\begin{bmatrix} 2 & 0 & 0 \\ 0 & -1 & 0 \\ 0 & 0 & 0 \end{bmatrix}$

**2.** $\begin{bmatrix} 0 & 3 & 1 & 4 \\ 0 & 0 & 0 & 0 \\ 0 & 0 & 2 & 5 \\ 0 & 0 & 0 & -1 \end{bmatrix}$

**3.** $\begin{bmatrix} 5 & 0 & 0 & 0 \\ 0 & -1 & 0 & 0 \\ 0 & 0 & 0 & 0 \\ 6 & 1 & 4 & 0 \end{bmatrix}$

**4.** $\begin{bmatrix} 5 & 0 & 0 \\ 0 & 5 & 0 \\ 0 & 0 & 5 \end{bmatrix}$

**5.** $\begin{bmatrix} 2 & 0 & 0 & 0 \\ 0 & 0 & 0 & 0 \\ 0 & 0 & 0 & 0 \end{bmatrix}$

**6.** $\begin{bmatrix} 0 & 0 \\ 0 & 0 \end{bmatrix}$

**7.** If $D = \text{diag}(2, -1, 0)$ find $DA$, where $A$ is the given matrix

(a) $\begin{bmatrix} 1 & -1 & 2 \\ 0 & 3 & -2 \\ 4 & 4 & -1 \end{bmatrix}$

(b) $\begin{bmatrix} 4 & -1 \\ 5 & 2 \\ 3 & 6 \end{bmatrix}$

**8.** If $D = \text{diag}(-2, 0, 3)$ find $BD$, where $B$ is the given matrix

(a) $\begin{bmatrix} 3 & 5 & -1 \\ 0 & 1 & 3 \\ -4 & -4 & 2 \end{bmatrix}$

(b) $\begin{bmatrix} 4 & 1 & 2 \\ -3 & 6 & -1 \end{bmatrix}$

**9.** Prove property (2.49b).

10. Prove the corollary to Theorem 2.4.

11. If $A$ and $B$ are both diagonal matrices of the same size, show that $A$ and $B$ commute.

12. Find the transpose of the given matrix.

(a) $\begin{bmatrix} 2 & -1 \\ 4 & 5 \end{bmatrix}$ 　　　(b) $\begin{bmatrix} 3 & 0 & -2 \\ -1 & 4 & 0 \end{bmatrix}$

(c) $\begin{bmatrix} 6 & 2 & 4 \\ 1 & 0 & 3 \\ 2 & 1 & 1 \end{bmatrix}$ 　　　(d) $\begin{bmatrix} 5 \\ 1 \\ 3 \end{bmatrix}$

13. Prove the properties (2.50).

14. (a) Give an example of a $3 \times 3$ symmetric matrix, none of whose elements is zero.

(b) Show that if $A$ is any square matrix then $AA^T$ and $A^TA$ are symmetric.

(c) If $A$ and $B$ are both symmetric matrices of the same size, is $AB$ necessarily symmetric?

(d) Give an example of a $3 \times 3$ skew-symmetric matrix that is not the zero matrix.


## 2.5   DETERMINANTS

Let $A$ be a square matrix of order $n$. Associated with such a matrix is a number, called the *determinant* of $A$ and denoted by det $A$. If

$$
A = \begin{bmatrix}
a_{11} & a_{12} & \cdots & a_{1n} \\
a_{21} & a_{22} & \cdots & a_{2n} \\
\vdots & \vdots & & \vdots \\
a_{n1} & a_{n2} & \cdots & a_{nn}
\end{bmatrix}, \tag{2.51}
$$

we write

$$
\det A = \begin{vmatrix}
a_{11} & a_{12} & \cdots & a_{1n} \\
a_{21} & a_{22} & \cdots & a_{2n} \\
\vdots & \vdots & & \vdots \\
a_{n1} & a_{n2} & \cdots & a_{nn}
\end{vmatrix}, \tag{2.52}
$$

when we wish to display the elements of $A$. If $A$ is of order $n$, we say that det $A$ is a determinant of order $n$.

If $A$ is of order 1, with a single element $a_{11}$, we define det $A = a_{11}$. If $A$ is of order 2,

$$A = \begin{bmatrix} a_{11} & a_{12} \\ a_{21} & a_{22} \end{bmatrix},$$

we define

$$\det A = a_{11}a_{22} - a_{12}a_{21}. \qquad (2.53)$$

Thus

$$\begin{vmatrix} 2 & -3 \\ 1 & 5 \end{vmatrix} = (2)(5) - (-3)(1) = 13.$$

We shall presently define the determinant of a square matrix of arbitrary order $n$. First, however, we must develop some preliminary ideas. Consider a set[4] $\{j_1, j_2, \ldots, j_n\}$ whose distinct elements $j_1, j_2, \ldots, j_n$ are positive integers. Each possible ordering of the elements of the set is called a *permutation* of the set. We use parentheses to denote an *ordered* set. For example, the possible permutations of the set $\{1, 2, 5\}$ are $(1, 2, 5)$, $(1, 5, 2)$, $(2, 1, 5)$, $(2, 5, 1)$, $(5, 1, 2)$, and $(5, 2, 1)$. The number of possible permutations of $n$ integers (or of any $n$ objects) is $n!$.

Let $(j_1, j_2, \ldots, j_n)$ be a permutation of a set of $n$ positive integers. Let $\alpha_1$ be the number of integers following $j_1$ that are smaller than $j_1$, let $\alpha_2$ be the number of integers following $j_2$ that are smaller than $j_2$, and so on. Note that $\alpha_n$ is always zero. The sum $\alpha_1 + \alpha_2 + \cdots + \alpha_{n-1}$ is called the number of *inversions* in the permutation $(j_1, j_2, \ldots, j_n)$. For example, in the permutation $(2, 4, 1, 3)$ of $\{1, 2, 3, 4\}$, we have $\alpha_1 = 1$, $\alpha_2 = 2$, and $\alpha_3 = 0$, so the number of inversions is three. A permutation is said to have *even or odd parity* according to whether the number of its inversions is even or odd. We define

$$\delta(j_1, j_2, \ldots, j_n) \qquad (2.54)$$

to be one if the parity of $(j_1, j_2, \ldots, j_n)$ is even and minus one if the parity is odd.

We are now in a position to define the determinant of the $n$th-order matrix (2.51). We form all possible products of the form

$$a_{1j_1} a_{2j_2} a_{3j_3} \cdots a_{nj_n} \qquad (2.55)$$

---

[4] One way to describe a set is to list the members of the set, enclosed in braces.

in which there occurs exactly one element from each row and each column of $A$. Thus $(j_1, j_2, j_3, \ldots, j_n)$ is a permutation of $\{1, 2, 3, \ldots, n\}$. The determinant of $A$ is defined by the formula

$$\det A = \sum \delta(j_1, j_2, \ldots, j_n)\, a_{1j_1} a_{2j_2} \cdots a_{nj_n}, \qquad (2.56)$$

where the sum is taken over all possible permutations $(j_1, j_2, \ldots, j_n)$ of $\{1, 2, \ldots, n\}$.

It can be verified that this definition agrees with those previously given for the cases $n = 1$ and $n = 2$. Let us apply the definition to find the determinant of a $3 \times 3$ matrix,

$$A = \begin{bmatrix} a_{11} & a_{12} & a_{13} \\ a_{21} & a_{22} & a_{23} \\ a_{31} & a_{32} & a_{33} \end{bmatrix}. \qquad (2.57)$$

The products of the form (2.55) are

$$a_{11}\,a_{22}\,a_{33}, \quad a_{11}\,a_{23}\,a_{32}, \quad a_{12}\,a_{21}\,a_{33}$$
$$a_{12}\,a_{23}\,a_{31}, \quad a_{13}\,a_{21}\,a_{32}, \quad a_{13}\,a_{22}\,a_{31}.$$

Finding the proper signs and summing, we have

$$\det A = a_{11}\,a_{22}\,a_{23} + a_{13}\,a_{21}a_{32} + a_{12}\,a_{23}\,a_{31}$$
$$- a_{13}\,a_{22}\,a_{31} - a_{12}\,a_{21}a_{33} - a_{11}a_{23}\,a_{32}. \qquad (2.58)$$

This formula is not easy to remember. However the device

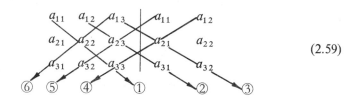

$$(2.59)$$

is useful. We have written down the elements of the matrix $A$ in array form and then repeated the first two columns. We form the six products associated with the six arrows, each product having as factors the three elements pierced by that arrow. We assign a plus sign to the products determined by arrows ①, ②, and ③ and a minus sign to the products associated with arrows

④, ⑤, and ⑥. The sum of the six signed products is det $A$. As an example, we evaluate the determinant

$$\begin{vmatrix} 1 & 2 & 3 \\ -1 & 3 & 0 \\ 2 & -4 & 5 \end{vmatrix}.$$

Forming the array

$$\begin{array}{ccc|cc} 1 & 2 & 3 & 1 & 2 \\ -1 & 3 & 0 & -1 & 3 \\ 2 & -4 & 5 & 2 & -4 \end{array}$$

and using the procedure described above, we find that the value of the determinant is

$$(1)(3)(5) + (2)(0)(2) + (3)(-1)(-4)$$
$$-(2)(-1)(5) - (1)(0)(-4) - (3)(3)(2) = 19.$$

The reader should be warned that the scheme just described works only for third-order determinants and not for higher order ones. A practical method for evaluating a determinant of any order will be developed in the next section.

A square matrix whose determinant is zero is said to be *singular*. A square matrix that is not singular is called *nonsingular*.

We consider one more topic, the derivative of the determinant of a matrix function. Consider an ordered set of real differentiable functions $a_{ij}$, $1 \le i \le n$, $1 \le j \le n$, with a common domain. Associated with each number $x$ in the domain is the matrix

$$A(x) = \begin{bmatrix} a_{11}(x) & a_{12}(x) & \cdots & a_{1n}(x) \\ a_{21}(x) & a_{22}(x) & \cdots & a_{2n}(x) \\ \vdots & \vdots & & \vdots \\ a_{n1}(x) & a_{n2}(x) & \cdots & a_{nn}(x) \end{bmatrix}.$$

For each $x$ we can inquire about $(d/dx)$ det $A(x)$ Recalling that

$$\det A(x) = \sum \pm a_{1j_1}(x)\, a_{2j_2}(x) \cdots a_{nj_n}(x),$$

we see that

$$(\det A)' = \sum \pm a'_{1j_1} a_{2j_2} \cdots a_{nj_n} + \sum \pm a_{1j_1} a'_{2j_2} \cdots a_{nj_n}$$
$$+ \cdots + \sum \pm a_{1j_1} a_{2j_2} \cdots a'_{nj_n}$$

or

$$(\det A)' = \begin{vmatrix} a'_{11} & a'_{12} & \cdots & a'_{1n} \\ a_{21} & a_{22} & \cdots & a_{2n} \\ \vdots & \vdots & & \vdots \\ a_{n1} & a_{n2} & \cdots & a_{nn} \end{vmatrix} + \begin{vmatrix} a_{11} & a_{12} & \cdots & a_{1n} \\ a'_{21} & a'_{22} & \cdots & a'_{2n} \\ \vdots & \vdots & & \vdots \\ a_{n1} & a_{n2} & \cdots & a_{nn} \end{vmatrix}$$

$$+ \cdots + \begin{vmatrix} a_{11} & a_{12} & \cdots & a_{1n} \\ a_{21} & a_{22} & \cdots & a_{2n} \\ \vdots & \vdots & & \vdots \\ a'_{n1} & a'_{n2} & \cdots & a'_{nn} \end{vmatrix} . \tag{2.60}$$

Thus the derivative of det $A$ is the sum of $n$ determinants that are obtained by successively differentiating the rows of $A$. As will be shown in the next section the derivative of det $A$ can also be expressed as the sum of $n$ determinants that are obtained by successively differentiating the columns of $A$.

For purposes of illustration, we observe that

$$\frac{d}{dx} \begin{vmatrix} x & x^2 & x^3 \\ e^x & 1 & 0 \\ \sin x & 0 & 0 \end{vmatrix}$$

$$= \begin{vmatrix} 1 & 2x & 3x^2 \\ e^x & 1 & 0 \\ \sin x & 0 & 0 \end{vmatrix} + \begin{vmatrix} x & x^2 & x^3 \\ e^x & 0 & 0 \\ \sin x & 0 & 0 \end{vmatrix} + \begin{vmatrix} x & x^2 & x^3 \\ e^x & 1 & 0 \\ \cos x & 0 & 0 \end{vmatrix} .$$

## Exercises for Section 2.5

1. Find all the permutations of the set $\{1, 2, 3, 4\}$.

2. Find the number of inversions in each of the following permutations:

(a) $(3, 1, 2)$         (b) $(3, 2, 1)$         (c) $(1, 2, 3)$
(d) $(1, 3, 2, 4)$      (e) $(4, 2, 3, 1)$      (f) $(2, 3, 4, 1)$
(g) $(1, 3, 5, 2, 4)$   (h) $(4, 1, 5, 3, 2)$

3. With $\delta(j_1, j_2, \ldots, j_n)$ defined as in the text, find the value of each of the following:

(a) $\delta(2, 1, 3)$        (b) $\delta(1, 2, 3)$        (c) $\delta(2, 4, 1, 3)$
(d) $\delta(4, 3, 2, 1)$     (e) $\delta(1, 3, 5, 2, 4)$

4. Derive the formula (2.53) for the value of a second-order determinant from the general formula (2.56).

5. Find the value of each of the second-order determinants by using the formula (2.53).

(a) $\begin{vmatrix} 2 & -3 \\ 4 & 5 \end{vmatrix}$      (b) $\begin{vmatrix} 2 & -3 \\ -4 & 6 \end{vmatrix}$

(c) $\begin{vmatrix} -5 & 2 \\ 1 & 3 \end{vmatrix}$      (d) $\begin{vmatrix} 0 & 3 \\ 2 & 4 \end{vmatrix}$

6. Find the value of each of the third-order determinants.

(a) $\begin{vmatrix} 1 & 3 & 2 \\ 3 & 1 & -1 \\ -2 & 4 & 5 \end{vmatrix}$      (b) $\begin{vmatrix} 2 & 0 & -2 \\ 1 & 1 & 5 \\ 3 & 4 & 5 \end{vmatrix}$

(c) $\begin{vmatrix} -4 & -1 & 3 \\ 2 & 2 & 1 \\ 3 & 5 & 0 \end{vmatrix}$      (d) $\begin{vmatrix} 1 & 2 & 3 \\ 3 & 2 & 1 \\ 1 & 1 & 1 \end{vmatrix}$

7. If $A$ is a square matrix of order $n$, prove that $\det(-A) = (-1)^n \det A$.

8. If $A$ is a square matrix of order $n$ and $c$ is a number, prove that

$$\det(cA) = c^n \det A .$$

9. Show, by means of an example, that $\det(A + B) \neq \det A + \det B$ in general.

10. If $D = \operatorname{diag}(d_1, d_2, \ldots, d_n)$, show that $\det D = d_1 d_2 \cdots d_n$.

11. Evaluate the derivative of the determinant

$$\begin{vmatrix} x^3 + 1 & x^2 \\ x^2 & 2x \end{vmatrix}$$

in two ways: first, by evaluating the determinant and then taking the derivative; second, by applying formula (2.60).

## 2.6 PROPERTIES OF DETERMINANTS

Listed below are some elementary properties of determinants. Each property is illustrated by an example. A proof of only the first property is given in the text. Proofs of others are left as exercises. The reader who encounters difficulty with these will find proofs in most college algebra texts and in some books on calculus. Property 8 is more difficult to prove, and we ask for a proof only for $2 \times 2$ matrices.

**Property 1**   $\det A^T = \det A$:

$$\begin{vmatrix} 1 & -1 & 2 \\ 3 & 0 & 1 \\ 2 & 1 & 5 \end{vmatrix} = \begin{vmatrix} 1 & 3 & 2 \\ -1 & 0 & 1 \\ 2 & 1 & 5 \end{vmatrix} .$$

**Property 2**  If every element in a row (column) of $A$ is zero, then $\det A = 0$:

$$\begin{vmatrix} 2 & 0 & 3 \\ 1 & 0 & 1 \\ -1 & 0 & 4 \end{vmatrix} = 0.$$

**Property 3**  If every element in one row (column) of $A$ is multiplied by the number $c$, the determinant of the resulting matrix is equal to $c \det A$:

$$\begin{vmatrix} -6 & 3 & -9 \\ 2 & 1 & 0 \\ 1 & 1 & 2 \end{vmatrix} = -3 \begin{vmatrix} 2 & -1 & 3 \\ 2 & 1 & 0 \\ 1 & 1 & 2 \end{vmatrix}.$$

**Property 4**  If two rows (columns) of $A$ are interchanged, the determinant of the resulting matrix is equal to $-\det A$:

$$\begin{vmatrix} 1 & -1 & 3 \\ 0 & 4 & 1 \\ 2 & 2 & 5 \end{vmatrix} = -\begin{vmatrix} 2 & 2 & 5 \\ 0 & 4 & 1 \\ 1 & -1 & 3 \end{vmatrix}.$$

**Property 5**  If two rows (columns) of $A$ are identical, then $\det A = 0$:

$$\begin{vmatrix} 2 & 1 & 2 \\ 1 & -4 & 1 \\ 3 & 5 & 3 \end{vmatrix} = 0.$$

**Property 6**  If any column vector of $A$, say the $i$th, is the sum of two column vectors, so that $\mathbf{a}_i = \mathbf{a}'_i + \mathbf{a}''_i$, then $\det A = \det[\mathbf{a}_1, \ldots, \mathbf{a}'_i, \ldots, \mathbf{a}_n]$ $+ \det[\mathbf{a}_1, \ldots, \mathbf{a}''_i, \ldots, \mathbf{a}_n]$. An analogous property holds for rows:

$$\begin{vmatrix} 1 & 2+3 & 3 \\ 0 & 1-4 & 5 \\ 2 & -2+0 & 6 \end{vmatrix} = \begin{vmatrix} 1 & 2 & 3 \\ 0 & 1 & 5 \\ 2 & -2 & 6 \end{vmatrix} + \begin{vmatrix} 1 & 3 & 3 \\ 0 & -4 & 5 \\ 2 & 0 & 6 \end{vmatrix}.$$

**Property 7**  If to every element in a row (column) of $A$ is added $c$ times the corresponding element of a different row (column), the determinant of the resulting matrix is equal to $\det A$:

$$\begin{vmatrix} 2 & -1 & 0 \\ 1 & 2 & -3 \\ 4 & 5 & 6 \end{vmatrix} = \begin{vmatrix} 4 & 3 & -6 \\ 1 & 2 & 3 \\ 4 & 5 & 6 \end{vmatrix}.$$

(Here, twice the second row has been added to the first row.)

**Property 8** If $A$ and $B$ are both matrices of order $n$, then $\det(AB) = \det A \cdot \det B$:

$$\begin{vmatrix} 1 & -1 \\ 2 & 1 \end{vmatrix} \cdot \begin{vmatrix} 2 & 1 \\ -1 & 3 \end{vmatrix} = \begin{vmatrix} 3 & -2 \\ 3 & 5 \end{vmatrix}.$$

The following lemma is needed in the derivations of some of the properties described above.

**Lemma** If two adjacent elements in a permutation are interchanged, the parity of the permutation is changed.

PROOF Suppose that $j_\alpha$ and $j_\beta$ are interchanged in the permutation $(j_1, j_2, \ldots, j_\alpha, j_\beta, \ldots, j_n)$. If $j_\alpha < j_\beta$, the interchange of $j_\alpha$ and $j_\beta$ introduces one new inversion. If $j_\alpha > j_\beta$, the number of inversions is decreased by one. In each case the parity is changed.

PROOF OF PROPERTY 1 Let $a_{ij}^{\mathrm{T}}$ be the element in the $i$th row and $j$th column of $A^{\mathrm{T}}$. Then $a_{ij}^{\mathrm{T}} = a_{ji}$. The determinant of $A^{\mathrm{T}}$ is the sum of terms of the form

$$\delta(j_1, j_2, \ldots, j_n) a_{1j_1}^{\mathrm{T}} a_{2j_2}^{\mathrm{T}} \cdots a_{nj_n}^{\mathrm{T}} = \delta(j_1, j_2, \ldots, j_n) a_{j_1 1} a_{j_2 2} \cdots a_{j_n n}.$$

Now

$$a_{j_1 1} a_{j_2 2} \cdots a_{j_n n} = a_{1k_1} a_{2k_2} \cdots a_{nk_n}, \tag{2.61}$$

where $(k_1, k_2, \ldots, k_n)$ is some permutation of $\{1, 2, \ldots, n\}$. We can think of the product on the right-hand side in Eq. (2.61) as being formed from the product on the left-hand side by successively interchanging adjacent factors. Each interchange changes the parity of the ordered set of first subscripts and simultaneously changes the parity of the ordered set of second subscripts. Consequently the parity of $(k_1, k_2, \ldots, k_n)$ must be the same as that of $(j_1, j_2, \ldots, j_n)$. Thus $\delta(k_1, k_2, \ldots, k_n) = \delta(j_1, j_2, \ldots, j_n)$, so that

$$\det A^T = \sum \delta(k_1, k_2, \ldots, k_n) a_{1k_1} a_{2k_2} \cdots a_{nk_n} = \det A.$$

The rows of $A^T$ are the same as the columns of $A$. Consequently, Property 1 allows us to convert theorems about rows of a determinant into corresponding theorems about columns. For example, the derivative of $\det A$ can be found by differentiating $\det A^T$ by rows; this amounts to differentiating $\det A$ by columns.

We shall presently give a practical method for evaluating a determinant of any order. Our method is based on the following result.

**Theorem 2.5**    If $A$ is an $n \times n$ triangular (upper or lower) matrix, then

$$\det A = a_{11} a_{22} \cdots a_{nn}. \tag{2.62}$$

PROOF    The determinant of $A$ is the sum of terms of the form

$$\pm a_{1j_1} a_{2j_2} \cdots a_{nj_n}. \tag{2.63}$$

If one factor, say $a_{pj_p}$, is such that $p < j_p$, there must be another factor, say $a_{qj_q}$, such that $q > j_q$, and conversely. This is because

$$1 + 2 + \cdots + n = j_1 + j_2 + \cdots + j_n$$

for every permutation $(j_1, j_2, \ldots, j_n)$. Thus if $A$ is triangular, every term of the the form (2.63) is zero except the one where $j_1 = 1, j_2 = 2, \ldots, j_n = n$. Hence $\det A$ is given by formula (2.62).

By using the elementary properties listed at the beginning of this section, we can reduce the problem of evaluating any determinant to one of evaluating the determinant of a triangular matrix. The essential features of the reduction are shown in the following example. Consider the determinant

$$\det A = \begin{vmatrix} 0 & 2 & 1 & -1 \\ 2 & -2 & 4 & 0 \\ -1 & 2 & 0 & 1 \\ -2 & 1 & 1 & 3 \end{vmatrix}.$$

We wish to place a nonzero element in the first row and column. Interchanging the first and third rows, we have

$$\det A = - \begin{vmatrix} -1 & 2 & 0 & 1 \\ 2 & -2 & 4 & 0 \\ 0 & 2 & 1 & -1 \\ -2 & 1 & 1 & 3 \end{vmatrix}.$$

We place zeros in every position of the first column below the first by adding appropriate multiples of the first row to the second and fourth rows. Thus

$$\det A = - \begin{vmatrix} -1 & 2 & 0 & 1 \\ 0 & 2 & 4 & 2 \\ 0 & 2 & 1 & -1 \\ 0 & -3 & 1 & 1 \end{vmatrix} = -2 \begin{vmatrix} -1 & 2 & 0 & 1 \\ 0 & 1 & 2 & 1 \\ 0 & 2 & 1 & -1 \\ 0 & -3 & 1 & 1 \end{vmatrix}.$$

We place zeros in every position of the second column below the second by adding appropriate multiples of the second row to the third and fourth rows. We find that

$$\det A = -2 \begin{vmatrix} -1 & 2 & 0 & 1 \\ 0 & 1 & 2 & 1 \\ 0 & 0 & -3 & -3 \\ 0 & 0 & 7 & 4 \end{vmatrix} = 6 \begin{vmatrix} -1 & 2 & 0 & 1 \\ 0 & 1 & 2 & 1 \\ 0 & 0 & 1 & 1 \\ 0 & 0 & 7 & 4 \end{vmatrix}.$$

We add $(-7)$ times the third row to the fourth to obtain the triangular form

$$\det A = 6 \begin{vmatrix} -1 & 2 & 0 & 1 \\ 0 & 1 & 2 & 1 \\ 0 & 0 & 1 & 1 \\ 0 & 0 & 0 & -3 \end{vmatrix}.$$

Using Theorem 2.5, we have

$$\det A = (6)(-1)(1)(1)(-3) = 18.$$

## Exercises for Section 2.6

1. Derive Properties 2 and 3 of determinants that are given at the beginning of this section.

2. Prove Property 4 (Suggestion: use the lemma of this section). Then use Property 4 to prove Property 5.

3. Derive Properties 6 and 7.

4. Prove that $\det(AB) = (\det A) \cdot (\det B)$ when $A$ and $B$ are matrices of order two.

In Exercises 5–9, evaluate the determinant by using Theorem 2.5 and elementary properties of determinants.

5. (a) $\begin{vmatrix} 1 & 3 & 2 \\ 2 & -1 & 5 \\ -2 & 4 & -4 \end{vmatrix}$   (b) $\begin{vmatrix} 5 & 2 & 3 \\ 2 & -1 & 0 \\ 3 & 4 & 7 \end{vmatrix}$

6. (a) $\begin{vmatrix} -3 & 5 & 7 \\ -5 & -4 & 3 \\ 2 & 5 & 6 \end{vmatrix}$   (b) $\begin{vmatrix} 6 & 5 & 2 \\ 4 & 3 & 2 \\ 7 & 3 & 7 \end{vmatrix}$

7. (a) $\begin{vmatrix} 2 & 1 & 4 & 7 \\ 3 & 0 & 1 & 5 \\ -4 & -3 & 3 & 4 \\ 2 & 2 & -1 & 0 \end{vmatrix}$   (b) $\begin{vmatrix} -2 & 1 & 4 & 2 \\ -3 & 0 & 1 & 6 \\ 1 & 2 & 3 & 4 \\ -4 & 3 & -2 & 1 \end{vmatrix}$

**8. (a)**
$$\begin{vmatrix} 4 & 0 & 2 & 0 \\ 0 & 1 & 0 & 3 \\ 5 & 0 & 7 & 0 \\ 0 & 8 & 0 & 6 \end{vmatrix}$$

**(b)**
$$\begin{vmatrix} 6 & 2 & 8 & 0 \\ 1 & 3 & 5 & 2 \\ 2 & 1 & 0 & -3 \\ 2 & -5 & -2 & -7 \end{vmatrix}$$

**9.**
$$\begin{vmatrix} 2 & -1 & 0 & 4 & 1 \\ 1 & 5 & 2 & 0 & -2 \\ -1 & 3 & -3 & 1 & 0 \\ 0 & 1 & 1 & 2 & -2 \\ 2 & 2 & 1 & 0 & -1 \end{vmatrix}$$

**10.** Let $C = AB$, where $A$ and $B$ are square matrices of the same order. If $\det C = 0$, show that either $\det A$ or $\det B$, or both, is zero.

**11.** If $A$ is a square matrix such that $A^2 = A$, what can be said about $\det A$?

**12.** Let $f_1, f_2, \ldots, f_n$ be functions that are defined and possess at least $n$ derivatives on an interval. If

$$A(x) = \begin{bmatrix} f_1(x) & f_2(x) & \cdots & f_n(x) \\ f'_1(x) & f'_2(x) & \cdots & f'_n(x) \\ \vdots & \vdots & & \vdots \\ f_1^{(n-1)}(x) & f_2^{(n-1)}(x) & \cdots & f_n^{(n-1)}(x) \end{bmatrix},$$

show that

$$\frac{d}{dx} \det A(x) = \begin{vmatrix} f_1(x) & f_2(x) & \cdots & f_n(x) \\ f'_1(x) & f'_2(x) & \cdots & f'_n(x) \\ \vdots & \vdots & & \vdots \\ f_1^{(n-2)}(x) & f_2^{(n-2)}(x) & \cdots & f_n^{(n-2)}(x) \\ f_1^{(n)}(x) & f_2^{(n)}(x) & \cdots & f_n^{(n)}(x) \end{vmatrix}.$$

## 2.7  COFACTORS

Let $A$ be an $m \times n$ matrix. Any matrix that is formed from $A$ by deleting rows of $A$ or columns of $A$, or both, is called a *submatrix* of $A$. In addition, it is sometimes convenient to regard $A$ as a submatrix of itself (deletion of no rows and no columns).

We now restrict attention to the case where $A$ is a square $n \times n$ matrix. If we delete the $i$th row and $k$th column of $A$ (the row and column containing the element $a_{ik}$), we obtain a square submatrix of order $n - 1$. The determinant of this submatrix is called the minor of the element $a_{ik}$; we denote the minor by $M_{ik}$. We may write

$$M_{ik} = \sum \delta(j_1, \cdots, j_{k-1}, j_{k+1}, \cdots, j_n) a_{1j_1} \cdots a_{k-1,j_{k-1}} a_{k+1,j_{k+1}} \cdots a_{nj_n},$$

where $i$ is excluded from the row subscripts and the sum is over all permutations of $1, 2, \ldots, k-1, k+1, \ldots, n$. The quantity

$$A_{ik} = (-1)^{i+k} M_{ik}$$

is called the *cofactor* of the element $a_{ik}$. For example, if

$$A = \begin{bmatrix} 1 & -2 & 0 \\ 3 & 1 & 4 \\ 2 & 2 & 1 \end{bmatrix}, \tag{2.64}$$

then

$$A_{11} = \begin{vmatrix} 1 & 4 \\ 2 & 1 \end{vmatrix} = -7, \qquad A_{12} = -\begin{vmatrix} 3 & 4 \\ 2 & 1 \end{vmatrix} = 5,$$

and so on.

The determinant of a matrix can be expressed in terms of the cofactors of the elements in any one row or column, as is shown by the following theorem.

**Theorem 2.6** If each element in any one row (column) of an $n \times n$ matrix $A$ is multiplied by its cofactor, the sum of the $n$ products so formed is equal to det $A$. Thus

$$\sum_{j=1}^{n} a_{ij} A_{ij} = \det A, \qquad 1 \le i \le n, \tag{2.65a}$$

$$\sum_{i=1}^{n} a_{ij} A_{ij} = \det A, \qquad 1 \le j \le n. \tag{2.65b}$$

PROOF   We first establish the relation (2.65a), starting with the formula

$$\det A = \sum \delta(j_1, j_2, \ldots, j_n) a_{1j_1} a_{2j_2} \cdots a_{nj_n}.$$

For fixed $i$, we collect the products that involve $a_{i1}, a_{i2}, \ldots, a_{in}$. Then

$$\det A = \sum_{j_i=1}^{n} a_{ij_i} \sum \delta(j_1, \ldots, j_i, \ldots, j_n) a_{1j_1} \cdots a_{i-1,j_{i-1}} a_{i+1,j_{i+1}} \cdots a_{nj_n},$$

where the inner sum is taken over all permutations $(j_1, \ldots, j_i, \ldots, j_n)$ of $\{1, 2, \ldots, n\}$ with $j_i$ fixed. Since $j_i$ is in the $i$th position,

$$\delta(j_1, \ldots, j_i, \ldots, j_n) = (-1)^{i-1} \delta(j_i, j_1, \ldots, j_{i-1}, j_{i+1}, \ldots, j_n).$$

In the symbol on the right, $j_i$ is followed by $j_i - 1$ smaller terms, so this is equal to

$$(-1)^{i-1}(-1)^{j_i-1} \delta(j_1, \ldots, j_{i-1}, j_{i+1}, \ldots, j_n).$$

Hence

$$\det A = \sum_{j_i=1}^{n} a_{ij_i} (-1)^{i+j_i} M_{ij_i} = \sum_{j_i=1}^{n} a_{ij_i} A_{ij_i}.$$

This establishes Eq. (2.65a).

Formula (2.65b) can be derived by expressing $\det A$ in terms of the cofactors of the elements of the $j$th row of $A^T$, and observing that the cofactor of $a_{ji}^T$ is the same as the cofactor of $a_{ij}$. We omit the details.

As an example, we consider the matrix (2.64). Applying formula (2.65a), with $i = 1$, we have

$$\det A = (1)\begin{vmatrix} 1 & 4 \\ 2 & 1 \end{vmatrix} - (-2)\begin{vmatrix} 3 & 4 \\ 2 & 1 \end{vmatrix} + (0)\begin{vmatrix} 3 & 1 \\ 2 & 2 \end{vmatrix}$$

$$= (1)(-7) - (-2)(-5) + 0$$

$$= -17.$$

**Theorem 2.7**   If the elements of the $i$th row (column) of an $n \times n$ matrix $A$ are multiplied by the cofactors of the corresponding elements of the $j$th row (column), the sum of the products is $\det A$ if $i = j$ and zero if $i \neq j$. In symbols,

$$\sum_{k=1}^{n} a_{ik} A_{jk} = \delta_{ij} \det A, \qquad (2.66a)$$

$$\sum_{k=1}^{n} a_{ki} A_{kj} = \delta_{ij} \det A, \qquad (2.66b)$$

where $\delta_{ij}$ is the Kronecker delta.

PROOF   The validity of the formulas (2.66) follows from Theorem 2.6 if $i = j$. For the case $i \neq j$, the sum on the left-hand side in Eq. (2.66a) can be regarded as the determinant of a matrix whose $j$th row is the same as its $i$th row. Hence its value must be zero. Similarly, the left-hand member of Eq. (2.66b) can be regarded as the determinant of a matrix whose $j$th column is the same as its $i$th column.

The formulas (2.66) will be used to derive some results of theoretical importance in the next section. Although formulas (2.65) can be used to evaluate determinants, the method described in the previous section requires fewer arithmetic operations to be performed, and is more efficient when the order of the matrix is large.

## Exercises for Section 2.7

1. Find the cofactor of each element of the given matrix.

    (a) $\begin{bmatrix} 2 & 3 \\ -1 & 0 \end{bmatrix}$
        
    (b) $\begin{bmatrix} a & b \\ c & d \end{bmatrix}$

    (c) $\begin{bmatrix} 2 & -1 & 3 \\ 1 & 0 & -2 \\ 3 & 1 & 1 \end{bmatrix}$
        
    (d) $\begin{bmatrix} 4 & 0 & 2 \\ 0 & 2 & -2 \\ 1 & 3 & 1 \end{bmatrix}$

2. Evaluate the given determinant by applying Theorem 2.6. Use any row or column.

    (a) $\begin{vmatrix} 1 & 2 & -1 \\ 4 & 1 & 2 \\ 1 & 1 & -3 \end{vmatrix}$
        
    (b) $\begin{vmatrix} 4 & 3 & -2 \\ -1 & 2 & 0 \\ 1 & -1 & 3 \end{vmatrix}$

3. Find det $D$ if

$$D = \begin{bmatrix} 0 & \cdots & 0 & 0 & 0 & d_1 \\ 0 & \cdots & 0 & 0 & d_2 & 0 \\ 0 & \cdots & 0 & d_3 & 0 & 0 \\ \vdots & & \vdots & \vdots & \vdots & \vdots \\ d_n & \cdots & 0 & 0 & 0 & 0 \end{bmatrix}$$

4. Let $A$ be a third-order matrix with elements $a_{ij}$. Verify that the sum

$$a_{11}A_{21} + a_{12}A_{22} + a_{13}A_{23}$$

    (whose terms are the products of the elements in the first row of $A$ with the cofactors of the corresponding elements of the second row) is the determinant of the matrix whose first and second rows are identical. (Hence the sum is equal to zero.)

5. Let $P_1$ and $P_2$ be distinct points with rectangular coordinates $(x_1, y_1)$ and $(x_2, y_2)$ in a plane. Show that the equation

$$\begin{vmatrix} x & y & 1 \\ x_1 & y_1 & 1 \\ x_2 & y_2 & 1 \end{vmatrix} = 0$$

is that of the straight line through $P_1$ and $P_2$.

**6.** Let $x_1, x_2, \ldots, x_n$ be distinct numbers. Show that the formula

$$P(x) = \begin{vmatrix} 1 & x & x^2 & \cdots & x^n \\ 1 & x_1 & x_1^2 & \cdots & x_1^n \\ 1 & x_2 & x_2^2 & \cdots & x_2^n \\ \vdots & \vdots & \vdots & & \vdots \\ 1 & x_n & x_n^2 & \cdots & x_n^n \end{vmatrix}$$

defines a polynomial $P$ whose zeros are $x_1, x_2, \ldots, x_n$.

## 2.8   CRAMER'S RULE

We consider a system of linear equations

$$\begin{aligned} a_{11}x_1 + a_{12}x_2 + \cdots + a_{1n}x_n &= b_1, \\ a_{21}x_1 + a_{22}x_2 + \cdots + a_{2n}x_n &= b_2, \\ &\cdots\cdots\cdots\cdots\cdots\cdots\cdots\cdots\cdots, \\ a_{n1}x_1 + a_{n2}x_2 + \cdots + a_{nn}x_n &= b_n \end{aligned} \tag{2.67}$$

with $n$ equations and $n$ unknowns. The $n \times n$ coefficient matrix of the system is denoted by $A$.

**Theorem 2.8**   If $\det A \neq 0$, the system (2.67) possesses exactly one solution.

PROOF   We first write down the $n \times (n + 1)$ matrix

$$\begin{bmatrix} a_{11} & a_{12} & \cdots & a_{1n} & b_1 \\ a_{21} & a_{22} & \cdots & a_{2n} & b_2 \\ \vdots & \vdots & & \vdots & \vdots \\ a_{n1} & a_{n2} & \cdots & a_{nn} & b_n \end{bmatrix}, \tag{2.68}$$

which is called the *augmented matrix* of the system (2.67). By interchanging two rows of this matrix, or by adding a constant multiple of one row to another, we obtain the augmented matrix of a system that is equivalent to the system (2.67). Furthermore, the determinant of the coefficient matrix of the new system is the same as $\det A$, except possibly for its sign.

Since $A$ is nonsingular, there is at least one element in the first column of the matrix (2.68) that is not zero. By interchanging rows, if necessary, we place a nonzero element in the first row and first column. Then by adding appropriate multiples of the first row to the other rows we introduce zeros in all positions of the first column below the first. The result is an augmented matrix

$$\begin{bmatrix} a'_{11} & a'_{12} & \cdots & a'_{1n} & b'_1 \\ 0 & a'_{22} & \cdots & a'_{2n} & b'_2 \\ \vdots & \vdots & & \vdots & \vdots \\ 0 & a'_{n2} & \cdots & a'_{nn} & b'_n \end{bmatrix} \qquad (2.69)$$

of a system that is equivalent to the system (2.67). Since the coefficient matrix of this system is nonsingular, at least one element in the second column below the first position must be different from zero. We place a nonzero element in the second position of the second column by interchanging rows if necessary. Then we introduce zeros in all positions of the second column below the second position by adding appropriate multiples of the second row to the lower rows. The result is an augmented matrix of the form

$$\begin{bmatrix} a''_{11} & a''_{12} & a''_{13} & \cdots & a''_{1n} & b''_1 \\ 0 & a''_{22} & a''_{23} & \cdots & a''_{2n} & b''_2 \\ 0 & 0 & a''_{33} & \cdots & a''_{3n} & b''_3 \\ \vdots & \vdots & \vdots & & \vdots & \vdots \\ 0 & 0 & a''_{n3} & \cdots & a''_{nn} & b''_n \end{bmatrix}, \qquad (2.70)$$

where $a''_{11}$ and $a''_{22}$ are not zero. Since det $A \neq 0$, at least one element in the third column of the matrix (2.70) below the second position must be different from zero, so we can continue this process. Finally, we arrive at an augmented matrix of the form

$$\begin{bmatrix} \tilde{a}_{11} & \tilde{a}_{12} & \tilde{a}_{13} & \cdots & \tilde{a}_{1n} & \tilde{b}_1 \\ 0 & \tilde{a}_{22} & \tilde{a}_{23} & \cdots & \tilde{a}_{2n} & \tilde{b}_2 \\ 0 & 0 & \tilde{a}_{33} & \cdots & \tilde{a}_{3n} & \tilde{b}_3 \\ \vdots & \vdots & \vdots & & \vdots & \vdots \\ 0 & 0 & 0 & \cdots & \tilde{a}_{nn} & \tilde{b}_n \end{bmatrix}, \qquad (2.71)$$

where none of the elements $\tilde{a}_{11}, \tilde{a}_{22}, \ldots, \tilde{a}_{nn}$ is zero. The last row of this matrix corresponds to the equation $\tilde{a}_{nn} x_n = \tilde{b}_n$, so we can solve for $x_n$, finding $x_n = \tilde{b}_n / \tilde{a}_{nn}$. The $(n-1)$st row corresponds to the equation

$$\tilde{a}_{n-1, n-1} x_{n-1} + \tilde{a}_{n-1, n} x_n = \tilde{b}_{n-1}$$

and $x_{n-1}$ is now found from this equation. By working upward, we find all the solution components, which are uniquely determined. This concludes our proof.

We have established the existence of a unique solution of the system (2.67) by carrying out a Gauss reduction. For a specific case, this procedure provides an efficient method for solving the system.

We shall now derive a formula, known as Cramer's rule, for the solution components of the system (2.67). The use of this rule to find the solution is less efficient than the Gauss reduction method. Nevertheless, it is important for some theoretical purposes, as we shall illustrate.

Cramer's rule can be stated as follows.

**Theorem 2.9**   If det $A \neq 0$, the components of the solution of the system (2.67) are given by the formula,

$$x_k = \frac{\det B_k}{\det A}, \qquad 1 \le k \le n, \tag{2.72}$$

where the matrix $B_k$ is the same as $A$ except that the elements $a_{ik}$, $1 \le i \le n$, in the $k$th column of $A$ have been replaced by the terms $b_i$, $1 \le i \le n$, respectively.

We shall look at an example before proving the theorem. Consider the system

$$
\begin{aligned}
2x_1 + 2x_2 - x_3 &= 2, \\
-3x_1 - x_2 + 3x_3 &= -2, \\
4x_1 + 2x_2 - 3x_3 &= 0.
\end{aligned}
\tag{2.73}
$$

Calculation shows that the determinant of the coefficient matrix is 2. Since this matrix is nonsingular, Theorem 2.9 applies. Using formula (2.72), we have

$$x_1 = \frac{1}{2} \begin{vmatrix} 2 & 2 & -1 \\ -2 & -1 & 3 \\ 0 & 2 & -3 \end{vmatrix} = \frac{-14}{2} = -7,$$

$$x_2 = \frac{1}{2} \begin{vmatrix} 2 & 2 & -1 \\ -3 & -2 & 3 \\ 4 & 0 & -3 \end{vmatrix} = \frac{10}{2} = 5,$$

$$x_3 = \frac{1}{2} \begin{vmatrix} 2 & 2 & 2 \\ -3 & -1 & -2 \\ 4 & 2 & 0 \end{vmatrix} = \frac{-12}{2} = -6.$$

PROOF OF THEOREM 2.9   Let $(x_1, x_2, \ldots, x_n)$ be the solution of the system (2.67). Then

$$\sum_{j=1}^{n} a_{ij} x_j = b_i, \qquad 1 \le i \le n.$$

Let $k$ be any fixed positive integer such that $1 \le k \le n$. Multiplying through in the $i$th equation by $A_{ik}$ and adding equations, we have

$$\sum_{i=1}^{n} \sum_{j=1}^{n} a_{ij} A_{ik} x_j = \sum_{i=1}^{n} b_i A_{ik} .$$

Interchanging the order of summation on the left-hand side, we have

$$\sum_{j=1}^{n} x_j \sum_{i=1}^{n} a_{ij} A_{ik} = \sum_{i=1}^{n} b_i A_{ik} .$$

But by Theorem 2.7, the inner sum on the left-hand side is equal to $\delta_{jk} \det A$. Consequently this equation becomes

$$\det A \sum_{j=1}^{n} x_j \delta_{jk} = \sum_{i=1}^{n} b_i A_{ik}$$

or

$$(\det A)x_k = \sum_{i=1}^{n} b_i A_{ik} . \qquad (2.74)$$

Now $\det A$ can be expressed in terms of the cofactors of the $k$th column of $A$ as

$$\sum_{i=1}^{n} a_{ik} A_{ik} .$$

Hence the sum on the right-hand side of Eq. (2.74) is the determinant of a matrix $B_k$ that is obtained from $A$ by replacing $a_{ik}$ by $b_i$ for $i = 1, 2, \ldots, n$. Since $\det A \ne 0$, we can divide through in Eq. (2.74) to obtain the formula (2.72)
    Notice that the relation (2.74), which can be written as

$$(\det A)x_k = \det B_k , \quad 1 \le k \le n,$$

holds regardless whether or not $\det A = 0$. The only place where we used the assumption $\det A \ne 0$ was where we divided both sides of this equation by $\det A$. Thus if $\det A \ne 0$, the system (2.67) has a unique solution; if $\det A = 0$, the system can have a solution only if $B_k = 0$ for all $k$.
    The special case when the system (2.67) is homogeneous ($b_1 = b_2 = \cdots = b_n = 0$) is of some interest. A homogeneous system always possesses the

trivial solution, all of whose components are zero. If $A$ is nonsingular Theorem 2.8 says that this is the only solution. In order to see what happens when $\det A = 0$ we use the Gauss reduction method of solution that was described in Section 2.1. Application of this method shows that the system (2.67) is equivalent to a system of the form

$$
\begin{aligned}
\tilde{a}_{11}\, x_{i_1} + \tilde{a}_{12}\, x_{i_2} + \cdots + \tilde{a}_{1r}\, x_{i_r} + \cdots + \tilde{a}_{1n}\, x_{i_n} &= 0, \\
\tilde{a}_{22}\, x_{i_2} + \cdots + \tilde{a}_{2r}\, x_{i_r} + \cdots + \tilde{a}_{2n}\, x_{i_n} &= 0, \\
\cdots\cdots\cdots\cdots\cdots\cdots\cdots\cdots\cdots\cdots\cdots\cdots\cdots\cdots\cdots& \\
\tilde{a}_{rr}\, x_{i_r} + \cdots + \tilde{a}_{rn}\, x_{i_n} &= 0, \\
0 &= 0, \\
\cdots& \\
0 &= 0,
\end{aligned}
$$

where $r \le n$ and $\tilde{a}_{ii} \ne 0$ for $1 \le i \le r$. Because of the types of operations used in deriving this system, the determinant of its coefficient matrix is also zero. Hence we must have $r < n$ and the system possesses an $n - r$ parameter family of solutions. We summarize as follows.

**Theorem 2.10**   The homogeneous system with $n$ equations and $n$ unknowns,

$$
\begin{aligned}
a_{11}x_1 + \cdots + a_{1n}x_n &= 0, \\
\cdots\cdots\cdots\cdots\cdots\cdots\cdots& \\
a_{n1}x_1 + \cdots + a_{nn}x_n &= 0,
\end{aligned}
$$

possesses a nontrivial solution if and only if $\det A = 0$.

As an application of Cramer's rule, let the quantities $a_{ij}$ and $b_i$, $1 \le i \le n$, $1 \le j \le n$, be *continuous functions* defined on an interval $\mathscr{I}$. If $\det A(x)$ does not vanish for any $x$ in $\mathscr{I}$, the system

$$
\begin{aligned}
a_{11}(x)f_1(x) + \cdots + a_{1n}(x)f_n(x) &= b_1(x), \\
\cdots\cdots\cdots\cdots\cdots\cdots\cdots\cdots\cdots\cdots\cdots& \\
a_{n1}(x)f_1(x) + \cdots + a_{nn}(x)f_n(x) &= b_n(x),
\end{aligned}
$$

determines a set of functions $f_1, f_2, \ldots, f_n$ each defined on $\mathscr{I}$. Cramer's rule allows us to conclude that each of these functions is *continuous*, since each can be expressed as the quotient of quantities that are the sums of products of continuous functions. It is not necessary to actually solve the system to determine this important property of the solution functions.

## Exercises for Section 2.8

1. Let $A$ be of size $m \times n$. Under what conditions can Cramer's rule be used to find the solutions of the equation $Ax = b$?

In Exercises 2–9, solve the system by Cramer's rule if the rule applies. If it does not, find all solutions that exist by another method.

2.
$$3x_1 - 2x_2 = 1$$
$$-2x_1 + 2x_2 = 5$$

3.
$$4x_1 + 5x_2 = 8$$
$$2x_1 + x_2 = -7$$

4.
$$2x_1 - 6x_2 = 1$$
$$-x_1 + 3x_2 = 4$$

5.
$$-3x_1 + x_2 = 6$$
$$9x_1 - 3x_2 = -18$$

6.
$$2x_1 - x_2 + 3x_3 = 1$$
$$x_2 + 2x_3 = -3$$
$$x_1 + x_3 = 0$$

7.
$$-4x_1 + x_2 = 3$$
$$2x_1 + 2x_2 + x_3 = -2$$
$$3x_1 + 4x_3 = 2$$

8.
$$3x_1 + x_3 = -2$$
$$x_1 + 2x_2 - x_3 = 0$$
$$x_1 - 4x_2 + 3x_3 = 1$$

9.
$$-2x_1 - x_2 = 3$$
$$x_1 + 3x_2 - x_3 = 0$$
$$5x_2 - 2x_3 = 3$$

10. Let $A$ be a singular matrix. Show that the equation $Ax = b$ is either inconsistent or else possesses infinitely many solutions. (Use the Gauss reduction method.)

11. Consider the system

$$a_{11}x_1 + a_{12}x_2 = b_1,$$
$$a_{21}x_1 + a_{22}x_2 = b_1,$$

where neither $a_{11}$ nor $a_{21}$ is zero. Clearly, we can eliminate $x_1$ from the second equation by multiplying through in the first equation by $a_{21}/a_{11}$ and subtracting the resulting equation from the second equation. Alternatively, we could eliminate $x_1$ from the first equation by multiplying through in the second equation by $a_{11}/a_{21}$ and subtracting the resulting equation from the first equation. In a practical problem the numbers $a_{ij}$ and $b_i$ would probably be rounded and not exact. If $|a_{11}/a_{21}| < 1$, explain why it would be better to eliminate $x_1$ from the first equation rather than the second.

## 2.9  THE INVERSE OF A MATRIX

Let us consider the equation

$$Ax = b, \tag{2.75}$$

where $A$ is a nonsingular $n \times n$ matrix. If we can find an $n \times n$ matrix $B$ with the property that

$$BA = I,\qquad\qquad(2.76)$$

then we can solve the equation easily. Upon multiplying both sides of Eq. (2.75) by $B$ we have

$$BA\mathbf{x} = B\mathbf{b}$$

or

$$\mathbf{x} = B\mathbf{b}.\qquad\qquad(2.77)$$

If there does exist a matrix $B$ such that $BA = I$, then $A$ and $B$ must both be nonsingular. For det $I = 1$ and hence $\det(BA) = \det B \cdot \det A = 1$.

If $BA = I$ then it is also true that $AB = I$. To prove this, let $C = AB$. Then

$$BC = BAB = IB = B,$$

so that

$$B(C - I) = 0.$$

Every column vector of $C - I$ is a solution of the equation $B\mathbf{x} = \mathbf{0}$, according to formula (2.41). Since $B$ is nonsingular, Cramer's rule asserts that $C - I = 0$ or $C = AB = I$.

Suppose that $A$ is nonsingular. Let $\mathbf{e}_1, \mathbf{e}_2, \ldots, \mathbf{e}_n$ be the column vectors of $I$ and let $B$ be an $n \times n$ matrix with column vectors $\mathbf{b}_1, \mathbf{b}_2, \ldots, \mathbf{b}_n$. Then $B$ satisfies the equation $AB = I$ if and only if

$$A[\mathbf{b}_1, \mathbf{b}_2, \ldots, \mathbf{b}_n] = [\mathbf{e}_1, \mathbf{e}_2, \ldots, \mathbf{e}_n]$$

or

$$A\mathbf{b}_1 = \mathbf{e}_1, \quad A\mathbf{b}_2 = \mathbf{e}_2, \ldots, \quad A\mathbf{b}_n = \mathbf{e}_n.\qquad\qquad(2.78)$$

Since $A$ is nonsingular, there exists exactly one vector $\mathbf{b}_i$ such that $A\mathbf{b}_i = \mathbf{e}_i$, $1 \le i \le n$. We summarize our results thus far as follows.

**Theorem 2.11**  If $A$ is a nonsingular $n \times n$ matrix, there exists one and only one $n \times n$ matrix $B$ such that

$$AB = BA = I.$$

The column vectors of $B$ are the solutions of Eqs. (2.78).

The matrix $B$ is called the *inverse* of the matrix $A$ and is denoted by $A^{-1}$.
As an example, let us find the inverse of the (nonsingular) matrix

$$A = \begin{bmatrix} 0 & 1 & 0 \\ 2 & 0 & -1 \\ 1 & 3 & 1 \end{bmatrix}. \tag{2.79}$$

To do this, we need to find vectors $\mathbf{b}_1$, $\mathbf{b}_2$, and $\mathbf{b}_3$ such that

$$A\mathbf{b}_1 = \begin{bmatrix} 1 \\ 0 \\ 0 \end{bmatrix}, \qquad A\mathbf{b}_2 = \begin{bmatrix} 0 \\ 1 \\ 0 \end{bmatrix}, \qquad A\mathbf{b}_3 = \begin{bmatrix} 0 \\ 0 \\ 1 \end{bmatrix}. \tag{2.80}$$

Then the inverse of $A$ will be given by the formula

$$A^{-1} = [\mathbf{b}_1, \mathbf{b}_2, \mathbf{b}_3].$$

We can solve the three equations (2.80) more or less simultaneously by using the following procedure. First we write down the $3 \times 6$ matrix

$$\begin{bmatrix} 0 & 1 & 0 & 1 & 0 & 0 \\ 2 & 0 & -1 & 0 & 1 & 0 \\ 1 & 3 & 1 & 0 & 0 & 1 \end{bmatrix}$$

formed by adjoining the identity matrix to $A$. By successively performing *elementary row operations* (interchanging two rows, multiplying through in a row by a number other than zero, adding a multiple of one row to another) we arrive at the matrix

$$\begin{bmatrix} 1 & 0 & 0 & -1 & \frac{1}{3} & \frac{1}{3} \\ 0 & 1 & 0 & 1 & 0 & 0 \\ 0 & 0 & 1 & -2 & -\frac{1}{3} & \frac{2}{3} \end{bmatrix}. \tag{2.81}$$

Hence the Eqs. (2.80) are are also equivalent to the equations

$$I\mathbf{b}_1 = \begin{bmatrix} -1 \\ 1 \\ -2 \end{bmatrix}, \qquad I\mathbf{b}_2 = \begin{bmatrix} \frac{1}{3} \\ 0 \\ -\frac{1}{3} \end{bmatrix}, \qquad I\mathbf{b}_3 = \begin{bmatrix} \frac{1}{3} \\ 0 \\ \frac{2}{3} \end{bmatrix},$$

respectively. The last three columns of the matrix (2.81) are therefore the column vectors of $A^{-1}$. Hence

$$A^{-1} = \begin{bmatrix} -1 & \frac{1}{3} & \frac{1}{3} \\ 1 & 0 & 0 \\ -2 & -\frac{1}{3} & \frac{2}{3} \end{bmatrix}. \tag{2.82}$$

In general, if $A$ is an $n \times n$ nonsingular matrix, we can form an $n \times 2n$ matrix by adjoining the $n \times n$ identity matrix to $A$. We write this matrix as

$$[A : I]. \tag{2.83}$$

This matrix can be reduced to the form

$$[I : C] \tag{2.84}$$

by elementary row operations (Exercise 17). Since the equation $A\mathbf{b}_k = \mathbf{e}_k$ is equivalent to the equation $I\mathbf{b}_k = \mathbf{c}_k$, $1 \le k \le n$, it is clear that $C = A^{-1}$. Thus the matrix (2.84) is

$$[I : A^{-1}]. \tag{2.85}$$

Our next theorem gives a formula for the elements of $A^{-1}$. While not very practical for computational purposes, the formula is occasionally useful for theoretical purposes.

**Theorem 2.12**  Let the element in the $i$th row and $j$th column of $A^{-1}$ be denoted by $\tilde{a}_{ij}$. Then

$$\tilde{a}_{ij} = \frac{A_{ji}}{\det A}, \tag{2.86}$$

where $A_{ji}$ is the cofactor of the element $a_{ji}$ of $A$. Thus to find $A^{-1}$ we replace each element $a_{ij}$ of $A$ by its cofactor $A_{ij}$, form the transpose of the resulting matrix, and divide each element by $\det A$.

PROOF   We must show that if $\tilde{a}_{ij}$ is given by formula (2.86) then

$$\sum_{k=1}^{n} a_{ik} \tilde{a}_{kj} = \delta_{ij}, \quad 1 \le i \le n, \quad 1 \le j \le n.$$

Using Theorem 2.7, we see that

$$\sum_{k=1}^{n} a_{ik}\, \tilde{a}_{kj} = \frac{1}{\det A} \sum_{k=1}^{n} a_{ik}\, A_{jk}$$

$$= \frac{1}{\det A}\, \delta_{ij}\, \det A$$

$$= \delta_{ij},$$

which we wished to show.

We apply the theorem to find the inverse of the matrix (2.79). Calculating cofactors, we have

$$\begin{bmatrix} A_{11} & A_{12} & A_{13} \\ A_{21} & A_{22} & A_{23} \\ A_{31} & A_{32} & A_{33} \end{bmatrix} = \begin{bmatrix} 3 & -3 & 6 \\ -1 & 0 & 1 \\ -1 & 0 & -2 \end{bmatrix}.$$

Taking the transpose and dividing by $\det A = -3$, we find that

$$A^{-1} = -\frac{1}{3}\begin{bmatrix} 3 & -1 & -1 \\ -3 & 0 & 0 \\ 6 & 1 & -2 \end{bmatrix} = \begin{bmatrix} -1 & \frac{1}{3} & \frac{1}{3} \\ 1 & 0 & 0 \\ -2 & -\frac{1}{3} & \frac{2}{3} \end{bmatrix}.$$

A more practical use of Theorem 2.12 is illustrated by the following result, which will be of theoretical importance in our study of linear differential equations.

**Theorem 2.13** Let each of the functions $a_{ij}$, $1 \le i \le n$, $1 \le j \le n$, be defined and continuous on an interval $\mathscr{I}$. Suppose that the matrix $A(x)$ with elements $a_{ij}(x)$ is nonsingular for each $x$ in $\mathscr{I}$. If the elements of $[A(x)]^{-1}$ are denoted by $\tilde{a}_{ij}(x)$ then the functions $\tilde{a}_{ij}$ are continuous on $\mathscr{I}$.

PROOF The functions $A_{ij}$ and $\det A$ are continuous, being the sums of products of continuous functions. According to Theorem 2.12, the functions $\tilde{a}_{ij}$ are the quotients of continuous functions and therefore are continuous.

### Exercises for Section 2.9

1. If $A$ is singular, why can there exist no matrix $B$ such that $AB = I$ or $BA = I$?

In Exercises 2–11, find the inverse of each nonsingular matrix (a) by using the method of the example (2.79); (b) by using Theorem 2.12.

2. $\begin{bmatrix} -2 & 2 \\ -4 & 3 \end{bmatrix}$

3. $\begin{bmatrix} 2 & 1 \\ -5 & -4 \end{bmatrix}$

4. $\begin{bmatrix} 2 & -1 \\ -4 & 2 \end{bmatrix}$

5. $\begin{bmatrix} 3 & 1 \\ 1 & 1 \end{bmatrix}$

6. $\begin{bmatrix} 1 & -2 & 0 \\ 2 & -1 & 1 \\ 0 & 4 & 2 \end{bmatrix}$

7. $\begin{bmatrix} 1 & 0 & 2 \\ 2 & -3 & 4 \\ 0 & 2 & 1 \end{bmatrix}$

8. $\begin{bmatrix} 3 & 1 & 0 \\ 2 & 1 & 1 \\ 1 & 0 & 1 \end{bmatrix}$

9. $\begin{bmatrix} 2 & -1 & 1 \\ 1 & 2 & -2 \\ 3 & 1 & 0 \end{bmatrix}$

10. $\begin{bmatrix} 2 & 3 & 2 \\ 3 & 1 & -2 \\ -1 & 0 & 1 \end{bmatrix}$

11. $\begin{bmatrix} 1 & 1 & -1 & 2 \\ 0 & 2 & 0 & -1 \\ -1 & 2 & 2 & -2 \\ 0 & -1 & 0 & 1 \end{bmatrix}$

12. If $A$ is nonsingular, show that $(A^T)^{-1} = (A^{-1})^T$.

13. Let $A$ and $B$ be nonsingular matrices of the same order. If $C = AB$, show that $C^{-1} = B^{-1}A^{-1}$.

14. Let $A$ be nonsingular. Show that the inverse of $A^{-1}$ is $A$.

15. The matrix $I$ is its own inverse, since $I \cdot I = I$.

    (a) Find at least two second-order matrices (other than $I$) that have this property.

    (b) If $A = A^{-1}$, show that det $A = \pm 1$.

16. A real nonsingular matrix $A$ is said to be *orthogonal* if $A^{-1} = A^T$.

    (a) If $A$ is orthogonal, show that det $A = \pm 1$.

    (b) Show that $A$ is orthogonal if and only if the row (column) vectors of $A$ are mutually orthogonal unit vectors.

17. If $A$ is nonsingular, show that $A$ can be changed into the identity matrix $I$ by a finite sequence of elementary row operations. Suggestion: modify the reduction procedure used in Section 2.8.

# III

## Vector Spaces and Linear Transformations

### 3.1 VECTOR SPACES

In Section 2.2 we defined $R^n$ to be the set of all ordered $n$-tuples of real numbers. We used the notation

$$\mathbf{u} = (u_1, u_2, \ldots, u_n), \qquad \mathbf{v} = (v_1, v_2, \ldots, v_n)$$

for elements, or vectors, in $R^n$. If $\mathbf{u}$ is in $R^n$ and $c$ is a real number, the product of the number $c$ and the vector $\mathbf{u}$ is defined as

$$c\mathbf{u} = (cu_1, cu_2, \ldots, cu_n).$$

This product is again an element of $R^n$. We indicate this fact by saying that $R^n$ is *closed under the operation of multiplication by a number.* The sum $\mathbf{u} + \mathbf{v}$ of two elements of $R^n$ is defined as

$$\mathbf{u} + \mathbf{v} = (u_1 + v_1, u_2 + v_2, \ldots, u_n + v_n).$$

Since this sum is an element of $R^n$, we say that $R^n$ is *closed under the operation of addition.* We now alter our definition of $R^n$ slightly. We define $R^n$ to be the set of all ordered $n$-tuples of real numbers, *together with two operations,* "multiplication by a number" and "addition." Thus $R^n$ is not merely a set, but a set with which two operations are associated. The set is closed under

*113*

these operations; that is, performance of either operation always leads to an element of the set. Notice that $R^n$ possesses a special element,

$$\mathbf{0} = (0, 0, \ldots, 0)$$

with the property that

$$\mathbf{u} + \mathbf{0} = \mathbf{u}$$

for every element $\mathbf{u}$ in $R^n$. The element $\mathbf{0}$ is called the *zero element* of $R^n$. Associated with every element $\mathbf{u}$ of $R^n$ is an element

$$-\mathbf{u} = (-1)\mathbf{u} = (-u_1, -u_2, \ldots, -u_n)$$

with the property that

$$\mathbf{u} + (-\mathbf{u}) = \mathbf{0}.$$

Many other sets of objects that are frequently encountered in mathematics possess properties similar to those of $R^n$. Of course, two operations ("multiplication by a number" and "addition") must be associated with the set.

For example, let us consider the set of all real-valued functions defined on an interval $\mathscr{I}$. If $f$ is such a function and $c$ is a real number, then the product of $c$ and $f$, $cf$, is defined to be the function whose value at each point $x$ in $\mathscr{I}$ is $cf(x)$. The sum $f + g$ of two functions is defined to be the function with the value $f(x) + g(x)$ at each $x$ in $\mathscr{I}$. Thus the set of functions defined on $\mathscr{I}$ is closed under the operations of addition and multiplication by a number. The zero function 0 (we use the symbol 0 for both the number zero and the zero function) whose values are all zero, has the property that

$$f + 0 = f$$

for every function $f$. Also, if we define

$$-f = (-1)f,$$

then

$$f + (-f) = 0$$

for every function $f$.

The similarities between the set of vectors in $R^n$ and the set of functions arise not so much because of the nature of the *elements* of the sets, but because of the nature of the *operations* defined on the sets. It is therefore more efficient to consider, in an abstract way, sets of objects whose elements are not defined, but on which are defined operations of "multiplication by a number" and "addition." Sets possessing the properties described above, in addition to certain others, are called *vector spaces* or *linear spaces*. We now formulate a precise definition.

Let $V$ be a nonempty collection or set of objects. (We denote elements of $V$ by **u**, **v**, **w**, and so on. We denote real numbers by $a$, $b$, $c$, and so on.) For every real number $a$ and every element **u** of $V$, let the "product" $a$**u** be defined and be an element of $V$. For every ordered pair (**u**, **v**) of elements of $V$ let the "sum" **u** + **v** be defined and be an element of $V$. Let the operations of "multiplication by a number" and "addition" be such that the following properties hold. (Here the symbol = means "is the same as.")

**Property 1**   For all **u**, **v**, and **w** in $V$,

$$\mathbf{v} + \mathbf{u} = \mathbf{u} + \mathbf{v}, \qquad (\mathbf{u} + \mathbf{v}) + \mathbf{w} = \mathbf{u} + (\mathbf{v} + \mathbf{w}).$$

**Property 2**   There exists an element **0** of $V$, called the *zero element*, with the property that

$$\mathbf{u} + \mathbf{0} = \mathbf{u}$$

for every element **u** of $V$.

**Property 3**   For every element **u** of $V$ there exists an element $-$**u** of $V$, called the *additive inverse* of **u**, such that

$$\mathbf{u} + (-\mathbf{u}) = \mathbf{0}.$$

**Property 4**   For all **u** and **v** in $V$ and for all real numbers $a$, $b$, and $c$,

$$c(\mathbf{u} + \mathbf{v}) = c\mathbf{u} + c\mathbf{v}, \qquad (a + b)\mathbf{u} = a\mathbf{u} + b\mathbf{u},$$
$$a(b\mathbf{u}) = (ab)\mathbf{u}, \qquad\qquad 1\mathbf{u} = \mathbf{u}.$$

Then $V$, together with the operations of "multiplication by a number" and "addition," is called a *vector space*, or a *linear space*, *over the real numbers*. More briefly, we refer to $V$ as a *real vector space*, or a *real linear space*.

If $\mathbf{v}_1$ and $v_2$ are in $V$ and if $c_1$ and $c_2$ are real numbers, then

$$c_1 \mathbf{v}_1 + c_2 \mathbf{v}_2$$

is in $V$. This is because $c_1 \mathbf{v}_1$ and $c_2 \mathbf{v}_2$ are in $V$, and the sum of two elements of $V$ is in $V$. If $\mathbf{v}_1$, $\mathbf{v}_2$, and $\mathbf{v}_3$ are in $V$, then we see that

$$c_1 \mathbf{v}_1 + c_2 \mathbf{v}_2 + c_3 \mathbf{v}_3 = (c_1 \mathbf{v}_1 + c_2 \mathbf{v}_2) + c_3 \mathbf{v}_3$$

is in $V$. It can be shown, by mathematical induction, that if $\mathbf{v}_1, \mathbf{v}_2, \ldots, \mathbf{v}_m$ are elements of $V$ (not necessarily all different) and if $c_1, c_2, \ldots, c_m$ are real numbers, then

$$c_1 \mathbf{v}_1 + c_2 \mathbf{v}_2 + \cdots + c_m \mathbf{v}_m$$

is in $V$.

The important properties

$$0\mathbf{u} = \mathbf{0}, \tag{3.1}$$

$$(-\mathbf{u}) = (-1)\mathbf{u}, \tag{3.2}$$

which hold for every element $\mathbf{u}$ of $V$, follow from the definition of a vector space (Exercise 1).

It is easy to see that $R^n$ is a real vector space. Another example of a real vector space is the set of all real-valued functions defined on an interval (together with the operations defined previously). The elements of a vector space need not be " vectors " in the sense of elements of $R^n$. In the last example they are functions. In the exercises are examples of vector spaces whose elements are numbers, matrices, infinite sequences, and other mathematical objects.

By allowing the numbers $a$, $b$, $c$, and so on, in our formal definition of a vector space to. be complex, we can define a *vector space over the complex numbers*, or a *complex vector space*. As an important example of such a space, let us consider the set $C^n$ of all ordered $n$-tuples of complex numbers. If

$$\mathbf{u} = (u_1, u_2, \ldots, u_n), \qquad \mathbf{v} = (v_1, v_2, \ldots, v_n)$$

are elements of $C^n$ (the quantities $u_i$ and $v_i$ are complex numbers) we define

$$\mathbf{u} + \mathbf{v} = (u_1 + v_1, u_2 + v_2, \ldots, u_n + v_n)$$

and for every complex number $c$ we define

$$c\mathbf{u} = (cu_1, cu_2, \ldots, cu_n).$$

Then it may be verified that $C^n$, together with the indicated operations, is a complex vector space.

The set of real numbers and the set of complex numbers are examples of mathematical entities known as *fields*. In more advanced treatments of linear algebra, vector spaces over arbitrary fields are considered. The elements of a field we call *scalars*. We shall restrict our attention only to real and complex numbers as scalars since, in applications, the important vector spaces are those over the fields of real and complex numbers.

## Exercises for Section 3.1

1. Show that the properties (3.1) and (3.2) hold for every vector space.

In Exercises 2–13, determine if the given set constitutes a real vector space. In each case the operations of "multiplication by a number" and "addition" are understood to be the usual operations associated with the elements of the set.

2. The set of all geometric vectors of the form $\mathbf{v} = v_1\mathbf{i} + v_2\mathbf{j} + v_3\mathbf{k}$, where $\mathbf{i}$, $\mathbf{j}$, and $\mathbf{k}$ are mutually orthogonal unit vectors.

3. The set of all real numbers.

4. The set of all elements of $R^3$ with first component 0.

5. The set of all elements of $R^3$ with first component 1.

6. The set of all $2 \times 2$ matrices with real elements.

7. The set of all nonsingular $2 \times 2$ matrices with real elements.

8. The set of all singular $2 \times 2$ matrices with real elements.

9. The set of all continuous functions defined on an interval $\mathscr{I}$.

10. The set of all polynomials of degree $\leq 2$.

11. The set of all solutions of the homogeneous equation $A\mathbf{x} = \mathbf{0}$. (Here $A$ is an $m \times n$ matrix; $\mathbf{x}$ and $\mathbf{0}$ are column vectors.)

12. The set of all convergent infinite sequences of real numbers.

13. The set of all convergent infinite series of real numbers.

14. Is the set of all ordered $n$-tuples of real numbers a complex vector space?

15. Is the set of all ordered $n$-tuples of complex numbers a real vector space?

**16.** If $V$ is a vector space over the complex numbers, show that the set of elements of $V$ constitutes a vector space over the real numbers.

**17.** Show that a vector space has only one zero element. That is, show that if $\mathbf{v} + \mathbf{0} = \mathbf{v}$ and $\mathbf{v} + \mathbf{0}' = \mathbf{v}$ for every element $\mathbf{v}$, then $\mathbf{0}' = \mathbf{0}$.

**18.** If $\mathbf{u}$, $\mathbf{v}$, and $\mathbf{w}$ are elements of a vector space and $\mathbf{u} + \mathbf{v} = \mathbf{u} + \mathbf{w}$, show that $\mathbf{v} = \mathbf{w}$.

**19.** In a vector space, show that $c\mathbf{0} = \mathbf{0}$ for every scalar $c$.

## 3.2  SUBSPACES

In order to illustrate the main concept of this section, let us consider the set of all elements in $R^3$ whose first components are zero. If

$$\mathbf{x} = (0, x_2, x_3), \qquad \mathbf{y} = (0, y_2, y_3),$$

then

$$c\mathbf{x} = (0, cx_2, cx_3)$$

for every real number $c$ and

$$\mathbf{x} + \mathbf{y} = (0, x_2 + y_2, x_3 + y_3).$$

Also, the zero element $(0, 0, 0)$ of $R^3$ belongs to this set. It is now easy to verify that the set of all elements of $R^3$ with first component 0 forms a real vector space under the same operations associated with $R^3$.

More generally, let $V$ be any vector space and suppose that some subset of $V$ is a vector space under the same operations and the same field associated with $V$. Then this subset (together with the two operations) is called a *subspace* of $V$. In particular, $V$ is a subspace of itself. Also, the set whose single element is the zero element of $V$ is a subspace of $V$. In order to show that a subset of $V$ is a subspace, it suffices to show that the subset is closed under the two basic operations (see Exercise 1).

As another example, let $\mathscr{F}$ be the vector space of all real valued functions defined on some interval $\mathscr{I}$. Then the set of all *continuous* real valued functions defined on $\mathscr{I}$ forms a subspace of $\mathscr{F}$: if $f$ and $g$ are continuous, $f + g$ and $cf$ are also continuous.

Now let $V$ be an arbitrary vector space and let $\mathbf{v}, \mathbf{v}_1, \mathbf{v}_2, \ldots, \mathbf{v}_m$ be elements of $V$. If there exist scalars $c_1, c_2, \ldots, c_m$ such that

$$\mathbf{v} = c_1\mathbf{v}_1 + c_2\mathbf{v}_2 + \cdots + c_m\mathbf{v}_m,$$

we say that $\mathbf{v}$ is a *linear combination* of the elements $\mathbf{v}_1, \mathbf{v}_2, \ldots, \mathbf{v}_m$. We also use the summation notation

$$\mathbf{v} = \sum_{i=1}^{m} c_i \mathbf{v}_i.$$

If *every* element of $V$ is a linear combination of the elements $\mathbf{v}_1, \mathbf{v}_2, \ldots, \mathbf{v}_m$ of $V$, we say that $V$ is *spanned* by these elements. We also say that the elements *span* $V$.

Suppose that $\mathbf{u}_1, \mathbf{u}_2, \ldots, \mathbf{u}_k$ are any $k$ elements of a vector space $V$. Then the set of all linear combinations of these elements, that is, the set of all elements of $V$ of the form

$$c_1 \mathbf{u}_1 + c_2 \mathbf{u}_2 + \cdots + c_k \mathbf{u}_k \tag{3.3}$$

is a subspace of $V$. This follows because if

$$\mathbf{v} = a_1 \mathbf{u}_1 + a_2 \mathbf{u}_2 + \cdots + a_k \mathbf{u}_k, \qquad \mathbf{w} = b_1 \mathbf{u}_1 + b_2 \mathbf{u}_2 + \cdots + b_k \mathbf{u}_k,$$

then

$$c\mathbf{v} = (ca_1)\mathbf{u}_1 + (ca_2)\mathbf{u}_2 + \cdots + (ca_k)\mathbf{u}_k$$

and

$$\mathbf{v} + \mathbf{w} = (a_1 + b_1)\mathbf{u}_1 + (a_2 + b_2)\mathbf{u}_2 + \cdots + (a_k + b_k)u_k$$

are again of the form (3.3). The elements $\mathbf{u}_1, \mathbf{u}_2, \ldots, \mathbf{u}_k$ are in the subspace, because

$$\mathbf{u}_1 = 1\mathbf{u}_1 + 0\mathbf{u}_2 + \cdots + 0\mathbf{u}_k, \qquad \mathbf{u}_2 = 0\mathbf{u}_1 + 1\mathbf{u}_2 + \cdots + 0\mathbf{u}_k,$$

and so on. The subspace is evidently spanned by these $k$ elements. Also, any subspace that contains the $k$ elements must contain every linear combination of them. Thus we may speak of *the* subspace spanned by the $k$ elements.

As an example, consider the space $R^3$ and let $\mathbf{u}_1 = (1, 1, 0)$ and $\mathbf{u}_2 = (0, 1, 2)$. Then the subspace of $R^3$ that is spanned by $\mathbf{u}_1$ and $\mathbf{u}_2$ is the set of all vectors of the form $c_1 \mathbf{u}_1 + c_2 \mathbf{u}_2$ or $(c_1, c_1 + c_2, 2c_2)$. A geometric intepretation of this subspace (obtained by regarding the components $x_1, x_2, x_3$ of a vector $\mathbf{x}$ as the rectangular coordinates of a point) is the plane through the origin and the points $(1, 1, 0)$ and $(0, 1, 2)$.

Suppose that

$$\mathbf{v}_1, \mathbf{v}_2, \ldots, \mathbf{v}_m \tag{3.4}$$

are elements of a vector space $V$. Then these elements span a subspace $\tilde{V}$ of $V$, where $\tilde{V}$ is the set of all linear combinations of the elements (3.4). If one of these elements, say the $i$th, is multiplied by a nonzero scalar $c$, we obtain the set of elements

$$\mathbf{v}_1, \mathbf{v}_2, \ldots, c\mathbf{v}_i, \ldots, \mathbf{v}_m \tag{3.5}$$

of $V$ which also span a subspace of $V$. This subspace is the same as $\tilde{V}$. To see this, let $\mathbf{x}$ be any element of $\tilde{V}$. Then

$$\mathbf{x} = a_1\mathbf{v}_1 + \cdots + a_i\mathbf{v}_i + \cdots + a_m\mathbf{v}_m,$$
$$= a_1\mathbf{v}_1 + \cdots + a_i c^{-1}(c\mathbf{v}_i) + \cdots + a_m\mathbf{v}_m,$$

so $\mathbf{x}$ is in the subspace spanned by the elements (3.5). On the other hand, if $\mathbf{y}$ is in the space spanned by the elements (3.5) it is also in $\tilde{V}$. This follows because if

$$\mathbf{y} = b_1\mathbf{v}_1 + \cdots + b_i(c\mathbf{v}_i) + \cdots + b_m\mathbf{v}_m,$$

then

$$\mathbf{y} = b_1\mathbf{v}_1 + \cdots + (b_i c)\mathbf{v}_i + \cdots + b_m\mathbf{v}_m.$$

Hence the elements (3.4) and the elements (3.5) span the same subspace of $V$.

If we add $k$ times one element (say the $i$th) of the set (3.4) to another (say the $j$th), we obtain the set of elements

$$\mathbf{v}_1, \ldots, \mathbf{v}_i, \ldots, \mathbf{v}_j + k\mathbf{v}_i, \ldots, \mathbf{v}_m \tag{3.6}$$

of $V$. We leave it to the reader (Exercise 10) to show that the subspace spanned by the elements (3.6) is the same as $\tilde{V}$. We summarize these results as follows.

**Theorem 3.1**    Let $V$ be a vector space and let $\tilde{V}$ be the subspace of $V$ that is spanned by the elements $\mathbf{v}_1, \mathbf{v}_2, \ldots, \mathbf{v}_m$ of $V$. Then $\tilde{V}$ is the same as the subspace that is spanned by the elements

$$\mathbf{v}_1, \mathbf{v}_2, \ldots, c\mathbf{v}_i, \ldots, \mathbf{v}_m,$$

where $c \neq 0$, and also $\tilde{V}$ is the same as the subspace that is spanned by the elements

$$\mathbf{v}_1, \mathbf{v}_2, \ldots, \mathbf{v}_i, \ldots, \mathbf{v}_j + k\mathbf{v}_i, \ldots, \mathbf{v}_m,$$

where $i \neq j$ and $k$ is any scalar.

We conclude this section with two definitions. Let $V_1$ and $V_2$ be subspaces of a vector space $V$. By the *intersection* of $V_1$ and $V_2$, written $V_1 \cap V_2$, we mean the set of all elements of $V$ that are in both $V_1$ and $V_2$. By the *sum of* $V_1$ and $V_2$, written $V_1 + V_2$, we mean the set of all elements of $V$ that are of the form $\mathbf{v}_1 + \mathbf{v}_2$, where $\mathbf{v}_1$ is in $V_1$ and $\mathbf{v}_2$ is in $V_2$. It is left to the exercises to show that $V_1 \cap V_2$ and $V_1 + V_2$ are subspaces of $V$.

## Exercises for Section 3.2

1. Let $V$ be a vector space. Show that a subset of $V$ is a subspace of $V$ (under the same operations and the same field associated with $V$) if it has the properties of being closed under the operations associated with $V$. (In other words, show that if the subset is closed under the two operations, it automatically has Properties 1–4 of a vector space of Section 3.1.)

2. Show that the set of all elements of $R^2$ that are of the form $(a, -a)$, where $a$ is any real number, is a subspace of $R^2$.

3. Show that the set of all elements of $R^2$ that are of the form $(1, a)$, where $a$ is any real number, is not a subspace of $R^2$.

4. Show that the set of all elements of $R^2$ that are of the form

$$(a + b, a + 2b),$$

where $a$ and $b$ are any real numbers, is a subspace of $R^2$.

5. Show that the set of all elements of $R^3$ that are of the form

$$(a + b, -a, 2b),$$

where $a$ and $b$ are any real numbers, is a subspace of $R^3$.

6. Let $A$ be a $2 \times 2$ real matrix. Show that the set of all (real) solutions of the equation $A\mathbf{x} = \mathbf{0}$ is a subspace of $R^2$.

7. Let $\mathcal{F}$ be the space of all functions that are defined on an interval $\mathcal{I}$. Show that the set of all differentiable functions on $\mathcal{I}$ forms a subspace of $\mathcal{F}$.

8. Let $\mathcal{F}$ be the space of all functions defined on an interval $\mathcal{I}$. Show that

the set of all functions $f$ in $\mathscr{I}$ that satisfy the condition $f''(x) - 3f'(x) + 2f(x) = 0$ for all $x$ in $\mathscr{I}$ is a subspace of $\mathscr{F}$.

9. Let $\mathscr{F}$ be the space of all functions defined on an interval $\mathscr{I}$. Show that the set of all functions $f$ in $\mathscr{F}$ that satisfy the condition $e^x f'(x) - (\sin x)f(x) = 0$ for all $x$ in $\mathscr{I}$ constitutes a subspace of $\mathscr{F}$.

10. Show that the elements (3.6) of $V$ span the same subspace as the elements (3.4).

11. Let $A$ be an $m \times n$ real matrix, and $\mathbf{b}$ a real $m$ dimensional vector. Show that the equation $A\mathbf{x} = \mathbf{b}$ has a solution if and only if $\mathbf{b}$ belongs to the subspace of $R^n$ that is spanned by the column vectors of $A$.

12. Let each of the elements $\mathbf{v}_1, \mathbf{v}_2, \ldots, \mathbf{v}_m$ of a vector space $V$ be a linear combination of the elements $\mathbf{u}_1, \mathbf{u}_2, \ldots, \mathbf{u}_k$ of $V$. Show that the space spanned by the $\mathbf{v}_i$ is also spanned by the $\mathbf{u}_i$. Is the space spanned by the $\mathbf{u}_i$ necessarily spanned by the $\mathbf{v}_i$?

13. If $V_1$ and $V_2$ are subspaces of a vector space $V$, show that $V_1 \cap V_2$ and $V_1 + V_2$ are subspaces of $V$.

14. If $U$ is a subspace of $V$, does the set of all elements of $V$ that are not in $U$ (together with the zero element) constitute a subspace of $V$?

15. Let $U$ be a subspace of $R^n$. Let $U^\perp$ be the set of all elements of $R^n$ that are orthogonal to every element of $U$; that is, $\mathbf{v}$ is in $U^\perp$ if $\mathbf{v} \cdot \mathbf{u} = 0$ for every element $\mathbf{u}$ of $U$. Show that $U^\perp$ is a subspace of $R^n$. (The subspace $U^\perp$ is called the *orthogonal complement* of $U$.)

## 3.3   LINEAR DEPENDENCE

Let $V$ be a vector space. In what follows, the word "number" refers to a real or complex number, according to whether $V$ is real or complex. A finite set $\{\mathbf{v}_1, \mathbf{v}_2, \ldots, \mathbf{v}_m\}$ of elements of $V$ is said to be *linearly dependent* (we also say that the elements $\mathbf{v}_1, \mathbf{v}_2, \ldots, \mathbf{v}_m$ are linearly dependent) if there exist numbers $c_1, c_2, \ldots, c_m$ not all zero such that

$$c_1 \mathbf{v}_1 + c_2 \mathbf{v}_2 + \cdots + c_m \mathbf{v}_m = \mathbf{0}. \tag{3.7}$$

The condition (3.7) is always satisfied if the numbers $c_i$ are all zero. The restriction that these numbers are not all zero is essential.

Notice how we have used the properties of a vector space in the definition of linear dependence. Each of the products $c_i \mathbf{v}_i$ must be defined and be an element of $V$. The sum of the $m$ products must also be defined and be an element of $V$. We also need the presence of $\mathbf{0}$, the zero element.

If the set $\{v_1, v_2, \ldots, v_m\}$ of elements of $V$ is not linearly dependent, it is said to be *linearly independent*.

A linearly dependent set can be characterized in another way.

**Theorem 3.2**   The set $\{v_1, v_2, \ldots, v_m\}$ is linearly dependent if and only if at least one element of the set is a linear combination of the others.

PROOF   Suppose that $v_i$ is a linear combination of the other elements such that

$$v_i = a_1 v_1 + \cdots + a_{i-1} v_{i-1} + a_{i+1} v_{i+1} + \cdots + a_m v_m.$$

Then

$$a_1 v_1 + \cdots + a_{i-1} v_{i-1} + (-1)v_i + a_{i+1} v_{i+1} + \cdots + a_m v_m = 0,$$

and since the coefficient of $v_i$ is not zero, the set is linearly dependent. Next suppose that

$$c_1 v_1 + c_2 v_2 + \cdots + c_m v_m = 0$$

and that $c_i$, say, is not zero. Then we may write

$$v_i = -\frac{c_1}{c_i} v_1 - \cdots - \frac{c_{i-1}}{c_i} v_{i-1} - \frac{c_{i+1}}{c_i} v_{i+1} - \cdots - \frac{c_m}{c_i} v_m,$$

hence $v_i$ is a linear combination of the other elements.

We consider some examples.

**Example 1**   Let $v_1 = (1, -1, 3)$ and $v_2 = (2, 1, 0)$ be elements of $R^3$. Then the condition

$$c_1 v_1 + c_2 v_2 = 0$$

is satisfied if and only if

$$c_1 + 2c_2 = 0,$$
$$-c_1 + c_2 = 0,$$
$$3c_1 + 0c_2 = 0.$$

But this system for $c_1$ and $c_2$ is satisfied if and only if $c_1 = c_2 = 0$. Hence $v_1$

and $v_2$ are linearly independent. If $v_3 = (-2, 2, -6)$, then $v_1$ and $v_3$ are linearly dependent because $v_3 = -2v_1$ or $2v_1 + v_3 = 0$.

**Example 2**   Let $\mathscr{I}$ be the set of all real numbers and let $\mathscr{F}$ be the space of functions defined on $\mathscr{I}$. If $f$, $g$, and $h$ are the functions defined by the formulas

$$f(x) = 3x^2, \qquad g(x) = 2x, \qquad h(x) = 6x - 6x^2,$$

then $f$, $g$, and $h$ are in $\mathscr{F}$. The set $\{f, g, h\}$ is linearly dependent because

$$2f(x) - 3g(x) + h(x) = 0$$

for all $x$; that is, $2f - 3g + h = 0$. However, the set $\{f, g\}$ is linearly independent, because if

$$c_1 f + c_2 g = 0,$$

then

$$3c_1 x^2 + 2c_2 x = 0$$

for all $x$. Taking the values $x = 1$ and $x = -1$, we see that

$$3c_1 + 2c_2 = 0, \qquad 3c_1 - 2c_2 = 0.$$

But because $c_1 = c_2 = 0$, it follows that $f$ and $g$ are linearly independent. Linear dependence of functions will be considered further in the next section.

The next theorem gives a sufficient condition for the linear dependence of a set of elements of an arbitrary vector space.

**Theorem 3.3**   Let each of the elements $v_1, v_2, \ldots, v_n$ of a vector space $V$ be a linear combination of the elements $u_1, u_2, \ldots, u_m$. If $m < n$, then the elements $v_1, v_2, \ldots, v_n$ are linearly dependent.

PROOF   We have

$$v_i = \sum_{j=1}^{m} a_{ij} u_j, \quad 1 \le i \le n.$$

The condition

$$\sum_{i=1}^{n} c_i v_i = 0$$

is equivalent to the condition

$$\sum_{i=1}^{n} c_i \sum_{j=1}^{m} a_{ij} \mathbf{u}_j = \mathbf{0}$$

or

$$\sum_{j=1}^{m} \left( \sum_{i=1}^{n} a_{ij} c_i \right) \mathbf{u}_j = \mathbf{0}.$$

This is satisfied if

$$\sum_{i=1}^{n} a_{ij} c_i = 0, \quad 1 \le j \le m.$$

But this is a system of $m$ equations for the $n$ quantities $c_1, c_2, \ldots, c_n$. Since $m < n$ (fewer equations than unknowns), the system has a nontrivial solution (Section 2.1). Hence the elements $\mathbf{v}_1, \mathbf{v}_2, \ldots, \mathbf{v}_n$ are linearly dependent.

Let us consider a homogeneous system of equations

$$
\begin{aligned}
a_{11}x_1 + a_{12}x_2 + \cdots + a_{1n}x_n &= 0, \\
a_{21}x_1 + a_{22}x_2 + \cdots + a_{2n}x_n &= 0, \\
&\phantom{=}\cdots\cdots\cdots\cdots\cdots\cdots \\
a_{m1}x_1 + a_{m2}x_2 + \cdots + a_{mn}x_n &= 0.
\end{aligned}
\tag{3.8}
$$

Let $A$ be the $m \times n$ matrix with elements $a_{ij}$. Then the system (3.8) may be written as

$$x_1 \mathbf{a}_1 + x_2 \mathbf{a}_2 + \cdots + x_n \mathbf{a}_n = \mathbf{0}, \tag{3.9}$$

where $\mathbf{a}_1, \mathbf{a}_2, \ldots, \mathbf{a}_n$ are the column vectors of $A$. From this formulation Theorem 3.4 follows immediately.

**Theorem 3.4**   The homogeneous equation $A\mathbf{x} = \mathbf{0}$ possesses a nontrivial solution if and only if the column vectors of the matrix $A$ are linearly dependent.

This theorem yields a criterion for the linear dependence, or independence, of a set of $n$ vectors in $R^n$.

**Theorem 3.5**   Let $\mathbf{v}_1, \mathbf{v}_2, \ldots, \mathbf{v}_n$ be vectors in $R^n$. These vectors are linearly dependent if and only if the $n \times n$ matrix with these vectors as its column (row) vectors is singular.

PROOF   Let $A$ be the matrix whose column vectors are the given vectors. By Theorem 3.4, the vectors are linearly dependent if and only if the equation $A\mathbf{x} = \mathbf{0}$ has a nontrivial solution. By Theorem 2.10 this equation has a nontrivial solution if and only if $A$ is singular. Now let $B$ be the matrix whose row vectors are the given vectors. Since $B = A^T$ and $\det A^T = \det A$, we conclude that the vectors are linearly dependent if and only if $B$ is singular.

To illustrate the theorem, we observe that $\mathbf{u}_1 = (2, -1)$ and $\mathbf{u}_2 = (3, 2)$ are linearly independent elements of $R^2$ and that $\mathbf{v}_1 = (1, 3)$ and $\mathbf{v}_2 = (-2, -6)$ are linearly dependent because

$$\begin{vmatrix} 2 & 3 \\ -1 & 2 \end{vmatrix} \neq 0, \qquad \begin{vmatrix} 1 & 3 \\ -2 & -6 \end{vmatrix} = 0.$$

We shall presently deduce a necessary and sufficient condition for the linear dependence of any finite set of vectors in $R^m$. First we need one more preliminary result.

**Lemma**   Let the row vectors of the coefficient matrix $A$ of the homogeneous system

$$a_{11}x_1 + \cdots + a_{1n}x_n = 0,$$
$$\cdots\cdots\cdots\cdots\cdots\cdots\cdots$$
$$a_{m1}x_1 + \cdots + a_{mn}x_n = 0$$

be linearly dependent. Then at least one equation of the system is implied by the remaining equations. The system is thus equivalent to a system with one less equation, formed by dropping an appropriate one of the original equations.

PROOF   By Theorem 3.2, at least one row vector, say $\mathbf{a}_r$, is a linear combination of the other row vectors. This means that the $r$th equation of the system is automatically satisfied if the remaining equations are satisfied. In fact, we can obtain an equivalent system in which one equation is the identity $0 = 0$ by adding to the $r$th equation appropriate multiples of the other equations.

**Theorem 3.6**   Let $\mathbf{a}_1, \mathbf{a}_2, \ldots, \mathbf{a}_k$ be vectors in $R^m$ and let $A$ be the matrix with these vectors as its column (row) vectors. If $k > m$ the vectors are linearly dependent. If $k \leq m$, the vectors are linearly dependent if and only if every square submatrix of $A$ of order $k$ is singular.

PROOF   We consider first the case where the given vectors are the column vectors of $A$. The condition

$$c_1\mathbf{a}_1 + c_2\mathbf{a}_2 + \cdots + c_k\mathbf{a}_k = \mathbf{0} \tag{3.10}$$

corresponds to the system

$$
\begin{aligned}
c_1 a_{11} + c_2 a_{12} + \cdots + c_k a_{1k} &= 0, \\
c_1 a_{21} + c_2 a_{22} + \cdots + c_k a_{2k} &= 0, \\
&\cdots \cdots \cdots \cdots \cdots \cdots \cdots \cdots \\
c_1 a_{m1} + c_2 a_{m2} + \cdots + c_k a_{mk} &= 0.
\end{aligned}
\tag{3.11}
$$

If $k > m$, there are more unknowns than equations in the system and hence (see Section 2.1) it has a nontrivial solution. In this case the vectors $\mathbf{a}_i$ are linearly dependent.

Now suppose that $k \leq m$. If the vectors $\mathbf{a}_i$ are linearly dependent, the system (3.11) has a nontrivial solution and therefore every set of $k$ equations of the system has a nontrivial solution. But a homogeneous system of $k$ equations and $k$ unknowns has a nontrivial solution if and only if the determinant of its coefficient matrix is zero (Theorem 2.10). Hence every $k \times k$ submatrix of $A$ is singular.

Next suppose that the vectors $\mathbf{a}_i$ are linearly independent. We want to show that at least one $k \times k$ submatrix of $A$ is nonsingular. Suppose that every $k \times k$ submatrix is singular. Consider any set of $k$ equations of the system (3.11). Since the $k \times k$ coefficient matrix of this system is singular, the row vectors of the matrix are linearly dependent, according to Theorem 3.5. Therefore one equation is implied by the other $k - 1$ equations, according to the lemma. Discarding this equation from the system (3.11), we have $m - 1$ equations for $c_1, c_2, \ldots, c_k$. If $m - 1 < k$ we stop. Otherwise we continue this process, discarding equations until we arrive at a system of $k - 1$ equations for the $k$ unknowns $c_1, c_2, \ldots, c_k$. This system, with fewer equations than unknowns, must have a nontrivial solution. Then the system (3.11) will have a nontrivial solution, since the additional equations of this system are implied by the $k$ equations that remain. But this is impossible, since the vectors $a_i$ are linearly independent. Hence our assumption that every $k \times k$ submatrix of $A$ is singular must be false, so there exists at least one nonsingular $k \times k$ submatrix.

The part of the theorem concerning row vectors follows by considering the transpose of the matrix $A$. The details are left as Exercise 9.

**Example 3** The vectors $\mathbf{x} = (1, 2, 4)$ and $\mathbf{y} = (3, 1, 2)$ are linearly independent because the matrix

$$
\begin{bmatrix} 1 & 3 \\ 2 & 1 \\ 4 & 2 \end{bmatrix}
$$

has a nonsingular 2 × 2 submatrix. One such submatrix is

$$\begin{bmatrix} 1 & 3 \\ 2 & 1 \end{bmatrix}.$$

## Exercises for Section 3.3

1. Determine whether the given vectors are linearly dependent or linearly independent.

   (a)  $(2, -1)$,  $(-4, 2)$
   (b)  $(3, -1)$,  $(2, 1)$
   (c)  $(2, 1)$,  $(3, 0)$,  $(1, 4)$
   (d)  $(-1, 0, 3)$,  $(2, 0, -6)$
   (e)  $(1, 2, 3)$,  $(2, -1, 0)$
   (f)  $(2, -1, 1)$,  $(2, 0, 3)$,  $(1, 1, -2)$
   (g)  $(2, -1, 1)$,  $(2, -3, -2)$,  $(2, 3, 7)$
   (h)  $(1, 0, 2, -2)$,  $(2, 1, 0, 1)$
   (i)  $(2, -1, 0, 1)$,  $(4, -2, 0, 2)$
   (j)  $(0, 0, 0, 1)$,  $(4, 0, 0, 2)$,  $(1, 1, 0, 1)$

2. Let $\mathscr{F}$ be the space of functions that are defined on the set of all real numbers. Determine whether the given functions are linearly dependent or linearly independent.

   (a)  $2x + 1$, $x^2$
   (b)  $e^{ax}$, $e^{bx}$, $a \neq b$
   (c)  $2xe^x$, $-3xe^x$
   (d)  $x$, $|x|$
   (e)  $\sin 2x$, $\cos 2x$, $\sin(2x - \pi/3)$
   (f)  $3x - 2$, $2x + 4$, $2x + 1$

3. If $S$ is any finite set of elements of a vector space $V$ that contains the zero element of $V$, show that $S$ is linearly dependent.

4. Show that any (finite) set of more than $n$ elements of $R^n$ is linearly dependent.

5. (a)  Let $S$ be a (finite) set of linearly independent elements of a vector space. Show that every subset of $S$ is linearly independent.

   (b)  If $S$ is a (finite) set of linearly dependent elements of a vector space, is every subset of $S$ linearly dependent?

   (c)  Let $S$ be a (finite) set of linearly dependent elements of a vector space $V$. Show that any (finite) set of elements of $V$ that contains $S$ is linearly dependent.

   (d)  If $S$ is a (finite) set of linearly independent elements of $V$, is every finite set that contains $S$ linearly independent?

6. (a) Let **x** and **y** be linearly independent elements of a vector space. If $\mathbf{u} = a\mathbf{x} + b\mathbf{y}$ and $\mathbf{v} = c\mathbf{x} + d\mathbf{y}$, show that **u** and **v** are linearly independent if and only if $ad - bc \neq 0$.

(b) Let $\mathbf{x}_1, \mathbf{x}_2, \ldots, \mathbf{x}_n$ be linearly independent elements of a vector space and let

$$\mathbf{u}_i = \sum_{j=1}^{n} a_{ij}\mathbf{x}_j, \quad 1 \leq i \leq n.$$

Show that the elements $\mathbf{u}_1, \mathbf{u}_2, \ldots, \mathbf{u}_n$ are linearly independent if and only if $\det A \neq 0$.

7. Let $\mathbf{x}_1, \mathbf{x}_2, \ldots, \mathbf{x}_n$ be linearly independent elements of a vector space, and suppose that **y** can be expressed as a linear combination of these elements. Show that **y** can be expressed as a linear combination in only one way; that is, if

$$\mathbf{y} = \sum_{i=1}^{n} a_i \mathbf{x}_i \quad \text{and} \quad \mathbf{y} = \sum_{i=1}^{n} b_i \mathbf{x}_i,$$

than $a_i = b_i$, $1 \leq i \leq n$.

8. Let **x**, **y**, and **z** be elements of a vector space. If the sets $\{\mathbf{x}, \mathbf{y}\}$ and $\{\mathbf{y}, \mathbf{z}\}$ are linearly independent, is the set $\{\mathbf{x}, \mathbf{y}, \mathbf{z}\}$ necessarily linearly independent?

9. Let $B$ be an $m \times n$ matrix with row vectors $\mathbf{b}_1, \mathbf{b}_2, \ldots, \mathbf{b}_m$. Show that these vectors are linearly independent if and only if $B$ has a nonsingular submatix of order $m$. (If $n < m$, the vectors are linearly dependent.) Suggestion: apply Theorem 3.6 to $B^T$, observing that if $C$ is a submatrix of $B$, then $C^T$ is a submatrix of $B^T$.

10. Let $\mathbf{v}_1, \mathbf{v}_2, \ldots, \mathbf{v}_k$ be mutually orthogonal vectors in $R^n$, so that $\mathbf{v}_i \cdot \mathbf{v}_j = 0$ if $i \neq j$. If none of the vectors is the zero vector, show that they are linearly independent.

11. The sequences $(1, 0, 0, 0, \ldots)$ and $(0, 1, 0, 0, \ldots)$ are elements of the vector space of convergent sequences. Are they linearly independent?

## 3.4 WRONSKIANS

In Example 2 of the previous section we showed that the functions $f$ and $g$, where

$$f(x) = 3x^2, \qquad g(x) = 2x$$

for all $x$, were linearly independent elements of the vector space of functions defined on the set of all real numbers. We shall now establish the linear independence of $f$ and $g$ by another method that is of considerable theoretical importance.

If

$$c_1 f + c_2 g = 0,$$

then we must have

$$c_1 f(x) + c_2 g(x) = 0$$

and

$$c_1 f'(x) + c_2 g'(x) = 0$$

for all $x$. That is,

$$3c_1 x^2 + 2c_2 x = 0,$$

$$6c_1 x + 2c_2 = 0$$

for all $x$. For any fixed $x$, this is a system of equations for $c_1$ and $c_2$. The determinant of the system is

$$6x^2 - 12x^2 = -6x^2.$$

Since this determinant is not zero when $x = 1$ (or for any other value of $x$ except $x = 0$), $c_1$ and $c_2$ must both be zero. Consequently the functions $f$ and $g$ must be linearly independent.

We now generalize the above procedure. Let $\mathscr{I}$ be a fixed interval. Let $\mathscr{F}^n$, where $n$ is a nonnegative integer, denote the vector space of all functions that are defined and possess at least $n$ derivatives on $\mathscr{I}$. (Here $\mathscr{F}^0$ is simply the set of all functions defined on $\mathscr{I}$.)

Let $f_1, f_2, \ldots, f_m$ be functions that belong to the space $\mathscr{F}^{m-1}$. Associated with this set of functions at each point $x$ of the interval $I$ is the determinant

$$\begin{vmatrix} f_1(x) & f_2(x) & \cdots & f_m(x) \\ f_1'(x) & f_2'(x) & \cdots & f_m'(x) \\ \vdots & \vdots & & \vdots \\ f_1^{(m-1)}(x) & f_2^{(m-1)}(x) & \cdots & f_m^{(m-1)}(x) \end{vmatrix}. \tag{3.12}$$

This determinant is called the *Wronskian* of the functions $f_1, f_2, \ldots, f_m$ at the point $x$. We shall denote it by the symbol

$$W(x; f_1, f_2, \ldots, f_m) \tag{3.13}$$

or sometimes simply by $W(x)$ when it is clear what functions are involved. The Wronskian of a set of functions provides a test for the linear independence of the functions, as we shall now see.

**Theorem 3.7** Let the functions $f_1, f_2, \ldots, f_m$ belong to the space $\mathscr{F}^{m-1}$, relative to an interval $\mathscr{I}$. If the functions are linearly dependent on $\mathscr{I}$, then the Wronskian $W(x; f_1, f_2, \ldots, f_m)$ is zero at each point $x$ of $\mathscr{I}$. Hence if the Wronskian is not zero at even one point of $\mathscr{I}$, the functions are linearly independent.

PROOF   If the functions are linearly dependent there exist numbers $c_1, c_2, \ldots,$ $c_m$, not all zero, such that

$$c_1 f_1(x) + c_2 f_2(x) + \cdots + c_m f_m(x) = 0$$

for all $x$ in $\mathscr{I}$. Since the function $c_1 f_1 + c_2 f_2 + \cdots + c_m f_m$ is the zero function, its derivatives must also be the zero function. Hence we have the $m$ relations

$$
\begin{aligned}
c_1 f_1(x) + c_2 f_2(x) + \cdots + c_m f_m(x) &= 0, \\
c_1 f_1'(x) + c_2 f_2'(x) + \cdots + c_m f_m'(x) &= 0, \\
&\hspace{-3em}\cdots\cdots\cdots\cdots\cdots\cdots\cdots\cdots\cdots\cdots\cdots\cdots\cdots \\
c_1 f_1^{(m-1)}(x) + c_2 f_2^{(m-1)}(x) + \cdots + c_m f_m^{(m-1)}(x) &= 0
\end{aligned}
\tag{3.14}
$$

for all $x$ in $\mathscr{I}$. For each fixed $x$ the system (3.14) is a linear homogeneous system of equations that is satisfied by $c_1, c_2, \ldots, c_m$. These numbers are not all zero, so by Theorem 2.10 the determinant of the system must be zero for every $x$ in $\mathscr{I}$. But this determinant is the Wronskian of the functions. If the Wronskian does not vanish at some point, the functions cannot be linearly dependent. Hence they must be linearly independent.

As an example, we consider the three functions $f_1, f_2$, and $f_3$, where

$$f_1(x) = \cos x, \qquad f_2(x) = \sin x, \qquad f_3(x) = x$$

for all $x$. The Wronskian is

$$
W(x) = \begin{vmatrix}
\cos x & \sin x & x \\
-\sin x & \cos x & 1 \\
-\cos x & -\sin x & 0
\end{vmatrix} = x.
$$

Since $W(x)$ is not zero for *all* $x$, the functions are linearly independent.

Theorem 3.7 states that if a set of functions is linearly dependent, then the Wronskian of the functions is identically zero. The converse is not true. If the

Wronskian of a set of function is identically zero the functions need not be linearly dependent. This can be seen from the following example. Let

$$g_1(x) = x^2, \qquad g_2(x) = x\,|x| = \begin{cases} x^2, & x \geq 0, \\ -x^2, & x < 0. \end{cases}$$

We note that $g_2'(0)$ exists and is equal to zero. When $x \geq 0$, we have

$$W(x) = \begin{vmatrix} x^2 & x^2 \\ 2x & 2x \end{vmatrix} = 0$$

and when $x < 0$ we have

$$W(x) = \begin{vmatrix} x^2 & -x^2 \\ 2x & -2x \end{vmatrix} = 0.$$

Hence $W(x) = 0$ for all $x$. But the functions $g_1$ and $g_2$ are linearly independent because if

$$c_1 g_1(x) + c_2 g_2(x) = 0$$

for all $x$, we see by setting $x = 1$ and $x = -1$ that $c_1 + c_2 = 0$ and $c_1 - c_2 = 0$. Hence $c_1 = c_2 = 0$. Thus the functions $g_1$ and $g_2$ are linearly independent even though their Wronskian vanishes identically.

We now use Theorem 3.7 to establish the linear independence, on any interval, of a set of exponential functions of the form

$$f_1(x) = e^{r_1 x}, \quad f_2(x) = e^{r_2 x}, \dots, \quad f_n(x) = e^{r_n x}, \tag{3.15}$$

where $r_i \neq r_j$ when $i \neq j$. The fact that these functions are linearly independent will be of some interest to us later on in our study of linear differential equations.

The Wronskian of the functions (3.15) is

$$W(x) = e^{(r_1 + r_2 + \cdots + r_n)x} D_n, \tag{3.16}$$

where

$$D_n = \begin{vmatrix} 1 & 1 & \cdots & 1 \\ r_1 & r_2 & \cdots & r_n \\ r_1^2 & r_2^2 & \cdots & r_n^2 \\ \vdots & \vdots & & \vdots \\ r_1^{n-1} & r_2^{n-1} & \cdots & r_n^{n-1} \end{vmatrix}. \tag{3.17}$$

The determinant (3.17) is known as *Vandermonde's determinant*.
    When $n = 2$, we have

$$D_2 = \begin{vmatrix} 1 & 1 \\ r_1 & r_2 \end{vmatrix} = r_2 - r_1.$$

When $n = 3$, we leave it to the reader to show that

$$D_3 = \begin{vmatrix} 1 & 1 & 1 \\ r_1 & r_2 & r_2 \\ r_1^2 & r_2^2 & r_3^2 \end{vmatrix} = (r_2 - r_1)(r_3 - r_1)(r_3 - r_2). \tag{3.18}$$

In general it can be shown (Exercise 5) that

$$D_n = (r_2 - r_1)[(r_3 - r_1)(r_3 - r_2)][(r_4 - r_1)(r_4 - r_2)(r_4 - r_3)] \cdots$$
$$\times [(r_n - r_1)(r_n - r_2) \cdots (r_n - r_{n-1})] \tag{3.19}$$

and hence that $D_n \neq 0$. This means that the exponential functions (3.15) are
linearly independent.

## Exercises for Section 3.4

1.  If the Wronskian of a set of functions is zero at every point of an interval,
    then the functions must be linearly dependent on that interval. True or
    false?

2.  Compute the Wronskian of the given set of functions. Then determine
    whether the functions are linearly dependent or linearly independent.
    (a)  $e^{ax}$, $e^{bx}$, $a \neq b$, $x$ in any interval
    (b)  $\cos ax$, $\sin ax$, $a \neq 0$, $x$ in any interval
    (c)  $x^m$, $|x|^m$, $m$ a positive integer, all $x$
    (d)  $1$, $x$, $x^2$, all $x$
    (e)  $x + 1$, $x + 2$, $x + 3$, all $x$
    (f)  $x$, $x + 1$, $(x + 1)/x$, $x > 0$

3.  Derive formula (3.18).

4.  Use formula (3.19) to find the Wronskian of the functions

$$e^x, e^{2x}, e^{3x}, e^{-x}$$

**5.** Prove formula (3.19). Use mathematical induction. Observe that if

$$P(r) = \begin{vmatrix} 1 & 1 & \cdots & 1 & 1 \\ r_1 & r_2 & \cdots & r_k & r \\ \vdots & \vdots & & \vdots & \vdots \\ r_1^k & r_2^k & \cdots & r_k^k & r^k \end{vmatrix},$$

then $P$ is a polynomial of degree $k$ with zeros $r_1, r_2, \ldots, r_k$ and $P(r_{k+1}) = D_{k+1}$.

**6.** The functions $f$ and $g$ are said to be *orthogonal* on the interval $[a, b]$ if

$$\int_a^b f(x)\, g(x)\, dx = 0.$$

Let $f_1, f_2, \ldots, f_k$ be functions of the space of continuous functions on $[a, b]$ such that $f_i$ and $f_j$ are orthogonal if $i \neq j$. Show that the functions are linearly independent, provided that none of them is the zero function. (If $f$ is continuous and $f$ is not the zero function, then

$$\int_a^b [f(x)]^2\, dx > 0.)$$

**7.** Let $f_m(x) = x^m$ for all $x$ and for $m$ a nonnegative integer. For every fixed $n$ show that the functions $f_0, f_1, \ldots, f_n$ are linearly independent.

**8.** Use the result of Exercise 7 to show that a polynomial is the zero function if and only if all its coefficients are zero.

## 3.5   DIMENSION

Let $V$ be a vector space. If for some positive integer $n$ there exists a set of $n$ linearly independent elements of $V$ and if every set of more than $n$ elements is linearly dependent, then $V$ is said to be a *finite-dimensional vector space* and to have *dimension n*. (In addition, the vector space whose only element is a zero element is said to be finite-dimensional and to have dimension zero.)

If $V$ is a vector space of dimension $n$, any set of $n$ linearly independent elements of $V$ is called a *basis* for $V$.

Let $S$ be a collection of elements of a vector space. If there exists a finite subset of $k$ linearly independent elements of $S$ and if every finite set with more than $k$ elements is linearly dependent, we say that *the maximum number of linearly independent elements of S is k*. In particular, the maximum number of linearly independent elements of an $n$-dimensional vector space is $n$.

We shall characterize finite dimensional vector spaces in another way. To do this we need the following result.

**Theorem 3.8** Let $S$ be a collection of elements of a vector space, and let $k$ be the maximum number of linearly independent elements of $S$. If

$$\{\mathbf{v}_1, \mathbf{v}_2, \ldots, \mathbf{v}_k\}$$

is a linearly independent subset of $S$, then every element of $S$ is a linear combination of the elements of this subset.

PROOF Let $v$ be any element of $S$. Then the set $\{\mathbf{v}, \mathbf{v}_1, \mathbf{v}_2, \ldots, \mathbf{v}_k\}$ is linearly dependent since it has $k + 1$ elements. Hence there exist scalars $c_0, c_1, \ldots, c_k$, not all zero, such that

$$c_0 \mathbf{v} + c_1 \mathbf{v}_1 + \cdots + c_k \mathbf{v}_k = \mathbf{0}.$$

We claim that $c_0 \neq 0$. Otherwise, we would have

$$c_1 \mathbf{v}_1 + \cdots + c_k \mathbf{v}_k = \mathbf{0},$$

and since $\mathbf{v}_1, \mathbf{v}_2, \ldots, \mathbf{v}_k$ are linearly independent, this means that $c_1 = c_2 = \cdots = c_k = 0$ also. Hence we may write

$$\mathbf{v} = \sum_{i=1}^{k} \left( -\frac{c_i}{c_0} \right) \mathbf{v}_i;$$

therefore, $\mathbf{v}$ is a linear combination of the elements of the linearly independent subset of $k$ elements.

**Theorem 3.9** A vector space $V$ has dimension $n$ if and only if there exists a set of $n$ linearly independent elements of $V$ that spans $V$. If $V$ is finite dimensional, every basis for $V$ spans $V$.

PROOF Suppose first that there exists a set of $n$ linearly independent elements of $V$ that spans $V$. Then every element of $V$ is a linear combination of these $n$ elements. By Theorem 3.3, every finite set with more than $n$ elements is linearly dependent. Hence $V$ has dimension $n$.

Next suppose that $V$ has dimension $n$. Then there exists a set of $n$ linearly independent elements of $V$. Let $\{\mathbf{v}_1, \mathbf{v}_2, \ldots, \mathbf{v}_n\}$ be any such set, that is, any basis for $V$. Since the maximum number of linearly independent elements of $V$ is $n$, it follows from Theorem 3.8 that the basis spans $V$.

We shall use Theorem 3.9 to prove that $R^n$ is an $n$-dimensional vector space. To accomplish this, we have to exhibit a set of $n$ linearly independent elements of $R^n$ and then show that every element of $R^n$ is a linear combination of these $n$ elements. Let

$$\mathbf{e}_1 = (1, 0, 0, 0, \ldots, 0), \quad \mathbf{e}_2 = (0, 1, 0, 0, \ldots, 0), \ldots, \quad \mathbf{e}_n = (0, 0, 0, 0, \ldots, 1).$$

The $n \times n$ matrix with these vectors as its column vectors is the identity matrix $I$, which is nonsingular. By Theorem 3.5, the vectors are linearly independent. If $\mathbf{x} = (x_1, x_2, \ldots, x_n)$ is any element of $R^n$, we can write

$$\mathbf{x} = x_1 \mathbf{e}_1 + x_2 \mathbf{e}_2 + \cdots + x_n \mathbf{e}_n.$$

Hence $R^n$ is $n$-dimensional and $\{\mathbf{e}_1, \mathbf{e}_2, \ldots, \mathbf{e}_n\}$ is a basis for $R^n$. We call this particular basis the *standard basis* for $R^n$. In similar fashion it can be shown (Exercise 3) that the complex vector space $C^n$ is $n$-dimensional.

Having shown that $R^n$ is $n$-dimensional, we know that any set of $n$ linearly independent elements constitutes a basis for $R^n$. For $R^2$, the vectors $\mathbf{e}_1 = (1, 0)$ and $\mathbf{e}_2 = (0, 1)$ constitute a basis. But the vectors $(2, -1)$ and $(-2, 2)$ also form a basis, since they are linearly independent.

Geometrically, the vectors in a one-dimensional subspace of $R^2$ or $R^3$ correspond to the points on a line through the origin of a rectangular coordinate-system. A two-dimensional subspace of $R^3$ corresponds to a plane through the origin.

The set of all real valued functions defined on an interval $\mathscr{I}$ is a real vector space, but not a finite-dimensional space. In fact, we can show that for every positive integer $n$ there exists a set of $n$ linearly independent elements of the space. Thus there can be no largest set of linearly independent elements. Let $f_m(x) = e^{mx}$ for $x$ in $\mathscr{I}$ and for every positive integer $m$. Then for each positive integer $n$ the functions $f_1, f_2, \ldots, f_n$ are linearly independent, as was indicated in Section 3.4. The set of all functions of the form $af_1 + bf_2$, where $a$ and $b$ are any real numbers, is a two-dimensional subspace of the space considered here. For $f_1$ and $f_2$ are linearly independent elements of the subspace, and every element of the subspace is a linear combination of $f_1$ and $f_2$.

We shall present two more results that involve the idea of the dimension of a vector space.

**Theorem 3.10**    Let $V$ be a vector space that is spanned by the elements of the set $S = \{\mathbf{v}_1, \mathbf{v}_2, \ldots, \mathbf{v}_n\}$. Then $V$ has dimension $k$ if and only if the maximum number of linearly independent elements of $S$ is $k$.

PROOF    Suppose that $V$ has dimension $k$. Then every subset of $S$ with more than $k$ elements must be linearly dependent, by the definition of a $k$-dimensional space. But $S$ must have a subset of $k$ linearly independent elements. Otherwise every element of $S$ and $V$ is a linear combination of fewer than $k$ elements of $V$, and then $V$ cannot have dimension $k$.

Now suppose that $S$ has the properties of the theorem. Then there is a subset of $k$ linearly independent elements of $S$. By Theorem 3.8, every element of $S$, and hence of $V$, is a linear combination of these $k$ elements. By Theorem 3.9, $V$ has dimension $k$.

**Theorem 3.11** Let $V$ be an $n$-dimensional vector space. If $\mathbf{v}_1, \mathbf{v}_2, \ldots, \mathbf{v}_k$ are linearly independent elements of $V$ and if $k < n$, there exist elements $\mathbf{v}_{k+1}$, $\mathbf{v}_{k+2}, \ldots, \mathbf{v}_n$ such that $\{\mathbf{v}_1, \mathbf{v}_2, \ldots, \mathbf{v}_n\}$ is a basis for $V$.

PROOF   There must exist an element $\mathbf{v}_{k+1}$ such that $\{\mathbf{v}_1, \mathbf{v}_2, \ldots, \mathbf{v}_k, \mathbf{v}_{k+1}\}$ is linearly independent. Otherwise, by Theorem 3.8, every element of $V$ is a linear combination of $\mathbf{v}_1, \mathbf{v}_2, \ldots, \mathbf{v}_k$ and $V$ has dimension $k$. If $n = k + 1$, the proof is finished. If $n > k + 1$ we can continue to add elements $\mathbf{v}_{k+2}, \mathbf{v}_{k+3}, \ldots$, $\mathbf{v}_n$ such that $\{\mathbf{v}_1, \mathbf{v}_2, \ldots, \mathbf{v}_n\}$ is a basis for $V$.

In $R^n$ (or $C^n$), bases whose elements are mutually orthogonal are often convenient. If $\mathbf{v}_1, \mathbf{v}_2, \ldots, \mathbf{v}_k$ are nonzero elements of $R^n$ (or $C^n$) such that $\mathbf{v}_i \cdot \mathbf{v}_j = 0$ for $i \neq j$, then these elements are linearly independent. To see this, suppose that

$$c_1 \mathbf{v}_1 + c_2 \mathbf{v}_2 + \cdots + c_k \mathbf{v}_k = \mathbf{0}.$$

Taking the scalar product of both members of this equation with $\mathbf{v}_i$, where $i$ is any integer such that $1 \leq i \leq k$, we have

$$c_1(\mathbf{v}_1 \cdot \mathbf{v}_i) + c_2(\mathbf{v}_2 \cdot \mathbf{v}_i) + \cdots + c_k(\mathbf{v}_k \cdot \mathbf{v}_i) = 0.$$

Because of the orthogonality of the elements, this equation reduces to

$$c_i(\mathbf{v}_i \cdot \mathbf{v}_i) = 0.$$

But since $\mathbf{v}_i \neq \mathbf{0}$ then $\mathbf{v}_i \cdot \mathbf{v}_i > 0$ and hence $c_i = 0$ for $1 \leq i \leq k$. Thus the elements must be linearly independent.

A basis $\mathscr{B} = \{v_1, v_2, \ldots, v_n\}$ for $R^n$ (or $C^n$) is called an *orthogonal basis* if $\mathbf{v}_i \cdot \mathbf{v}_j = 0$ for $i \neq j$. The basis $\mathscr{B}$ is called an *orthonormal basis* if it is orthogonal and if in addition its elements are unit vectors. Thus $\mathscr{B}$ is an orthonormal basis if $\mathbf{v}_i \cdot \mathbf{v}_j = \delta_{ij}$. The standard basis is an example of an orthonormal basis for $R^n$. Another example of an orthonormal basis for $R^3$ is $\{\mathbf{v}_1, \mathbf{v}_2, \mathbf{v}_3\}$ where

$$\mathbf{v}_1 = \tfrac{1}{3}(2, 2, 1), \qquad \mathbf{v}_2 = \tfrac{1}{3}(2, -1, -2), \qquad \mathbf{v}_3 = \tfrac{1}{3}(-1, 2, -2).$$

### Exercises for Section 3.5

1.  If $V$ is a finite-dimensional vector space with dimension $n$, show that any subspace $U$ of $V$ is also finite-dimensional and that its dimension cannot exceed $n$. If $U$ has dimension $n$, show that $U = V$.

2.  If $V$ is an $n$-dimensional vector space, can a basis for $V$ have
    (a)  more than $n$ elements?
    (b)  fewer than $n$ elements?

3.  Show that the complex vector space $C^n$ is $n$-dimensional.

4.  Determine if the set of vectors $\{u, v, w\}$ constitutes a basis for $R^3$.
    (a)  $u = (1, 1, 0), \quad v = (2, 0, 1), \quad w = (1, -1, 1)$
    (b)  $u = (0, 1, -1), \quad v = (1, 0, 1), \quad w = (2, 2, 2)$

5.  Show that the set of all $2 \times 2$ matrixes forms a finite-dimensional vector space. Determine the dimension and find a basis for the space.

6.  Show that the vector space of all convergent infinite sequences is not finite-dimensional.

7.  Show that the set of all real polynomials of degree $<2$ is a finite-dimensional vector space. Determine the dimension and find at least two bases.

8.  Is the space of all real valued continuous functions defined on the interval $[0, 1]$ finite-dimensional?

9.  Let $v = (v_1, v_2)$ be an arbitrary element of $R^2$. Find numbers $a$ and $b$ such that $v = ax + by$, where $\{x, y\}$ is the indicated basis for $R^2$.
    (a)  $x = (2, -1), \quad y = (1, 1)$
    (b)  $y = (1, 3), \quad y = (2, 0)$

10. Let $\mathscr{P}^n$ be the space of all (real) polynomials of degree $<n$. Show that $\mathscr{P}^n$ has dimension $n$.

11. Let $V$ be an $n$-dimensional vector space and let $U$ be a $k$-dimensional subspace of $V$. Show that there exists a subspace $W$, with dimension $n - k$, such that $U + W = V$ and $U \cap W = \{0\}$. Suggestion: let

$$\{u_1, u_2, \ldots, u_k\}$$

be a basis for $U$ and use Theorem 3.11.

## 3.6   LINEAR TRANSFORMATIONS

Let $A$ and $B$ be arbitrary sets. A rule of correspondence that assigns to each element of $A$ exactly one element of $B$ is called a function, or mapping, from $A$ into $B$. Alternatively, a function from $A$ into $B$ may be defined as a collection of ordered pairs $(a, b)$, where $a$ is in $A$, $b$ is in $B$, and each element of $A$ occurs in exactly one ordered pair. The set $A$ is called the *domain* of the mapping. We say that the transformation maps $a$ into $b$. We denote transformations by capital letters $S$, $T$, and so on. If $T$ is a mapping from $A$ into $B$, we write

$$T: A \rightarrow B.$$

To indicate that $T$ maps $a$ into $b$, we write $b = Ta$ or $b = T(a)$. We call $b$ the *image* of $a$ under $T$.

We shall be concerned with mappings from a vector space into a vector space. A mapping from $R^1$ into $R^1$ is usually called a function of a real variable. A mapping from $R^2$ into $R^1$ is usually called a function of two real variables. A mapping from $R^2$ into $R^2$ can be described by an ordered pair of functions of two real variables. The equations

$$y_1 = f_1(x_1, x_2), \qquad y_2 = f_2(x_1, x_2) \tag{3.20}$$

describe such a mapping, in which $(x_1, x_2)$ is mapped into $(y_1, y_2)$.

Let $V$ and $W$ be vector spaces, both real or both complex, and let $T$ be a mapping from $V$ into $W$. The mapping $T$ is said to be a *linear transformation*, or a *linear mapping*, or a *linear operator* if it has the following properties.

**Property 1**  For every number $c$ and every element $\mathbf{v}$ of $V$,

$$T(c\mathbf{v}) = cT\mathbf{v}. \tag{3.21}$$

**Property 2**  For every pair $\{\mathbf{u}, \mathbf{v}\}$ of elements of $V$,

$$T(\mathbf{u} + \mathbf{v}) = T\mathbf{u} + T\mathbf{v}. \tag{3.22}$$

Notice the essential uses of the properties of a vector space in the definition of a linear transformation. If $\mathbf{u}$ and $\mathbf{v}$ are in $V$, the product $c\mathbf{v}$ and the sum $\mathbf{u} + \mathbf{v}$ are defined and are elements of $V$. Since $T\mathbf{u}$ and $T\mathbf{v}$ are in $W$, the product $cT\mathbf{v}$ and the sum $T\mathbf{u} + T\mathbf{v}$ are defined and are elements of $W$.

If $\mathbf{v}$ is any element of $V$, we see by setting $c = 0$ in Property 1 that

$$T\mathbf{0} = T(0\mathbf{v}) = 0T\mathbf{v} = \mathbf{0}.$$

Thus the zero element of $V$ is mapped into the zero element of $W$ by the linear transformation $T$.

The set of all elements of $V$ that are mapped into the zero element of $W$ by $T$ form a subspace of $V$. To see this, suppose that $T\mathbf{u} = \mathbf{0}$ and $T\mathbf{v} = \mathbf{0}$. Then

$$T(c\mathbf{u}) = cT\mathbf{u} = c\mathbf{0} = \mathbf{0}$$

and

$$T(\mathbf{u} + \mathbf{v}) = T\mathbf{u} + T\mathbf{v} = \mathbf{0} + \mathbf{0} = \mathbf{0}.$$

This subspace is called the *kernel*, or *null space*, of the linear transformation $T$.

The set of all elements $\mathbf{w}$ of $W$ such that $T\mathbf{v} = \mathbf{w}$ for at least one element $\mathbf{v}$ of $V$ is a subspace of $W$. For if $\mathbf{x} = T\mathbf{u}$ and $\mathbf{y} = T\mathbf{v}$, then $c\mathbf{x} = T(c\mathbf{u})$ and $\mathbf{x} + \mathbf{y} = T(\mathbf{u} + \mathbf{v})$. This subspace of $W$ is called the *image*, or *range*, of the linear transformation $T$.

To illustrate some of these concepts, we describe a linear transformation from $R^n$ into $R^m$. Let us represent the elements of $R^n$ and $R^m$ by their associated column vectors, and let $A$ be an $m \times n$ real matrix. If $\mathbf{x}$ is in $R^n$, then $A\mathbf{x}$ is in $R^m$ and we write

$$T\mathbf{x} = A\mathbf{x}.$$

It is not hard to see that $T$ is a linear transformation. For if $\mathbf{u}$ and $\mathbf{v}$ are in $R^n$, and $c$ is a real number, then

$$T(c\mathbf{u}) = A(c\mathbf{u}) = cA\mathbf{u} = cT(\mathbf{u})$$

and

$$T(\mathbf{u} + \mathbf{v}) = A(\mathbf{u} + \mathbf{v}) = A\mathbf{u} + A\mathbf{v} = T(\mathbf{u}) + T(\mathbf{v}).$$

The kernel of $T$ is the set of all solutions of the homogeneous equation

$$A\mathbf{x} = \mathbf{0}.$$

The image of $T$ is the set of all elements $\mathbf{y}$ in $R^m$ for which the equation

$$A\mathbf{x} = \mathbf{y}$$

has a solution.

If $T_1$ and $T_2$ are both linear transformations from a vector space $V$ into a vector space $W$, we say that $T_1$ is *equal* to $T_2$, written $T_1 = T_2$, if

$$T_1\mathbf{v} = T_2\mathbf{v}$$

for every element $\mathbf{v}$ of $V$.

If $T_1$ and $T_2$ are both linear transformations from $V$ into $W$, we define the *sum*, $T_1 + T_2$, to be the transformation $S$ such that

$$S\mathbf{v} = T_1\mathbf{v} + T_2\mathbf{v}$$

for every element $\mathbf{v}$ of $V$. It is easily verified (Exercise 8) that $T_1 + T_2$ is a linear transformation from $V$ into $W$, and that $T_2 + T_1 = T_1 + T_2$.

Let $V$ and $W$ be vector spaces over the real (complex) numbers, and let $T$ be a linear transformation from $V$ into $W$. If $c$ is a real (complex) number, we define the *product* $cT$ to be the transformation $P$ such that

$$P\mathbf{v} = cT\mathbf{v}$$

for every element $\mathbf{v}$ of $V$. It can be verified that $cT$ is a linear transformation.

Now let $U$, $V$, and $W$ be vector spaces, all real or all complex. Let $T_1$ and $T_2$ be linear transformations such that $T_1 : U \to V$ and $T_2 : V \to W$. The product $T_2 T_1$ is defined to be the transformation $S$ from $U$ into $W$ such that

$$S\mathbf{u} = T_2(T_1\mathbf{u})$$

for every element $\mathbf{u}$ of $U$. The situation is illustrated in Figure 3.1. Notice that $T_1\mathbf{u}$ is in $V$, so $T_2(T_1\mathbf{u})$ is defined and is an element of $W$. Verification that $T_2 T_1$ is a *linear* transformation from $U$ into $W$ is left as an exercise.

In case $T_1 : V \to V$ and $T_2 : V \to V$, then $T_2 T_1$ and $T_1 T_2$ are both defined and map $V$ into $V$. In general, $T_1 T_2 \neq T_2 T_1$, as we shall see in the following example, and in the examples of the next section.

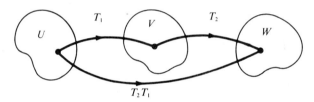

**Figure 3.1**

**Example**    Let $A$ and $B$ be $n \times n$ matrices. If we represent elements in $R^n$ by column vectors then the transformations $T_1$ and $T_2$ that are defined by the relations

$$T_1\mathbf{x} = A\mathbf{x}, \qquad T_2\mathbf{x} = B\mathbf{x}$$

are linear transformations from $R^n$ into $R^n$. Then

$$T_2 T_1\mathbf{x} = BA\mathbf{x}$$

and

$$T_1 T_2\mathbf{x} = AB\mathbf{x}$$

so that $T_2 T_1 \neq T_1 T_2$ unless $A$ and $B$ commute.

Let $T$ be a linear transformation from $V$ into $W$. The transformation $T$ is said to be a transformation from $V$ *onto* $W$ if the image of $T$ is $W$. Thus $T$ is onto if for every element $w$ in $W$ there is at least one element $v$ of $V$ such that $Tv = w$.

The transformation $T$ is said to be *one-to-one* if $Tx \neq Ty$ whenever $x \neq y$. Thus if $T$ is one-to-one, distinct elements in $V$ cannot map into the the same element of $W$.

If $T$ is both one-to-one and onto then there is associated with each element $w$ of $W$ exactly one element $v$ of $V$ such that $Tv = w$. This association defines a transformation from $W$ into $V$ that is called the *inverse* of $T$. If $w = Tv$ we write $v = T^{-1}w$. We see that

$$T^{-1}Tv = v$$

for every $v$ in $V$ and

$$TT^{-1}w = w$$

for every $w$ in $W$. It can be shown (Exercise 12) that the mapping $T^{-1}$ is a *linear* mapping if $T$ is linear.

### Exercises for Section 3.6

1.  Let $T$ be the linear transformation from $R^2$ into $R^2$ that is defined by the relations

$$Tx = Ax, \qquad A = \begin{bmatrix} 2 & -3 \\ -1 & 2 \end{bmatrix}.$$

Find $Tx$ if $x$ is the given point and determine if the point belongs to the kernel of $T$.

    (a)  (1, 2)     (b)  (0, 0)     (c)  (−3, 1)

2.  Let $T$ be the transformation of Exercise 1. Find all points, if any, that are mapped into the point (2, 3) by $T$.

3.  Let $T$ be the linear transformation from $R^2$ into $R^2$ that is defined by the relations

$$Tx = Ax, \qquad A = \begin{bmatrix} 3 & -2 \\ -6 & 4 \end{bmatrix}.$$

Which, if any, of the given points belong to the kernel of $T$?

    (a)  (1, 1)     (b)  (2, 3)     (c)  (−4, 6)

4.  If $T$ is the transformation of Exercise 3, which, if any, of the given points belong to the image of $T$?

    (a)  $(-4, 8)$      (b)  $(2, 2)$      (c)  $(1, -2)$

5.  Let $f$ be a function of a single real variable that maps $R^1$ into $R^1$. Show that $f$ is a linear transformation if and only if $f(x) = mx$ for some number $m$. Suggestion: let $m = f(1)$.

6.  If $T$ is a mapping from $U$ into $V$ such that $T\mathbf{u} = \mathbf{0}$ for every element $\mathbf{u}$ of $U$, show that $T$ is a linear transformation. ($T$ is called the *zero transformation* from $U$ into $V$.)

7.  Let $T$ be a mapping from a vector space $V$ into $V$ such that $T\mathbf{v} = \mathbf{v}$ for every element of $V$. Show that $T$ is a linear transformation (called the *identity transformation* on $V$).

8.  If $T_1$ and $T_2$ are linear transformations from $V$ into $W$, verify that $T_1 + T_2$ and $cT_1$ are linear transformations.

9.  Let $T_1$ and $T_2$ be linear transformations, with $T_1 : U \to V$ and $T_2 : V \to W$. Verify that $T_2 T_1$ is a linear transformation.

10. Let $T : R^2 \to R^2$ be defined by the relation $T(x_1, x_2) = (2x_1 - x_2, -4x_1 + 2x_2)$. Show that $T$ is linear. Find the kernel and image of $T$ and give a geometric interpretation of each.

11. Show that the set of all linear transformations from a vector space $V$ into a vector space $W$ forms a vector space. The operations of addition and multiplication by a number for linear transformations are as defined in the text. See Exercise 6.

12. If the linear transformation $T$ has an inverse, show that $T^{-1}$ is a linear transformation. Suggestion: to show that $T^{-1}(c\mathbf{w}) = cT^{-1}\mathbf{w}$, it suffices to show that $T(T^{-1}(c\mathbf{w})) = T(cT^{-1}\mathbf{w})$ because $T$ is one-to-one.

13. Show that a linear transformation $T$ is one-to-one if and only if the kernel of $T$ consists of the zero vector.

14. Let $T$ be a linear transformation from an $n$-dimensional space $V$ into an $m$-dimensional space $W$.

    (a)  If $m > n$, show that $T$ cannot be a mapping from $V$ onto $W$.

    (b)  If $m < n$, show that $T$ cannot be one-to-one.

15. Let $A$ be an $n \times n$ matrix and let $T$ be the linear transformation from $R^n$ into $R^n$ that is defined by the formula $T\mathbf{x} = A\mathbf{x}$. Show that $T$ has an inverse if and only if $A$ is nonsingular. If $A$ is nonsingular show that $T^{-1}\mathbf{y} = A^{-1}\mathbf{y}$.

## 3.7   DIFFERENTIAL OPERATORS

We now consider a class of linear transformations that are defined on spaces of functions. Let $\mathscr{I}$ be a fixed interval and let $\mathscr{F}^n$ denote the set of all functions that are defined and possess at least $n$ derivatives on $\mathscr{I}$. By $\mathscr{F}^0$ or $\mathscr{F}$, we mean the set of all functions defined on $\mathscr{I}$. For each $n$, the set $\mathscr{F}^n$ forms a vector space if the usual operations for functions are used. We shall therefore speak of the space $\mathscr{F}^n$.

First we introduce the derivative operator $D$, writing

$$Df = f', \qquad Df(x) = f'(x) \tag{3.23}$$

for every function $f$ in $\mathscr{F}^1$ and all $x$ in $\mathscr{I}$. Thus $D$ maps $\mathscr{F}^1$ into $\mathscr{F}$. Certainly $D$ is a linear operator, because

$$D(cf) = (cf)' = cf' = cDf$$

and

$$D(f + g) = (f + g)' = f' + g' = Df + Dg$$

for $f$ and $g$ in $\mathscr{F}^1$ and for every real number $c$.

The product $DD$ is defined on $\mathscr{F}^2$ and maps $\mathscr{F}^2$ into $\mathscr{F}$. If $h$ is in $\mathscr{F}^2$, then

$$DDh = D(Dh) = Dh' = h''.$$

We write

$$D^2 = DD,$$
$$D^3 = DD^2,$$
$$\cdots\cdots\cdots$$
$$D^n = DD^{n-1}.$$

It is easily seen that $D^n f = f^{(n)}$ for every function $f$ in $\mathscr{F}^n$. We also define the operator $D^0$ by means of the formula

$$D^0 f = f.$$

Thus $D^0$ is the *identity operator* that maps each element of $\mathscr{F}$ into itself.

We now define a more general class of linear operators. To begin with, let $a_0$, $a_1$, and $a_2$ be specific functions that are defined on an interval $\mathscr{I}$. We introduce an operator $L$,

$$L = a_0 D^2 + a_1 D + a_2, \tag{3.24}$$

where

$$Lf = a_0 D^2 f + a_1 Df + a_2 f = a_0 f'' + a_1 f' + a_2 f \qquad (3.25)$$

for every function $f$ in $\mathscr{F}^2$. Thus $L$ maps $\mathscr{F}^2$ into $\mathscr{F}$. To see that $L$ is a linear operator, we observe that

$$
\begin{aligned}
L(cf) &= a_0(cf)'' + a_1(cf') + a_2(cf) \\
&= c(a_0 f'' + a_1 f' + a_2 f) \\
&= cLf
\end{aligned}
$$

and that

$$
\begin{aligned}
L(f + g) &= a_0(f + g)'' + a_1(f + g)' + a_2(f + g) \\
&= (a_0 f'' + a_1 f' + a_2 f) + (a_0 g'' + a_1 g' + a_2 g) \\
&= Lf + Lg
\end{aligned}
$$

for $f$ and $g$ in $\mathscr{F}^2$ and for every real number $c$.

More generally, let $a_0, a_1, \ldots, a_n$ be specific functions defined on an interval $\mathscr{I}$. (It is assumed that $a_0$ is not the zero function.) We define an operator $L$,

$$L = a_0 D^n + a_1 D^{n-1} + a_2 D^{n-2} + \cdots + a_{n-1} D + a_n, \qquad (3.26)$$

on the space $\mathscr{F}^n$ by means of the formula

$$Lf = a_0 f^{(n)} + a_1 f^{(n-1)} + a_2 f^{(n-2)} + \cdots + a_{n-1} f' + a_n f, \qquad (3.27)$$

where $f$ is in $\mathscr{F}^n$. It can be shown that $L$ is a linear operator, and we say that $L$ is a *linear differential operator of order n*. This operator maps $\mathscr{F}^n$ into $\mathscr{F}$.

To consider some specific examples, let $L_1$ and $L_2$ be defined by the formulas

$$L_1 f(x) = xf'(x) - f(x),$$
$$L_2 f(x) = f'(x) + 2xf(x).$$

Here $L_1$ and $L_2$ are both first-order linear differential operators that are defined on the space $\mathscr{F}^1$. (The interval $\mathscr{I}$ may be taken as the set of all real numbers.) We may write

$$L_1 = xD - 1, \qquad L_2 = D + 2x.$$

Notice that

$$(L_1 + L_2)f(x) = xf'(x) - f(x) + f'(x) + 2xf(x)$$
$$= (x + 1)f'(x) + (2x - 1)f(x)$$

so that

$$L_1 + L_2 = (x + 1)D + (2x - 1).$$

Both $L_1 L_2$ and $L_2 L_1$ are defined on the space $\mathscr{F}^2$. If $g$ is in $\mathscr{F}^2$ we have

$$L_1 L_2 g = L_1(g' + 2xg)$$
$$= x(g' + 2xg)' - (g' + 2xg)$$
$$= xg'' + (2x^2 - 1)g'$$

and

$$L_2 L_1 g = L_2(xg' - g)$$
$$= (xg' - g)' + 2x(xg' - g)$$
$$= xg'' + 2x^2 g' - 2xg.$$

We see that $L_1 L_2 \neq L_2 L_1$ because if $g(x) = 1$ for all $x$, then

$$L_1 L_2\, g(x) = 0,$$

but

$$L_2 L_1 g(x) = -2x.$$

We recall that two operators $T_1$ and $T_2$, defined on the same space, are equal if and only if $T_1 \mathbf{u} = T_2 \mathbf{u}$ for every element $\mathbf{u}$ in the space. Let $\mathscr{I}$ be a fixed interval and let

$$L_1 = a_0 D^m + a_1 D^{m-1} + \cdots + a_{m-1}D + a_m, \qquad (3.28)$$

$$L_2 = b_0 D^n + b_1 D^{n-1} + \cdots + b_{n-1}D + b_n, \qquad (3.29)$$

where the $a_i$ and $b_i$ are functions defined on $\mathscr{I}$, and neither $a_0$ nor $b_0$ is the zero function. If $k$ is the largest of $m$ and $n$, that is, if $k = \max(m, n)$, then $L_1$ and $L_2$ are linear operators on $\mathscr{F}^k$. The following theorem characterizes equality for linear differential operators.

**Theorem 3.12** The operators $L_1$ and $L_2$ that are defined by formulas (3.28) and (3.29), respectively, are equal if and only if $m = n$ and $a_i = b_i$ for $1 \le i \le m$.

PROOF If $m = n$ and $a_i = b_i$ for all $i$, it is evident that $L_1 f = L_2 f$ for every function $f$ in $\mathscr{F}^n$ and hence $L_1 = L_2$.

Now suppose that $L_1 = L_2$. Let $f_0(x) = 1$ and $f_k(x) = x^k$ for $x$ in $\mathscr{I}$ and for every positive integer $k$. Since $L_1 f_0 = a_m$ and $L_2 f_0 = b_n$, the condition $L_1 f_0 = L_2 f_0$ requires that $a_m = b_n$. Next,

$$L_1 f_1(x) = a_{m-1}(x) + x a_m(x), \qquad L_2 f_1(x) = b_{n-1}(x) + x b_n(x),$$

so that the condition $L_1 f_1 = L_2 f_1$ says that $a_{m-1} = b_{n-1}$. Since

$$L_1 f_2(x) = 2 a_{m-2}(x) + 2 x a_{m-1}(x) + x^2 a_m(x),$$

$$L_2 f_2(x) = 2 b_{n-2}(x) + 2 x b_{n-1}(x) + x^2 b_n(x),$$

we must have $a_{m-2} = b_{n-2}$.

Let $a_j = 0$ if $j < 0$ and $b_j = 0$ if $j < 0$. Then, by mathematical induction, it can be shown that $a_{m-k} = b_{n-k}$ for $k = 0, 1, 2, \ldots$, and hence that $m = n$. The equation

$$Ly = 0, \tag{3.30}$$

where $L$ is as in Eq. (3.26) and $y$ is an unknown function, is called a linear homogeneous differential equation of order $n$. A function of the space $\mathscr{F}^n$ that satisfies the equation is called a solution of the differential equation. A function is a solution if and only if it belongs to the kernel of the operator $L$. Differential equations of the type (3.30) will be discussed in Chapter 5.

### Exercises for Section 3.7

1. Let $a_0(x) = 2$, $a_1(x) = x^2$, $a_2(x) = -x$ for all $x$, and let

$$L = a_0 D^2 + a_1 D + a_2.$$

Find $Lf$, where $f$ is as given
(a) $f(x) = 4x$ (b) $f(x) = e^{3x}$ (c) $f(x) = \sin x$

2. Let $a_0$ and $a_1$ be functions that are defined on an interval $\mathscr{I}$, and let

$$L = a_0 D + a_1.$$

Verify that $L$ is a linear operator from $\mathscr{F}^1$ into $\mathscr{F}$.

3. Let $L = D^2 - 3D + 2$. Which, if any, of the following functions belongs to the kernel of $L$?

    (a) $f(x) = e^{2x}$    (b) $g(x) = e^{3x}$    (c) $h(x) = 2e^x - e^{2x}$

4. Let $C[a, b]$ be the space of all continuous functions on the interval $[a, b]$. For $f$ in $C[a, b]$ let

$$Tf = \int_a^b f(x)\, dx.$$

    Show that $T$ is a linear transformation from $C[a, b]$ into $R^1$.

5. Find $L_1 L_2$ and $L_2 L_1$ if

    (a) $L_1 = D + x$,    $L_2 = D + 2$

    (b) $L_1 = xD$,    $L_2 = xD + 1$

    (c) $L_1 = D - 1$,    $L_2 = D^2 - xD + 1$

    (d) $L_1 = D - \tan x$,    $L_2 = D + \tan x$

6. Let $L_1 = a_1 D + b_1$ and $L_2 = a_2 D + b_2$, where $a_1$, $b_1$, $a_2$, and $b_2$ are constants. Show that $L_1$ and $L_2$ commute; that is, that $L_1 L_2 = L_2 L_1$.

7. Let $a_1, b_1, a_2, b_2$ be differentiable functions on an interval $\mathscr{I}$, such that $a_1$ and $a_2$ are never zero. If

$$L_1 = a_1 D + b_1, \qquad L_2 = a_2 D + b_2,$$

    show that $L_1 L_2 = L_2 L_1$ if and only if $a_2 = ka_1$ and $b_2 = kb_1 + K$, where $k$ and $K$ are constants.

8. Let $L_1$ and $L_2$ be linear differential operators.

    (a) If the function $f$ belongs to the kernel of $L_2$, show that $f$ belongs to the kernel of $L_1 L_2$.

    (b) If the function $g$ belongs to the kernel of $L_1$, does $g$ necessarily belong to the kernel of $L_1 L_2$?

9. If $L_1, L_2$, and $L_3$ are linear differential operators such that $L_1 L_2$ and $L_1 L_3$ are defined, show that $L_1(L_2 + L_3) = L_1 L_2 + L_1 L_3$.

10. (a) Describe the kernel of the derivative operator $D$ that maps $\mathscr{F}^1$ into $\mathscr{F}$.

    (b) Describe the kernel of the transformation $D^n$ that maps $\mathscr{F}^n$ into $\mathscr{F}$.

11. Show that the kernel of the operator $L = D - 1$, which maps $\mathscr{F}^1$ into $\mathscr{F}$, is the set of all functions $y$ of the form $y(x) = ce^x$, where $c$ is a constant. Suggestion: multiply through in the equation $y'(x) - y(x) = 0$ by $e^{-x}$.

## 3.8 RANK

Let $V$ and $W$ be vector spaces over the same field and let $T$ be a linear transformation from $V$ into $W$. We recall that the image of $T$ is the set of all elements $\mathbf{w}$ of $W$ such that $\mathbf{w} = T(\mathbf{v})$ for some $\mathbf{v}$ in $V$. The kernel of $T$ is the set of all elements in $V$ that map into the zero element of $W$. As shown in Section 3.6, the image of $T$ is a subspace of $W$ and the kernel of $T$ is a subspace of $V$. If $V$ is a finite dimensional space, so is the kernel of $T$. The image of $T$ is also finite-dimensional, because if $\{\mathbf{v}_1, \mathbf{v}_2, \ldots, \mathbf{v}_n\}$ is a basis for $V$ then the image of $T$ is spanned by the set $\{T\mathbf{v}_1, T\mathbf{v}_2, \ldots, T\mathbf{v}_n\}$.

The dimension of the image of $T$ is called the *rank* of $T$. The dimension of the kernel of $T$ is called the *nullity* of $T$. The rank and nullity are related in the following way.

**Theorem 3.13** If $T$ has rank $r$ and if the dimension of $V$ is $n$, then the nullity of $T$ is $n - r$.

PROOF Let $k$ denote the nullity of $T$ and let $\{\mathbf{v}_1, \mathbf{v}_2, \ldots, \mathbf{v}_k\}$ be a basis for the kernel of $T$. Then, by Theorem 3.11, there exist elements $\mathbf{v}_{k+1}, \mathbf{v}_{k+2}, \ldots, \mathbf{v}_n$ of $V$ such that $\{\mathbf{v}_1, \mathbf{v}_2, \ldots, \mathbf{v}_n\}$ is a basis for $V$. If $\mathbf{y}$ is in the image of $T$, then $\mathbf{y} = T(\mathbf{x})$ for some $\mathbf{x}$ in $V$. Let

$$\mathbf{x} = \sum_{i=1}^{n} a_i \mathbf{v}_i.$$

Then

$$\mathbf{y} = T(\mathbf{x}) = \sum_{i=1}^{n} a_i T(\mathbf{v}_i) = \sum_{i=k+1}^{n} a_i T(\mathbf{v}_i)$$

because $T(\mathbf{v}_i) = \mathbf{0}$ for $1 \leq i \leq k$. Hence the elements

$$T(\mathbf{v}_{k+1}), T(\mathbf{v}_{k+2}), \ldots, (T\mathbf{v}_n)$$

span the image of $T$. We shall show that these elements are linearly independent. Then, according to Theorem 3.9, the dimension of the image must be $n - k$, so that $r = n - k$ or $k = n - r$.

Suppose that $c_{k+1}, c_{k+2}, \ldots, c_n$ are scalars such that

$$\sum_{i=k+1}^{n} c_i T(\mathbf{v}_i) = \mathbf{0}.$$

Then

$$T\left( \sum_{i=k+1}^{n} c_i \mathbf{v}_i \right) = \mathbf{0},$$

so the sum is in the null space of $T$. Therefore

$$\sum_{i=k+1}^{n} c_i \mathbf{v}_i = \sum_{i=1}^{k} d_i \mathbf{v}_i$$

or

$$-d_1 \mathbf{v}_1 - d_2 \mathbf{v}_2 - \cdots - d_k \mathbf{v}_k + c_{k+1} \mathbf{v}_{k+1} + \cdots + c_n \mathbf{v}_n = \mathbf{0}.$$

But this is possible only if

$$c_{k+1} = c_{k+2} = \cdots = c_n = d_1 = d_2 = \cdots = d_k = 0.$$

Hence the vectors $T(\mathbf{v}_i)$, $k + 1 \le i \le n$, are linearly independent. This concludes our proof.

We consider an important special case. Let $A$ be an $m \times n$ real matrix.[1] Then the formula

$$T(\mathbf{x}) = A\mathbf{x}, \tag{3.31}$$

where $\mathbf{x}$ is an $n$-dimensional column vector, defines a linear transformation from $R^n$ into $R^m$. (In Section 3.10 we shall show that any real linear transformation from an $n$-dimensional space into an $m$-dimensional space can be described by such a formula.) Writing

$$A = [\mathbf{a}_1, \mathbf{a}_2, \ldots, \mathbf{a}_n],$$

we see from Eq. (3.31) that if $\mathbf{y} = T(\mathbf{x})$ then

$$\mathbf{y} = x_1 \mathbf{a}_1 + x_2 \mathbf{a}_2 + \cdots + x_n \mathbf{a}_n. \tag{3.32}$$

Thus $\mathbf{y}$ belongs to the image of $T$ if and only if $\mathbf{y}$ is a linear combination of the column vectors of $A$. The rank of $T$ is the dimension of the subspace of $R^n$ that is spanned by the column vectors of $A$. We define the *rank of a matrix* to be the dimension of the space spanned by its column vectors. According to Theorem 3.10, the rank of a matrix is equal to the maximum number of linearly independent vectors in the set of column vectors of the matrix.

The kernel of the transformation defined by Eq. (3.31) is the set of all solutions of the equation $A\mathbf{x} = \mathbf{0}$. We define the *kernel of a matrix A* to be the

---

[1] If $A$ is permitted to be complex, the spaces $R^n$ and $R^m$ in the ensuing discussion must be replaced by $C^n$ and $C^m$, respectively.

set of all vectors **x** such that $A\mathbf{x} = \mathbf{0}$. In order to describe the set of all solutions of the equation $A\mathbf{x} = \mathbf{0}$ we need to find a basis for the kernel of $A$. The following theorem is an immediate consequence of Theorem 3.13.

**Theorem 3.14** If $A$ has rank $r$, the set of all solutions of the equation $A\mathbf{x} = \mathbf{0}$ is a subspace of $R^n$ that has dimension $n - r$. Thus, if $\mathbf{u}_1, \mathbf{u}_2, ..., \mathbf{u}_{n-r}$, are linearly independent solutions, the set of all solutions consists of all vectors in $R^n$ that are of the form

$$c_1 \mathbf{u}_1 + c_2 \mathbf{u}_2 + \cdots + c_{n-r} \mathbf{u}_{n-r},$$

where $c_1, c_2, ..., c_{n-r}$ are arbitrary numbers.

The rank of a matrix $A$ shows up when we solve the equation $A\mathbf{x} = \mathbf{0}$, as we shall see. First we need the following lemma.

**Lemma** Let $(i_1, i_2, ..., i_n)$ be a permutation of $\{1, 2, ..., n\}$, where $n$ is a positive integer. The permutation associates with each vector

$$\mathbf{x} = (x_1, x_2, ..., x_n)$$

in $R^n$ the vector $\hat{\mathbf{x}} = (x_{i_1}, x_{i_2}, ..., x_{i_n})$. Then the vectors $\mathbf{x}_1, \mathbf{x}_2, ..., \mathbf{x}_k$ are linearly dependent if and only if the vectors $\hat{\mathbf{x}}_1, \hat{\mathbf{x}}_2, ..., \hat{\mathbf{x}}_k$ are linearly dependent.

PROOF   The condition

$$c_1 \mathbf{x}_1 + c_2 \mathbf{x}_2 + \cdots + c_k \mathbf{x}_k = \mathbf{0} \tag{3.33}$$

corresponds to the system

$$c_1 x_{11} + c_2 x_{12} + \cdots + c_k x_{1k} = 0,$$
$$c_1 x_{21} + c_2 x_{22} + \cdots + c_k x_{2k} = 0,$$
$$\cdots\cdots\cdots\cdots\cdots\cdots\cdots\cdots\cdots\cdots\cdots \tag{3.34}$$
$$c_1 x_{n1} + c_2 x_{n2} + \cdots + c_k x_{nk} = 0,$$

where $x_{ij}$ is the $i$th component of $x_j$. But the system that corresponds to the condition

$$c_1 \hat{\mathbf{x}}_1 + c_2 \hat{\mathbf{x}}_2 + \cdots + c_k \hat{\mathbf{x}}_k = \mathbf{0} \tag{3.35}$$

is equivalent to the system (3.34); in fact, it consists of the same equations with the order changed. Thus an ordered set of numbers $(c_1, c_2, ..., c_k)$

satisfies the condition (3.33) if and only if it satisfies the condition (3.35).
We can now prove the main result.

**Theorem 3.15**  Let $A$ be an $m \times n$ matrix. Suppose that the system of equations that corresponds to the equation $A\mathbf{x} = \mathbf{0}$ is equivalent to a system of the form

$$
\begin{aligned}
\tilde{a}_{11}x_{i_1} + \tilde{a}_{12}x_{i_2} + \cdots + \tilde{a}_{1r}x_{i_r} + \cdots + \tilde{a}_{1n}x_{i_n} &= 0, \\
\tilde{a}_{22}x_{i_2} + \cdots + \tilde{a}_{2r}x_{i_r} + \cdots + \tilde{a}_{2n}x_{i_n} &= 0, \\
&\cdots\cdots\cdots\cdots\cdots\cdots\cdots \\
\tilde{a}_{rr}x_{i_r} + \cdots + \tilde{a}_{rn}x_{i_n} &= 0,
\end{aligned}
\tag{3.36}
$$

where $\tilde{a}_{11}, \tilde{a}_{22}, \ldots, \tilde{a}_{rr}$ are not zero. Then $A$ has rank $r$ and the equation $A\mathbf{x} = \mathbf{0}$ has $n - r$ linearly independent solutions.

PROOF   We first show that the coefficient matrix of the system (3.36) has rank $r$. Its rank cannot be greater than $r$, because its column vectors are elements of $R^r$. But the $r$ column vectors

$$
\begin{bmatrix} \tilde{a}_{11} \\ 0 \\ 0 \\ \vdots \\ 0 \end{bmatrix},
\begin{bmatrix} \tilde{a}_{12} \\ \tilde{a}_{22} \\ 0 \\ \vdots \\ 0 \end{bmatrix}, \ldots,
\begin{bmatrix} \tilde{a}_{1r} \\ \tilde{a}_{2r} \\ \tilde{a}_{3r} \\ \vdots \\ \tilde{a}_{rr} \end{bmatrix}
$$

are linearly independent, since their determinant is $\tilde{a}_{11}\tilde{a}_{22}\cdots\tilde{a}_{rr} \neq 0$. Thus the coefficient matrix of the system (3.36) has rank $r$ and the maximum number of linearly independent solutions of this system is $n - r$. By the lemma, the maximum number of linearly independent solutions of the equation $A\mathbf{x} = \mathbf{0}$ is $n - r$, therefore, $A$ has rank $r$.

In order to obtain solutions of the system (3.36) we can assign values at will to the $n - r$ unknowns $x_{i_j}$ with $j = r + 1, r + 2, \ldots, n$. The remaining unknowns are then completely determined. A definite procedure for finding $n - r$ linearly independent solutions is as follows. First, we let $x_{i_{r+1}} = 1$ and $x_{i_j} = 0$ for $j > r + 1$ and determine the corresponding solution of the system (3.36). Next we find the solution that corresponds to the choice $x_{i_{r+2}} = 1$, $x_{i_j} = 0$ for $j = r + 1$ and for $j > r + 2$. Continuing, we obtain $n - r$ solutions. An application of Theorem 3.6 shows that the solutions are linearly independent.

As an example we consider the system

$$
\begin{aligned}
x_1 - x_2 + 2x_3 &= 0, \\
2x_1 - 2x_2 + 4x_3 &= 0, \\
-3x_1 + 3x_2 - 6x_3 &= 0, \\
4x_1 - 4x_2 + 8x_3 &= 0,
\end{aligned}
\tag{3.37}
$$

whose coefficient matrix is

$$A = \begin{bmatrix} 1 & -1 & 2 \\ 2 & -2 & 4 \\ -3 & 3 & -6 \\ 4 & -4 & 8 \end{bmatrix}.$$

The Gauss reduction procedure (see Section 2.1) can be used to reduce the system (3.37) to the equivalent system

$$
\begin{aligned}
x_1 - x_2 + 2x_3 &= 0, \\
0 &= 0, \\
0 &= 0, \\
0 &= 0.
\end{aligned}
\tag{3.38}
$$

This system is of the form (3.36), with $r = 1$. The rank of $A$ is therefore 1, so the maximum number of linearly independent solutions of the system (3.37) is $3 - 1 = 2$. From the system (3.38) we see that $(1, 1, 0)$ and $(-2, 0, 1)$ are linearly independent solutions. Hence the complete solution consists of all vectors of the form

$$c_1(1, 1, 0) + c_2(-2, 0, 1),$$

where $c_1$ and $c_2$ are arbitrary. We may also write

$$
\begin{aligned}
x_1 &= c_1 - 2c_2, \\
x_2 &= c_1 \\
x_3 &= c_2.
\end{aligned}
$$

Classical results that concern the rank of a matrix are presented in the next three theorems. Their proofs are left as exercises.

**Theorem 3.16** A matrix has rank $r$ if and only if it has at least one $r \times r$ nonsingular submatrix and every $(r + 1) \times (r + 1)$ submatrix (if there are any) is singular.

The dimension of the space spanned by the column vectors of a matrix is sometimes called the *column rank* of the matrix. Similarly, the dimension of the space spanned by the row vectors of a matrix is called the *row rank* of the matrix. The next theorem asserts the equality of these two ranks.

**Theorem 3.17**   The column rank of a matrix is equal to its row rank.

The next theorem concerns the effect of performing an elementary row or column operation (see Section 2.9) on a matrix.

**Theorem 3.18**   If a matrix $B$ is obtained from a matrix $A$ by the performance of an elementary row or column operation on $A$, then the rank of $B$ is equal to the rank of $A$.

We shall use Theorems 3.16 and 3.18 to determine the rank of the matrix

$$\begin{bmatrix} 2 & -1 & -3 & 4 \\ 4 & 1 & -1 & 2 \\ 2 & 5 & 7 & -8 \end{bmatrix}.$$

Subtracting twice the first row from the second, and then subtracting the first row from the third, we obtain the matrix

$$\begin{bmatrix} 2 & -1 & -3 & 4 \\ 0 & 3 & 5 & -6 \\ 0 & 6 & 10 & -12 \end{bmatrix}.$$

Subtracting twice the second row from the third yields the matrix

$$\begin{bmatrix} 2 & -1 & -3 & 4 \\ 0 & 3 & 5 & -6 \\ 0 & 0 & 0 & 0 \end{bmatrix}.$$

If we multiply through in the first column by $\frac{1}{2}$ and add appropriate multiples of the first column to the other columns, we obtain the matrix

$$\begin{bmatrix} 1 & 0 & 0 & 4 \\ 0 & 3 & 5 & -6 \\ 0 & 0 & 0 & 0 \end{bmatrix}.$$

Every $3 \times 3$ submatrix of this matrix is singular, but there is a $2 \times 2$ nonsingular submatrix. Thus the rank of this and the original matrix is 2. The row vectors

$$[2 \quad -1 \quad -3 \quad 4], \qquad [4 \quad 1 \quad -1 \quad 2]$$

of the original matrix are linearly independent, by Theorem 3.6, and form a basis for the two-dimensional space spanned by the row vectors of the matrix.

The column vectors

$$\begin{bmatrix} -3 \\ -1 \\ 7 \end{bmatrix}, \quad \begin{bmatrix} 4 \\ 2 \\ 8 \end{bmatrix}$$

are linearly independent and form a basis for the space spanned by the column vectors of the matrix.

### Exercises for Section 3.8

1. What can be said about a linear transformation whose rank is zero?

2. Let $T$ be a linear transformation from an $n$-dimensional space $V$ into a space $W$. If $\mathscr{B} = \{v_1, v_2, \ldots, v_n\}$ is a basis for $V$, show that the rank of $T$ is equal to the maximum number of linearly independent vectors in the set $\{Tv_1, Tv_2, \ldots, Tv_n\}$.

3. Let $\mathscr{P}^n$ be the vector space of all polynomials of degree $<n$, and let $D$ be the derivative operator that maps $\mathscr{P}^n$ into $\mathscr{P}^{n-1}$. Determine the rank and nullity of $D$.

4. Let $\mathscr{P}^2$ be the space of polynomials of degree $<2$ and let $T : \mathscr{P}^2 \to R^1$ be the mapping defined by

$$Tf = \int_0^1 f(t)\, dt,$$

for $f$ a function in $\mathscr{P}^2$. Find a basis for the kernel and a basis for the image of $T$.

5. Let $A$ be a $5 \times 6$ matrix with rank 2.

(a) What is the maximum number of linearly independent column vectors of $A$?

(b) What is the maximum number of linearly independent solutions of the equation $Ax = 0$?

In Exercises 6–13, find for the given matrix (a) its rank; (b) its nullity; (c) a basis for its kernel.

6. $[1 \quad -2 \quad 1]$

7. $\begin{bmatrix} 1 & 0 & -1 \\ 0 & 1 & 1 \\ 1 & 1 & 0 \end{bmatrix}$

8. $\begin{bmatrix} 1 & -1 & 1 & 0 \\ 1 & 0 & 1 & 1 \\ 0 & 0 & 0 & 1 \\ 1 & 1 & 1 & 2 \end{bmatrix}$

9. $\begin{bmatrix} 1 & 0 & 1 & -1 \\ -2 & 0 & -2 & 2 \end{bmatrix}$

10. $\begin{bmatrix} 0 & 0 \\ 0 & 0 \end{bmatrix}$

11. $\begin{bmatrix} 2 & 0 & -1 \\ -2 & 0 & 1 \\ 4 & 0 & -2 \end{bmatrix}$

12. $\begin{bmatrix} 2 & -1 \\ 3 & 2 \end{bmatrix}$

13. $\begin{bmatrix} 1 & -3 \\ -2 & 6 \end{bmatrix}$

14. Let $S$ and $T$ be linear mappings from a finite-dimensional space $V$ into $V$, such that $ST = I$, where $I$ is the identity mapping: $I(\mathbf{v}) = \mathbf{v}$ for all $\mathbf{v}$ in $V$. Show that $T$ has an inverse and that $S = T^{-1}$. Suggestion: show that the kernel of $T$ consists of the zero element, and use Theorem 3.13.

15. Let $T_2 : U \to V$ and $T_1 : V \to W$, where $U$ is finite-dimensional.
    (a)  Show that the nullity of $T_1 T_2$ is at least as large as the nullity of $T_2$.
    (b)  Show that the rank of $T_1 T_2$ is no larger than the rank of $T_2$. Suggestion: suppose that $T_2$ has rank $r$, let $\mathbf{v}_1, \mathbf{v}_2, \ldots, \mathbf{v}_{r+1}$ be elements in the image of $T_1 T_2$, and let $\mathbf{u}_1, \mathbf{u}_2, \ldots, \mathbf{u}_{r+1}$ be elements such that $T_1 \mathbf{u}_i = \mathbf{v}_i$.

16.  Prove Theorem 3.16. Suggestion: apply Theorem 3.6 to the submatrices.

17.  Prove Theorem 3.17. Suggestion: use Theorems 3.6 and 3.16.

18.  Prove Theorem 3.18. Suggestion: use Theorem 3.1.

19.  Let $A$ be an $m \times n$ matrix of rank $r$. Let $B$ be a nonsingular matrix of order $m$ and let $C$ be a nonsingular matrix of order $n$.

    (a)  Show that $BA$ has rank $r$. Suggestion: if $\mathbf{a}_1, \mathbf{a}_2, \ldots, \mathbf{a}_n$ are the column vectors of $A$, then the column vectors of $BA$ are

$$B\mathbf{a}_1, B\mathbf{a}_2, \ldots, B\mathbf{a}_n.$$

    (b)  Show that $AC$ has rank $r$. Suggestion: use part (a) and Theorem 3.17.

20.  Use Theorem 3.18 to determine the ranks of the matrices in Exercises 6–13.

21.  The matrix $[\mathbf{a}_1, \mathbf{a}_2, \ldots, \mathbf{a}_n, \mathbf{b}]$ is called the *augmented matrix* of the equation $A\mathbf{x} = \mathbf{b}$. Show that the equation has a solution if and only if the rank of the augmented matrix is equal to the rank of the coefficient matrix $A$.

## 3.9    COORDINATES

Let $V$ be a finite-dimensional vector space of dimension $n$ over the real numbers. We shall show that many important questions about such a space can be answered by a study of the particular space $R^n$.

Let $\mathscr{B}$ be an *ordered basis* for $V$. That is, let $\mathscr{B}$ be an *ordered* set

$$(\mathbf{v}_1, \mathbf{v}_2, \ldots, \mathbf{v}_n)$$

of $n$ linearly independent elements of $V$. If $\mathbf{v}$ is any element of $V$, we know that there exist numbers $a_1, a_2, \ldots, a_n$ such that

$$\mathbf{v} = a_1\mathbf{v}_1 + a_2\mathbf{v}_2 + \cdots + a_n\mathbf{v}_n.$$

These numbers $a_1, a_2, \ldots, a_n$ are unique, because if we also have

$$\mathbf{v} = b_1\mathbf{v}_1 + b_2\mathbf{v}_2 + \cdots + b_n\mathbf{v}_n,$$

then

$$(a_1 - b_1)\mathbf{v}_1 + (a_2 - b_2)\mathbf{v}_2 + \cdots + (a_n - b_n)\mathbf{v}_n = 0.$$

Since the basis elements are linearly independent, we have $a_i - b_i = 0$ or $a_i = b_i$ for $1 \le i \le n$.

The numers $a_1, a_2, \ldots, a_n$, taken in order, are called the *coordinates* of $\mathbf{v}$ relative to the ordered basis $\mathscr{B}$. The element $\mathbf{a} = (a_1, a_2, \ldots, a_n)$ of $R^n$ is called the *coordinate vector* of $\mathbf{v}$ relative to the ordered basis $\mathscr{B}$. If a different ordered basis for $V$ were selected, the coordinates and coordinate vector of $v$ relative to that basis might very well be different.

**Example 1** Let $V$ be the three-dimensional space $R^3$. If we choose the standard basis $(\mathbf{e}_1, \mathbf{e}_2, \mathbf{e}_3)$, where

$$\mathbf{e}_1 = (1, 0, 0), \qquad \mathbf{e}_2 = (0, 1, 0), \qquad \mathbf{e}_3 = (0, 0, 1)$$

for our ordered basis, then the coordinate vector of any element $\mathbf{v}$ of $V$ is $\mathbf{v}$ itself, because if $\mathbf{v} = (v_1, v_2, v_3)$ then

$$\mathbf{v} = v_1\mathbf{e}_1 + v_2\mathbf{e}_2 + v_3\mathbf{e}_3.$$

**Example 2** Again let $V$ be $R^3$, but choose as an ordered basis $(\mathbf{u}_1, \mathbf{u}_2, \mathbf{u}_3)$, where

$$\mathbf{r}_1 = (1, 1, 0), \qquad \mathbf{u}_2 = (2, -1, 1), \qquad \mathbf{u}_3 = (0, 1, -1).$$

If $\mathbf{v} = (v_1, v_2, v_3)$ is an element of $V$, then

$$\mathbf{v} = a_1\mathbf{u}_1 + a_2\mathbf{u}_2 + a_3\mathbf{u}_3$$

or

$$a_1 + 2a_2 \quad\ = v_1,$$
$$a_1 - \ a_2 + a_3 = v_2,$$
$$a_2 - a_3 = v_3.$$

Solving, we find that the coordinates of **v** relative to the given basis are

$$a_1 = v_2 + v_3, \qquad a_2 = \tfrac{1}{2}(v_1 - v_2 - v_3), \qquad a_3 = \tfrac{1}{2}(v_1 - v_2 - 3v_3).$$

For instance, the coordinate vector of $\mathbf{v} = (1, 1, 1)$ is $\mathbf{a} = (2, -1/2, -3/2)$.

**Example 3**   We consider the two-dimensional space $\mathscr{P}^2$ of all polynomials of degree $< 2$. An ordered basis for $\mathscr{P}^2$ is $(f_1, f_2)$, where $f_1(t) = 1$ and $f_2(t) = t$. The coordinate vector of $f$, where $f(t) = a + bt$, is $(a, b)$. If $g_1(t) = t + 2$ and $g_2(t) = -3t + 1$, the coordinate vector of $f$ relative to the basis $(g_1, g_2)$ is $(\tfrac{1}{7}(3a - 5b), \tfrac{1}{7}(a - 2b))$.

Once an ordered basis $\mathscr{B} = (\mathbf{u}_1, \mathbf{u}_2, \ldots, \mathbf{u}_n)$ for $V$ has been chosen, the element

$$\mathbf{a} = (a_1, a_2, \ldots, a_n)$$

of $R^n$ is associated with the element $\mathbf{v} = a_1\mathbf{u}_1 + a_2\mathbf{u}_2 + \cdots + a_n\mathbf{u}_n$ of $V$. On the other hand, if

$$\mathbf{b} = (b_1, b_2, \ldots, b_n)$$

is an element of $R^n$, then **b** is the coordinate vector of an element **w** of $V$, where

$$\mathbf{w} = b_1\mathbf{u}_1 + b_2\mathbf{u}_2 + \cdots + b_n\mathbf{u}_n.$$

We write

$$\mathbf{v} \leftrightarrow \mathbf{a}, \qquad \mathbf{w} \leftrightarrow \mathbf{b}$$

to indicate this correspondence. This correspondence is preserved under the fundamental operations of "multiplication by a number" and "addition." If $c$ is a number, then

$$c\mathbf{v} = ca_1\mathbf{u}_1 + ca_2\mathbf{u}_2 + \cdots + ca_n\mathbf{u}_n$$

and

$$\mathbf{v} + \mathbf{w} = (a_1 + b_1)\mathbf{u}_1 + (a_2 + b_2)\mathbf{u}_2 + \cdots + (a_n + b_n)\mathbf{u}_n.$$

Thus the coordinate vectors of $cv$ and $\mathbf{v} + \mathbf{w}$ are $c\mathbf{a}$ and $\mathbf{a} + \mathbf{b}$, respectively. That is,

$$cv \leftrightarrow c\mathbf{a}, \qquad \mathbf{v} + \mathbf{w} \leftrightarrow \mathbf{a} + \mathbf{b}.$$

Notice that the zero element $\mathbf{0}$ of $V$ corresponds to the zero element

$$(0, 0, \ldots, 0)$$

of $R^n$ because

$$\mathbf{0} = 0\mathbf{u}_1 + 0\mathbf{u}_2 + \cdots + 0\mathbf{u}_n.$$

The proof of the next theorem relies on the preservation of the correspondence between elements of $V$ and elements of $R^n$ under the fundamental operations.

**Theorem 3.19** Let $\mathbf{v}_1, \mathbf{v}_2, \ldots, \mathbf{v}_m$ be elements of $V$ and let $\mathbf{a}_1, \mathbf{a}_2, \ldots, \mathbf{a}_m$ be the corresponding coordinate vectors relative to an ordered basis for $V$. Then the elements $\mathbf{v}_1, \mathbf{v}_2, \ldots, \mathbf{v}_m$ are linearly dependent if and only if the elements $\mathbf{a}_1, \mathbf{a}_2, \ldots, \mathbf{a}_m$ are linearly dependent.

PROOF   Consider the relation

$$c_1\mathbf{v}_1 + c_2\mathbf{v}_2 + \cdots + c_m\mathbf{v}_m = \mathbf{0}. \tag{3.39}$$

In accordance with the discussion preceding the statement of this theorem, the coordinate vector of the element

$$c_1\mathbf{v}_1 + c_2\mathbf{v}_2 + \cdots + c_m\mathbf{v}_m$$

of $V$ is

$$c_1\mathbf{a}_1 + c_2\mathbf{a}_2 + \cdots + c_m\mathbf{a}_m$$

and the coordinate vector of $\mathbf{0}$ in $V$ is $\mathbf{0}$ in $R^n$. Hence the condition (3.39) is satisfied if and only if the condition

$$c_1\mathbf{a}_1 + c_2\mathbf{a}_2 + \cdots + c_m\mathbf{a}_m = \mathbf{0}$$

is satisfied. The $\mathbf{v}_i$ are linearly dependent if and only if the $\mathbf{a}_i$ are linearly dependent.

In the preceding discussion, we have assumed that $V$ is a vector space over the real numbers. The coordinate vectors are then ordered $n$-tuples of real numbers. If $V$ is complex, the coordinate vectors will be elements of $C^n$, the set of all ordered $n$-tuples of complex numbers. With this minor change, all the theory of this section applies to complex vector spaces.

Let us now investigate what happens to the coordinate vector of an element of $V$ when the basis is changed. Let

$$\mathscr{B} = (\mathbf{v}_1, \mathbf{v}_2, \ldots, \mathbf{v}_n), \qquad \mathscr{B}' = (\mathbf{v}'_1, \mathbf{v}'_2, \ldots, \mathbf{v}'_n)$$

be ordered bases for an $n$-dimensional vector space $V$. Let $\mathbf{w}$ be any element of $V$, with coordinate vectors

$$\mathbf{a} = (a_1, a_2, \ldots, a_n), \qquad \mathbf{a}' = (a'_1, a'_2, \ldots, a'_n)$$

relative to $\mathscr{B}$ and $\mathscr{B}'$, respectively. Thus

$$\mathbf{w} = a_1 \mathbf{v}_1 + a_2 \mathbf{v}_2 + \cdots + a_n \mathbf{v}_n = a'_1 \mathbf{v}'_1 + a'_2 \mathbf{v}'_2 + \cdots + a'_n \mathbf{v}'_n.$$

Every element in $V$ can be expressed as a linear combination of the elements of $\mathscr{B}'$. In particular, there exists numbers $c_{ij}$, $1 \le i \le n$, $1 \le j \le n$, such that

$$
\begin{aligned}
\mathbf{v}_1 &= c_{11}\mathbf{v}'_1 + c_{21}\mathbf{v}'_2 + \cdots + c_{n1}\mathbf{v}'_n, \\
\mathbf{v}_2 &= c_{12}\mathbf{v}'_1 + c_{22}\mathbf{v}'_2 + \cdots + c_{n2}\mathbf{v}'_n, \\
&\phantom{=}\ \cdots\cdots\cdots\cdots\cdots\cdots\cdots\cdots\cdots \\
\mathbf{v}_n &= c_{1n}\mathbf{v}'_1 + c_{2n}\mathbf{v}'_2 + \cdots + c_{nn}\mathbf{v}'_n
\end{aligned}
\tag{3.40}
$$

or

$$\mathbf{v}_j = \sum_{i=1}^{n} c_{ij}\mathbf{v}'_i, \quad 1 \le j \le n. \tag{3.41}$$

The relationship between the coordinate vectors $\mathbf{a}$ and $\mathbf{a}'$ is given by the following theorem.

**Theorem 3.20**   Let $\mathbf{a}$ and $\mathbf{a}'$ be the coordinate vectors of the same element of the $n$-dimensional vector space $V$ relative to the bases

$$\mathscr{B} = (\mathbf{v}_1, \mathbf{v}_2, \ldots, \mathbf{v}_n), \qquad \mathscr{B}' = (\mathbf{v}'_1, \mathbf{v}'_2, \ldots, \mathbf{v}'_n),$$

respectively. Let the elements of the two bases be related by Eq. (3.40). Then the column vector representations of $\mathbf{a}$ and $\mathbf{a}'$ are related by the equation

$$\mathbf{a}' = C\mathbf{a} \tag{3.42}$$

where the $n \times n$ nonsingular matrix $C$ with elements $c_{ij}$ is the transpose of the matrix associated with the system (3.40). The matrix $C$ is called the *matrix of the coordinate transformation* from $\mathscr{B}$ to $\mathscr{B}'$.

PROOF The matrix $C$ is nonsingular, since the elements of $\mathscr{B}$ are linearly independent. (See Exercise 6, Section 3.3.)

Making use of formula (3.41), we may write

$$\mathbf{w} = \sum_{j=1}^{n} a_j \mathbf{v}_j$$

$$= \sum_{j=1}^{n} a_j \sum_{i=1}^{n} c_{ij} \mathbf{v}'_i$$

$$= \sum_{i=1}^{n} \left( \sum_{j=1}^{n} c_{ij} a_j \right) \mathbf{v}'_i .$$

But we also have

$$\mathbf{w} = \sum_{i=1}^{n} a'_i \mathbf{v}'_i$$

so

$$a'_i = \sum_{j=1}^{n} c_{ij} a_j, \quad 1 \le i \le n .$$

This equation says that the column vectors $\mathbf{a}$ and $\mathbf{a}'$, which are $n \times 1$ matrices, satisfy the relation (3.42).

Since $C$ is nonsingular, we may write $\mathbf{a} = C^{-1}\mathbf{a}'$. Thus the matrix of the coordinate transformation from $\mathscr{B}'$ to $\mathscr{B}$ is $C^{-1}$.

**Example 4**  Let $\mathscr{B} = (\mathbf{v}_1, \mathbf{v}_2, \mathbf{v}_3)$ and $\mathscr{B}' = (\mathbf{v}'_1, \mathbf{v}'_2, \mathbf{v}'_3)$ in $R^3$, where

$$\mathbf{v}_1 = (2, 4, 0), \qquad \mathbf{v}'_1 = (1, 0, 1),$$
$$\mathbf{v}_2 = (0, -2, 0), \qquad \mathbf{v}'_2 = (2, 0, 0),$$
$$\mathbf{v}_3 = (0, 4, -2), \qquad \mathbf{v}'_3 = (2, 1, 1).$$

There are numbers $a$, $b$, and $c$ such that

$$\mathbf{v}_1 = a\mathbf{v}'_1 + b\mathbf{v}'_2 + c\mathbf{v}'_3$$

or

$$2 = a + 2b + 2c,$$
$$4 = \qquad\qquad c,$$
$$0 = a \qquad + c.$$

Solving for $a$, $b$, and $c$, we find that $a = -4$, $b = -1$, $c = 4$, and thus

$$\mathbf{v}_1 = -4\mathbf{v}_1' - \mathbf{v}_2' + 4\mathbf{v}_3'.$$

Treating $\mathbf{v}_2$ and $\mathbf{v}_3$ in similar fashion, we find that

$$\mathbf{v}_1 = -4\mathbf{v}_1' - \mathbf{v}_2' + 4\mathbf{v}_3',$$
$$\mathbf{v}_2 = 2\mathbf{v}_1' + \mathbf{v}_2' - 2\mathbf{v}_3',$$
$$\mathbf{v}_3 = -6\mathbf{v}_1' - \mathbf{v}_2' + 4\mathbf{v}_3'.$$

The matrix $C$ of the coordinate transformation is the transpose of the matrix

$$\begin{bmatrix} -4 & -1 & 4 \\ 2 & 1 & -2 \\ -6 & -1 & 4 \end{bmatrix}.$$

Thus

$$C = \begin{bmatrix} -4 & 2 & -6 \\ -1 & 1 & -1 \\ 4 & -2 & 4 \end{bmatrix}.$$

If $\mathbf{v}$ is the element of $R^3$ such that

$$\mathbf{v} = 2\mathbf{v}_1 - 3\mathbf{v}_2 + \mathbf{v}_3,$$

then

$$a = \begin{bmatrix} 2 \\ -3 \\ 1 \end{bmatrix}$$

and the coordinate vector $\mathbf{a}'$ of $\mathbf{v}$ relative to $\mathscr{B}'$ is given by

$$\mathbf{a}' = C\mathbf{a} = \begin{bmatrix} -20 \\ -6 \\ 18 \end{bmatrix}.$$

Thus

$$\mathbf{v} = -20\mathbf{v}_1' - 6\mathbf{v}_2' + 18\mathbf{v}_3'.$$

## Exercises for Section 3.9

1. Find the coordinates of the vector $\mathbf{v} = (2, -3)$ with respect to the given basis $\mathscr{B} = (\mathbf{u}_1, \mathbf{u}_2)$ for $R^2$.
   (a) $\mathbf{u}_1 = (1, 1)$, $\mathbf{u}_2 = (3, 4)$    (b) $\mathbf{u}_1 = (2, -4)$, $\mathbf{u}_2 = (-1, 3)$
   (c) $\mathbf{u}_1 = (0, 1)$, $\mathbf{u}_2 = (1, 0)$    (d) $\mathbf{u}_1 = (0, 3)$, $\mathbf{u}_2 = (2, -1)$

2. Find the coordinates of the vector $\mathbf{v} = (-1, 3, -2)$ with respect to the given basis $\mathscr{B} = (\mathbf{u}_1, \mathbf{u}_2, \mathbf{u}_3)$ for $R^3$.
   (a) $\mathbf{u}_1 = (0, 1, 1)$, $\mathbf{u}_2 = (1, 0, 1)$, $\mathbf{u}_3 = (1, 1, 0)$
   (b) $\mathbf{u}_1 = (1, 1, 2)$, $\mathbf{u}_2 = (2, 0, 0)$, $\mathbf{u}_3 = (-1, 3, -2)$

3. Let $\mathscr{P}^3$ be the space of all polynomials of degree $<3$. Show that $(g_1, g_2, g_3)$, where $g_1(x) = x - 1$, $g_2(x) = x^2 - x$, $g_3(x) = 2x$ for all $x$ is an ordered basis for $\mathscr{P}^3$. Find the coordinates, relative to this basis, for the function $f$, where $f(x) = 4x^2 - 6x + 2$.

4. (a) Let $\mathbf{u}$ be an element of $R^n$ and let $\mathbf{a}$ be the coordinate vector of $\mathbf{u}$ with respect to an ordered basis for $R^n$. Is it necessarily true that $\|\mathbf{u}\| = \|\mathbf{a}\|$? (Here $\|\mathbf{v}\|$ is the length of $\mathbf{v}$.)

   (b) Let $\mathbf{u}$ and $\mathbf{v}$ be orthogonal elements of $R^n$. Let $\mathbf{a}$ and $\mathbf{b}$ be the coordinate vectors of $\mathbf{u}$ and $\mathbf{v}$, respectively, with respect to an ordered basis for $R^n$. Are $\mathbf{a}$ and $\mathbf{b}$ necessarily orthogonal?

5. Let $\mathscr{B}$ be an orthonormal basis for $R^n$. Let $\mathbf{u}$ and $\mathbf{v}$ be elements of $R^n$ with coordinate vectors $\mathbf{a}$ and $\mathbf{b}$, respectively, relative to $\mathscr{B}$. Show that $\mathbf{u} \cdot \mathbf{v} = \mathbf{a} \cdot \mathbf{b}$, and that $\|\mathbf{u}\| = \|\mathbf{a}\|$.

6. Let $\mathscr{B} = (\mathbf{v}_1, \mathbf{v}_2)$ and $\mathscr{B}' = (\mathbf{v}_1', \mathbf{v}_2')$ be ordered bases for $R^2$ with

   $$\mathbf{v}_1 = (4, 1), \qquad \mathbf{v}_2 = (-2, -2), \qquad \mathbf{v}_1' = (-2, 1), \qquad \mathbf{v}_2' = (-3, 2).$$

   (a) Find the matrix of the coordinate transformation from $\mathscr{B}$ to $\mathscr{B}'$.

   (b) Find equations that relate the coordinates, relative to $\mathscr{B}$ and $\mathscr{B}'$. of an arbitrary vector in $R^2$.

7. Repeat Exercise 6 with

   $$\mathbf{v}_1 = (1, 0), \qquad \mathbf{v}_2 = (0, 1), \qquad \mathbf{v}_1' = (-1, 1), \qquad \mathbf{v}_2' = (-1, 3).$$

8. If the matrix of the coordinate transformation from $\mathscr{B} = (\mathbf{v}_1, \mathbf{v}_2, \mathbf{v}_3)$ to $\mathscr{B}' = (\mathbf{v}_1', \mathbf{v}_2', \mathbf{v}_3')$ is

   $$\begin{bmatrix} 2 & 1 & 0 \\ 0 & 1 & -2 \\ -1 & 0 & 3 \end{bmatrix}$$

   how are the elements of $\mathscr{B}$ and $\mathscr{B}'$ related?

9. Let $\mathscr{B}_1$ be the standard basis for $R^n$ and let $\mathscr{B}_2 = (\mathbf{u}_1, \mathbf{u}_2, \ldots, \mathbf{u}_n)$ be any ordered basis for $R^n$. Show that the matrix of the coordinate transformation from $\mathscr{B}_1$ to $\mathscr{B}_2$ is $[\mathbf{u}_1, \mathbf{u}_2, \ldots, \mathbf{u}_n]^{-1}$.

10. Let $\mathscr{B} = (\mathbf{u}_1, \mathbf{u}_2, \mathbf{u}_3)$, where $\mathbf{u}_1 = (2, 1, 0)$, $\mathbf{u}_2 = (0, 1, 1)$, $\mathbf{u}_3 = (-1, 1, 1)$. Find the matrix of the coordinate transformation in $R^3$ from the standard basis to $\mathscr{B}$. Suggestion: use the result of Problem 8.

11. Let $\mathscr{P}^3$ be the space of all polynomials of degree $< 3$. Let $f_1(t) = 1$, $f_2(t) = t$, $f_3(t) = t^2$, $g_1(t) = 2t$, $g_2(t) = t + 1$, $g_3(t) = t^2 - 1$. Show that $\mathscr{B} = (f_1, f_2, f_3)$ and $\mathscr{B}' = (g_1, g_2, g_3)$ are ordered bases for $\mathscr{P}^3$ and find the matrix of the coordinate transformation from $\mathscr{B}$ to $\mathscr{B}'$.

## 3.10   MATRICES AND LINEAR TRANSFORMATIONS

In this section, we restrict our attention to linear transformations from one finite dimensional space into another. Let $V$ and $W$ be vector spaces (both real or both complex) whose dimensions are $n$ and $m$, respectively, and let $T$ be a linear transformation from $V$ into $W$.

In order to describe $T$, let $\mathscr{B}_1 = (\mathbf{v}_1, \mathbf{v}_2, \ldots, \mathbf{v}_n)$ be an ordered basis for $V$ and let $\mathscr{B}_2 = (\mathbf{w}_1, \mathbf{w}_2, \ldots, \mathbf{w}_m)$ be an ordered basis for $W$. Since

$$T\mathbf{v}_1, T\mathbf{v}_2, \ldots, T\mathbf{v}_n$$

are elements of $W$, there exist numbers $c_{ij}$, $1 \le i \le m$, $1 \le j \le n$, such that

$$
\begin{aligned}
T\mathbf{v}_1 &= c_{11}\mathbf{w}_1 + c_{21}\mathbf{w}_2 + \cdots + c_{m1}\mathbf{w}_m, \\
T\mathbf{v}_2 &= c_{12}\mathbf{w}_1 + c_{22}\mathbf{w}_2 + \cdots + c_{m2}\mathbf{w}_m, \\
&\phantom{= c_{12}\mathbf{w}_1 + c_{22}} \cdots \\
T\mathbf{v}_n &= c_{1n}\mathbf{w}_1 + c_{2n}\mathbf{w}_2 + \cdots + c_{mn}\mathbf{w}_m
\end{aligned}
\tag{3.43}
$$

or

$$T\mathbf{v}_j = \sum_{i=1}^{m} c_{ij}\mathbf{w}_i, \qquad 1 \le j \le n. \tag{3.44}$$

If we know what happens to each element of $\mathscr{B}_1$ under $T$, we know what happens to every element of $V$, because if $\mathbf{x}$ is in $V$ and

$$\mathbf{x} = a_1\mathbf{v}_1 + a_2\mathbf{v}_2 + \cdots + a_n\mathbf{v}_n,$$

then

$$T\mathbf{x} = a_1 T\mathbf{v}_1 + a_2 T\mathbf{v}_2 + \cdots + a_n T\mathbf{v}_n.$$

The relationship between the coordinate vectors of **x** and $T$**x** may be described as follows

**Theorem 3.21** Let $T$ map $V$ into $W$, and let

$$\mathscr{B}_1 = (\mathbf{v}_1, \mathbf{v}_2, \ldots, \mathbf{v}_n), \qquad \mathscr{B}_2 = (\mathbf{w}_1, \mathbf{w}_2, \ldots, \mathbf{w}_m)$$

be bases for $V$ and $W$, respectively. Let **a** and **b** be the coordinate vectors of an element **x** in $V$ and its image $T$**x** in $W$, respectively. If the bases $\mathscr{B}_1$ and $\mathscr{B}_2$ are related by Eq. (3.43), then the column vectors **a** and **b** are related by the equation

$$\mathbf{b} = C\mathbf{a},$$

where $C$ is the $m \times n$ matrix with elements $c_{ij}$, that is, the transpose of the matrix of the system (3.43). The matrix $C$ is called the *matrix of the transformation T relative to the bases $\mathscr{B}_1$ and $\mathscr{B}_2$*.

PROOF   Since

$$\mathbf{x} = \sum_{j=1}^{n} a_j \mathbf{v}_j$$

we have

$$T\mathbf{x} = \sum_{j=1}^{n} a_j T\mathbf{v}_j.$$

Using the relations (3.44), we have

$$T\mathbf{x} = \sum_{j=1}^{n} a_j \sum_{i=1}^{m} c_{ij} \mathbf{w}_i = \sum_{i=1}^{m} \left( \sum_{j=1}^{n} c_{ij} a_j \right) \mathbf{w}_i.$$

Since we also have

$$T\mathbf{x} = \sum_{i=1}^{m} b_i \mathbf{w}_i,$$

it follows that

$$b_i = \sum_{j=1}^{n} c_{ij} a_j.$$

This is the relationship that we wished to establish.

From Eqs. (3.43) we see that the column vectors of $C$ (the row vectors of $C^T$) are the coordinate vectors of the elements $T\mathbf{v}_1, T\mathbf{v}_2, \ldots, T\mathbf{v}_n$ of $W$. It is easily shown (Exercise 7) that the rank of the matrix $C$ is equal to the rank of the transformation $T$. Thus the rank of the matrix of a linear transformation does not depend on the bases used, but only on the rank of the transformation. The elements of the matrix $C$ do depend on the bases chosen for $V$ and $W$. Once bases are chosen, the equation $\mathbf{b} = C\mathbf{a}$ associates with $T$ a linear transformation from $R^n$ into $R^m$ (or $C^n$ into $C^m$) since the coordinate vectors $\mathbf{a}$ and $\mathbf{b}$ are elements of these spaces, respectively.

**Example 1**    Let us consider a linear transformation $T$ from $R^3$ into $R^2$. A basis for $R^3$ is $\mathscr{B}_1 = (\mathbf{v}_1, \mathbf{v}_2, \mathbf{v}_3)$, where

$$\mathbf{v}_1 = (1, -1, 0), \qquad \mathbf{v}_2 = (0, 1, 2), \qquad \mathbf{v}_3 = (0, 1, 1).$$

A basis for $R^2$ is $\mathscr{B}_2 = (\mathbf{w}_1, \mathbf{w}_2)$, where

$$\mathbf{w}_1 = (2, 3), \qquad \mathbf{w}_2 = (-1, 2).$$

Suppose that $T$ is such that

$$T\mathbf{v}_1 = (5, 4), \qquad T\mathbf{v}_2 = (1, 5), \qquad T\mathbf{v}_3 = (3, -6).$$

This specification of the images of the elements of $\mathscr{B}_1$ completely determines $T$. It may be verified that

$$T\mathbf{v}_1 = 2\mathbf{w}_1 - \mathbf{w}_2,$$
$$T\mathbf{v}_2 = \mathbf{w}_1 + \mathbf{w}_2,$$
$$T\mathbf{v}_3 = -3\mathbf{w}_2.$$

According to Theorem 3.21, the matrix of $T$ relative to the indicated bases is

$$C = \begin{bmatrix} 2 & 1 & 0 \\ -1 & 1 & -3 \end{bmatrix}.$$

If $\mathbf{x} = a_1\mathbf{v}_1 + a_2\mathbf{v}_2 + a_3\mathbf{v}_3$ is an element of $R^3$ and if $T\mathbf{x} = b_1\mathbf{w}_1 + b_2\mathbf{w}_2$, then $\mathbf{b} = C\mathbf{a}$ or

$$b_1 = 2a_1 + a_2, \qquad b_2 = -a_1 + a_2 - 3a_3.$$

**Example 2**    We again consider a transformation $T$ from $R^3$ into $R^2$. This time we choose the standard bases $\mathscr{B}_1 = (\mathbf{e}_1, \mathbf{e}_2, \mathbf{e}_3)$, with

$$\mathbf{e}_1 = (1, 0, 0), \qquad \mathbf{e}_2 = (0, 1, 0), \qquad \mathbf{e}_3 = (0, 0, 1),$$

and $\mathscr{B}_2 = (\mathbf{e}_1', \mathbf{e}_2')$, where

$$\mathbf{e}_1' = (1, 0), \qquad \mathbf{e}_2' = (0, 1).$$

Then the coordinates of a vector are the same as the components of the vector. Suppose that

$$Te_1 = (a, b) = ae'_1 + be'_2,$$
$$Te_2 = (c, d) = ce'_1 + de'_2,$$
$$Te_3 = (e, f) = ee'_1 + fe'_2.$$

Then using Theorem 3.21, we see that the matrix of the transformation is

$$\begin{bmatrix} a & c & e \\ b & d & f \end{bmatrix} = [Te_1, Te_2, Te_3],$$

where we have used column vector representations of $Te_1$, $Te_2$, and $Te_3$.

In general, if $T$ maps $R^n$ into $R^m$ and if the standard bases

$$\mathscr{B}_1 = (e_1, e_2, \ldots, e_n), \qquad \mathscr{B}_2 = (e'_1, e'_2, \ldots, e'_m)$$

are used, then it can be shown that the $m \times n$ matrix of the transformation is

$$C = [Te_1, Te_2, \ldots, Te_n]. \tag{3.45}$$

Let us now consider the special case of a linear transformation $T$ from an $n$-dimensional vector space $V$ into $V$. In this case, $T$ is often called a linear transformation *on* $V$. Choosing a single basis

$$\mathscr{B} = (v_1, v_2, \ldots, v_n),$$

we write

$$x = a_1 v_1 + a_2 v_2 + \cdots + a_n v_n$$

and

$$Tx = b_1 v_1 + b_2 v_2 + \cdots + b_n v_n$$

for $x$ in $V$. If

$$Tv_j = \sum_{i=1}^{n} c_{ij} v_i, \qquad 1 \le j \le n,$$

the coordinate column vectors $\mathbf{a}$ and $\mathbf{b}$ are related by the equation

$$\mathbf{b} = C\mathbf{a}. \tag{3.46}$$

Let us now make a change of basis from $\mathscr{B}$ to a new basis

$$\mathscr{B}' = (v'_1, v'_2, \ldots, v'_n).$$

Then we may write

$$x = a'_1 v'_1 + a'_2 v'_2 + \cdots + a'_n v'_n, \qquad Tx = b'_1 v'_1 + b'_2 v'_2 + \cdots + b'_n v'_n.$$

Also, there is an $n \times n$ matrix $A$ with elements $a_{ij}$ such that

$$v_j = \sum_{i=1}^{n} a_{ij} v'_i, \qquad 1 \le j \le n.$$

According to Theorem 3.20, we must have

$$a' = Aa, \qquad b' = Ab$$

or

$$a = A^{-1}a', \qquad b = A^{-1}b'.$$

From Eq. (3.46) we have $A^{-1}b' = CA^{-1}a'$ or

$$b' = A^{-1}CAa'. \tag{3.47}$$

Thus if $C$ is the matrix of $T$ relative to the old basis, the matrix relative to the new basis is

$$A^{-1}CA. \tag{3.48}$$

If it is possible to find a nonsingular matrix $A$ such that $A^{-1}CA$ is a diagonal matrix, then the transformation on $R^n$ (or $C^n$) that is defined by Eq. (3.47) will have a particularly simple form. In fact, if $A^{-1}CA = \operatorname{diag}(d_1, d_2, \ldots, d_n)$, we have

$$b'_1 = d_1 a'_1, \quad b'_2 = d_2 a'_2, \ldots, \quad b'_n = d_n a'_n.$$

The possibility of choosing $A$ so that $A^{-1}CA$ is diagonal will be explored in the next chapter.

We consider one more topic of interest. Let $T$ be a linear transformation on $V$ and let $\mathscr{B} = (v_1, v_2, \ldots, v_n)$ be a basis for $V$. Let the image of $v_i$ be denoted by $v'_i$, so that $v'_i = Tv_i$, $1 \le i \le n$. If

$$v'_j = \sum_{i=1}^{n} c_{ij} v_i, \qquad 1 \le j \le n \tag{3.49}$$

and if **a** and **b** are the coordinate vectors, relative to $\mathscr{B}$, of elements **x** and $T\mathbf{x}$ of $V$, then according to Theorem 3.21, $\mathbf{b} = C\mathbf{a}$. If the vectors $\mathbf{v}_1'$, $\mathbf{v}_2'$, ..., $\mathbf{v}_n'$ are linearly independent, then $\mathscr{B}' = (\mathbf{v}_1', \mathbf{v}_2', ..., \mathbf{v}_n')$ is another basis for $V$ and we can inquire about the coordinate transformation from $\mathscr{B}$ to $\mathscr{B}'$. Let **p** and **p**′ be the coordinate vectors of an element of $V$, relative to $\mathscr{B}$ and $\mathscr{B}'$, respectively. Then from Eq. (3.49) and Theorem 3.20 (with the roles of $\mathbf{v}_i$ and $\mathbf{v}_i'$ interchanged in the statement of the theorem) we see that

$$\mathbf{p} = C\mathbf{p}' \quad \text{or} \quad \mathbf{p}' = C^{-1}\mathbf{p}.$$

Thus the matrix of the coordinate transformation from $\mathscr{B}$ to $\mathscr{B}'$ is the *inverse* of the matrix of the transformation $T$ relative to' $\mathscr{B}$ that takes $\mathscr{B}$ into $\mathscr{B}'$.

## Exercises for Section 3.10

1. Let $T$ be the linear transformation from $R^2$ into $R^2$ that maps $(1, -1)$ into $(1, 4)$ and $(2, 3)$ into $(2, 0)$. Find the matrix of $T$ relative to the basis $(\mathbf{u}_1, \mathbf{u}_2)$, where $\mathbf{u}_1$ and $\mathbf{u}_2$ are as given.

   (a) $\mathbf{u}_1 = (1, -1)$, $\mathbf{u}_2 = (2, 3)$    (b) $\mathbf{u}_1 = (1, 0)$, $\mathbf{u}_2 = (0, 1)$
   (c) $\mathbf{u}_1 = (2, 1)$,    $\mathbf{u}_2 = (1, 1)$    (d) $\mathbf{u}_1 = (1, 1)$, $\mathbf{u}_2 = (-1, 3)$

2. What is the rank of the transformation $T$ of Exercise 1 ?

3. Let $\mathscr{B}_1 = (\mathbf{e}_1, \mathbf{e}_2)$ and $\mathscr{B}_2 = (\mathbf{e}_1', \mathbf{e}_2', \mathbf{e}_3')$ be the standard bases for $R^2$ and $R^3$, respectively. Let $T$ be the linear transformation from $R^2$ into $R^3$ such that $T\mathbf{e}_1 = \mathbf{v}_1$ and $T\mathbf{e}_2 = \mathbf{v}_2$, where $\mathbf{v}_1$ and $\mathbf{v}_2$ are as given. Find the matrix of $T$ relative to $\mathscr{B}_1$ and $\mathscr{B}_2$, and determine the rank of $T$.

   (a) $\mathbf{v}_1 = (2, -1, 0)$, $\mathbf{v}_2 = (-4, 2, 0)$
   (b) $\mathbf{v}_1 = (1, 1, -3)$, $\mathbf{v}_2 = (2, 0, 2)$
   (c) $\mathbf{v}_1 = (0, 0, 0)$,    $\mathbf{v}_2 = (0, 0, 0)$

4. Let $\mathscr{P}^3$ denote the space of all polynomials of degree $< 3$. If $D$ is the derivative operator, then $D: \mathscr{P}^3 \to \mathscr{P}^3$. Find the matrix of the operator relative to the basis $(f_1, f_2, f_3)$, where

   (a) $f_1(x) = 1$, $f_2(x) = x$, $f_3(x) = x^2$
   (b) $f_1(x) = 2x - 1$, $f_2(x) = x^2 + 1$, $f_3(x) = 2x$

5. Let $S$ and $T$ be linear transformations from $V$ into $W$ with matrices $A$ and $B$, respectively, relative to a pair of bases. Find the matrix of
   (a) $cS$    (b) $S + T$.

6. Let $S$ be a linear transformation from $U$ into $V$ and let $T$ be a linear transformation from $V$ into $W$. If the matrix of $S$ is $A$, relative to $\mathscr{B}_1$

and $\mathscr{B}_2$, and if the matrix of $T$ is $B$, relative to $\mathscr{B}_2$ and $\mathscr{B}_3$, what is the matrix of $TS$, relative to $\mathscr{B}_1$ and $\mathscr{B}_3$? Justify your answer.

7. Let $T$ be a linear transformation from $V$ into $W$ and let $C$ be the matrix of $T$ with respect to a pair of bases. Show that the rank of $C$ is equal to the rank of $T$.

8. If $T: V \to V$ has the matrix $A$ relative to one basis and the matrix $B$ relative to another basis, show that det $B =$ det $A$.

9. Let $A$ be the matrix of a linear transformation $T$ with respect to bases $\mathscr{B}_1$ and $\mathscr{B}_2$. Suppose that $\mathscr{B}_1$ is replaced by a new basis $\mathscr{B}_1'$ and that $\mathscr{B}_2$ is replaced by $\mathscr{B}_2'$. If $P$ is the matrix of the coordinate transformation from $\mathscr{B}_1$ to $\mathscr{B}_1'$ and if $Q$ is the matrix of the coordinate transformation from $\mathscr{B}_2$ to $\mathscr{B}_2'$, what is the matrix of $T$ relative to $\mathscr{B}_1'$ and $\mathscr{B}_2'$?

10. Let $\mathscr{B}$ and $\mathscr{B}'$ be bases for a finite-dimensional vector space $V$. Show that the matrix of the coordinate transformation from $\mathscr{B}$ to $\mathscr{B}'$ is the matrix of the identity transformation from $V$ into $V$, relative to $\mathscr{B}$ and $\mathscr{B}'$.

11. Let $T$ be a linear transformation from an $n$-dimensional space $V$ into an $n$-dimensional space $W$. Let $A$ be the matrix of $T$, relative to bases $\mathscr{B}_1$ and $\mathscr{B}_2$. Show that $T$ has an inverse if and only if $A$ is nonsingular. If $A$ is nonsingular, show that $A^{-1}$ is the matrix for $T^{-1}$, relative to $\mathscr{B}_2$ and $\mathscr{B}_1$.

## 3.11   ORTHOGONAL TRANSFORMATIONS

Let $T$ be a linear transformation from $R^n$ into $R^n$. Then $T$ is called an *orthogonal transformation* if it has the property that

$$\|T\mathbf{v}\| = \|\mathbf{v}\| \tag{3.50}$$

for every element $\mathbf{v}$ of $R^n$. Thus an orthogonal transformation preserves the lengths of vectors in $R^n$. Orthogonal transformations can also be characterized in the following way.

**Theorem 3.22**   The linear transformation $T$ is orthogonal if and only if it preserves inner products; that is, if and only if

$$T\mathbf{u} \cdot T\mathbf{v} = \mathbf{u} \cdot \mathbf{v} \tag{3.51}$$

for all vectors $\mathbf{u}$ and $\mathbf{v}$ in $R^n$. (Recall that in two and three dimensions, the angle $\theta$ between two vectors $\mathbf{u}$ and $\mathbf{v}$ is given by the relation $\cos \theta = \mathbf{u} \cdot \mathbf{v}/ \|\mathbf{u}\| \|\mathbf{v}\|$. Thus an orthogonal transformation preserves angles as well as lengths.)

PROOF   If condition (3.51) holds, we see upon setting $\mathbf{u} = \mathbf{v}$ that

$$T\mathbf{v} \cdot T\mathbf{v} = \mathbf{v} \cdot \mathbf{v}$$

or

$$\|T\mathbf{v}\|^2 = \|\mathbf{v}\|^2.$$

The equality (3.50) follows upon taking square roots. Now suppose that condition (3.50) holds. Consider the identities

$$2\mathbf{u} \cdot \mathbf{v} = \|\mathbf{u} + \mathbf{v}\|^2 - \|\mathbf{u}\|^2 - \|\mathbf{v}\|^2, \tag{3.52}$$

$$2T\mathbf{u} \cdot T\mathbf{v} = \|T\mathbf{u} + T\mathbf{v}\|^2 - \|T\mathbf{u}\|^2 - \|T\mathbf{v}\|^2, \tag{3.53}$$

whose verification is left as an exercise. Since

$$\|T\mathbf{u} + T\mathbf{v}\| = \|T(\mathbf{u} + \mathbf{v})\| = \|\mathbf{u} + \mathbf{v}\| ,$$

$$\|T\mathbf{u}\| = \|\mathbf{u}\| , \qquad \|T\mathbf{v}\| = \|\mathbf{v}\| ,$$

we see that $T\mathbf{u} \cdot T\mathbf{v} = \mathbf{u} \cdot \mathbf{v}$ for all $\mathbf{u}$ and $\mathbf{v}$ in $V$. Hence preservation of lengths implies preservation of inner products.

The matrix of an orthogonal transformation, relative to an orthonormal basis, has certain special properties that we now investigate. Let $\mathscr{B} = (\mathbf{e}_1, \mathbf{e}_2, \ldots, \mathbf{e}_n)$ be an orthonormal basis for $R^n$ and let $A$ be the matrix of an orthogonal transformation $T$, relative to $\mathscr{B}$. If $\mathbf{u}$ and $\mathbf{v}$ are elements of $R^n$ such that

$$\mathbf{u} = x_1\mathbf{e}_1 + x_2\mathbf{e}_2 + \cdots + x_n\mathbf{e}_n, \qquad \mathbf{v} = y_1\mathbf{e}_1 + y_2\mathbf{e}_2 + \cdots + y_n\mathbf{e}_n,$$

then $\mathbf{u} \cdot \mathbf{v} = \mathbf{x} \cdot \mathbf{y}$, where $\mathbf{x}$ and $\mathbf{y}$ are the coordinate vectors of $\mathbf{u}$ and $\mathbf{v}$, respectively. The coordinate vectors of $T\mathbf{u}$ and $T\mathbf{v}$ are $A\mathbf{x}$ and $A\mathbf{y}$, respectively. Then $T\mathbf{u} \cdot T\mathbf{v} = \mathbf{u} \cdot \mathbf{v}$ if and only if

$$(A\mathbf{x})^T(A\mathbf{y}) = \mathbf{x}^T\mathbf{y}.$$

Making use of property (2.50c), we see that this equation can be written as

$$\mathbf{x}^T(A^T A)\mathbf{y} = \mathbf{x}^T\mathbf{y}.$$

Taking $\mathbf{x}$ to be the coordinate vector of $\mathbf{e}_i$ and $\mathbf{y}$ to be that of $\mathbf{e}_j$, we see from this last equation that the element in the $i$th row and $j$th column of $A^T A$ is $\delta_{ij}$. Thus

$$A^T A = I \tag{3.54}$$

or

$$A^{-1} = A^T. \tag{3.55}$$

A real matrix $A$ with the property (3.55) is called an *orthogonal matrix*. If $A = [\mathbf{a}_1, \mathbf{a}_2, \ldots, \mathbf{a}_n]$, then $A$ satisfies the condition (3.54) if and only if

$$\mathbf{a}_i \cdot \mathbf{a}_j = \delta_{ij}.$$

Thus $A$ is orthogonal if and only if its column vectors form an orthonormal set. (See Section 2.2.) Since

$$AA^{-1} = AA^T = I$$

we can also say that $A$ is orthogonal if and only if its row vectors constitute an orthonormal set. Finally, from the condition (3.54) we see that $(\det A)^2 = 1$ so that

$$\det A = \pm 1 \tag{3.56}$$

if $A$ is orthogonal. We say that $A$ is a *proper* or an *improper* orthogonal matrix according as $\det A = 1$ or $\det A = -1$. The matrix of $T$ relative to any other basis has the form $K^{-1}AK$, according to the expression (3.48). By Property 8, Section 2.6, the determinant of this matrix is the same as $\det A$. We may therefore speak of a *proper* or an *improper orthogonal transformation*, according as the determinant of the matrix of the transformation relative to any basis is positive or negative. The geometric significance of the two cases will be explained for $n = 2$ and $n = 3$. Notice that the orthgonal matrices

$$\begin{bmatrix} \frac{3}{5} & -\frac{4}{5} \\ \frac{4}{5} & \frac{3}{5} \end{bmatrix}, \quad \begin{bmatrix} \frac{3}{5} & \frac{4}{5} \\ \frac{4}{5} & -\frac{3}{5} \end{bmatrix}$$

are proper and improper, respectively.

We consider first the two dimensional case. Let $\mathscr{B} = (\mathbf{e}_1, \mathbf{e}_2)$ be an orthonormal basis for $R^2$. Let $\mathbf{e}_1'$ and $\mathbf{e}_2'$ be unit vectors obtained by rotating $\mathbf{e}_1$ and $\mathbf{e}_2$ through an angle $\theta$, as shown in Fig. 3.2. From the geometry of this figure

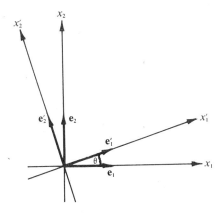

**Figure 3.2**

we see that

$$e'_1 = e_1 \cos \theta + e_2 \sin \theta,$$
$$e'_2 = -e_1 \sin \theta + e_2 \cos \theta. \tag{3.57}$$

The linear transformation $T$ that maps $e_1$ and $e_2$ into $e'_1$ and $e'_2$, respectively, has the matrix

$$A = \begin{bmatrix} \cos \theta & -\sin \theta \\ \sin \theta & \cos \theta \end{bmatrix} \tag{3.58}$$

with respect to the basis $\mathscr{B} = (e_1, e_2)$. Evidently $A$ is a proper orthogonal matrix. Since $T$ preserves lengths and angles, any vector with initial point at the origin is rotated rigidly through an angle $\theta$.

Let us also consider the coordinate transformation from the basis $\mathscr{B} = (e_1, e_2)$ to the basis $\mathscr{B}' = (e'_1, e'_2)$. If $v$ is any vector in $R^2$, we may write

$$\mathbf{v} = x_1 e_1 + x_2 e_2 = x'_1 e'_1 + x'_2 e'_2. \tag{3.59}$$

Then $\mathbf{x} = (x_1, x_2)$ and $\mathbf{x}' = (x'_1, x'_2)$ are the coordinate vectors of $\mathbf{v}$, relative to $\mathscr{B}$ and $\mathscr{B}'$. Taking scalar products with $e'_1$ and $e'_2$, we see that

$$x'_1 = x_1(e'_1 \cdot e_1) + x_2(e'_1 \cdot e_2),$$
$$x'_2 = x_1(e'_2 \cdot e_1) + x_2(e'_2 \cdot e_2)$$

or

$$x'_1 = x_1 \cos \theta + x_2 \sin \theta,$$
$$x'_2 = -x_1 \sin \theta + x_2 \cos \theta. \tag{3.60}$$

Thus the coordinate transformation is

$$\mathbf{x}' = A^T \mathbf{x}, \tag{3.61}$$

where $A$ is as in Eq. (3.58). Of course this formula also follows from the results at the end of the last section.

In the situation illustrated in Fig. 3.3, the $x_2$-axis has been reflected in the $x_1$-axis and then both axes have been rotated counterclockwise through an angle $\theta$. We consider the two steps separately. Let $T_1$ be the linear transformation that maps $\mathbf{e}_1$ into $\mathbf{e}_1$ and $\mathbf{e}_2$ into $-\mathbf{e}_2$. If $T_1\mathbf{e}_1 = \mathbf{e}_1'$ and $T_1\mathbf{e}_2 = \mathbf{e}_2'$ then

$$
\begin{aligned}
\mathbf{e}_1' &= \mathbf{e}_1, \\
\mathbf{e}_2' &= -\mathbf{e}_2.
\end{aligned}
\tag{3.62}
$$

The matrix of $T_1$, relative to the basis $\mathscr{B} = (\mathbf{e}_1, \mathbf{e}_2)$ is the improper orthgonal matrix

$$A_1 = \begin{bmatrix} 1 & 0 \\ 0 & -1 \end{bmatrix}. \tag{3.63}$$

Geometrically, $T_1$ maps every vector with initial point at the origin into its reflection with respect to the $x_1$ axis.

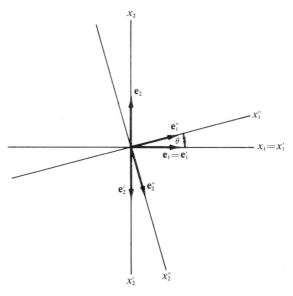

**Figure 3.3**

The mapping $T_2$ that takes $e_1'$ and $e_2'$ into $e_1''$ and $e_2''$, respectively, is simply a rotation through an angle $\theta$. We can use our earlier result to find the matrix $A_2$ of $T_2$ relative to the basis $\mathscr{B}' = (e_1', e_2')$. We see that $A_2$ is the proper orthogonal matrix

$$A_2 = \begin{bmatrix} \cos\theta & -\sin\theta \\ \sin\theta & \cos\theta \end{bmatrix}. \tag{3.64}$$

The linear transformation $T = T_2 T_1$ that takes $e_1$ into $e_1''$ and $e_2$ into $e_2''$ has the matrix (see Exercise 6, Section 3.10)

$$A = A_2 A_1 = \begin{bmatrix} \cos\theta & \sin\theta \\ \sin\theta & -\cos\theta \end{bmatrix}, \tag{3.65}$$

relative to $\mathscr{B} = (e_1, e_2)$. Note that $A$ is an improper orthogonal matrix. The orthogonal transformation considered here corresponds geometrically to a reflection followed by a rotation. As seen from Fig. 3.3, it can also be interpreted as a rotation followed by a reflection. In Exercise 6, the reader is asked to show that every $2 \times 2$ orthogonal matrix has one of the forms (3.58) or (3.65).

We consider somewhat more briefly the three-dimensional case. Let $\mathscr{B} = (e_1, e_2, e_3)$ and $\mathscr{B}' = (e_1', e_2', e_3')$ be orthonormal bases for $R^3$. One possible configuration of the two sets of unit vectors is illustrated in Fig. 3.4. There exist numbers $a_{ij}$ such that

$$\begin{aligned} e_1' &= a_{11} e_1 + a_{21} e_2 + a_{31} e_3, \\ e_2' &= a_{12} e_1 + a_{22} e_2 + a_{32} e_3, \\ e_3' &= a_{13} e_1 + a_{23} e_2 + a_{33} e_3 \end{aligned} \tag{3.66}$$

or

$$e_j' = \sum_{i=1}^{3} a_{ij} e_i, \qquad j = 1, 2, 3. \tag{3.67}$$

Taking the scalar product in this last equation with $e_k$, we see that

$$e_j' \cdot e_k = \sum_{i=1}^{3} a_{ij} e_i \cdot e_k = \sum_{i=1}^{3} a_{ij} \delta_{ik} = a_{kj}.$$

Thus $a_{kj}$ is the cosine of the angle between $e_k$ and $e_j'$. The matrix of the linear transformation $T$ that maps $e_1, e_2, e_3$ into $e_1', e_2', e_3'$ is the matrix $A$ with

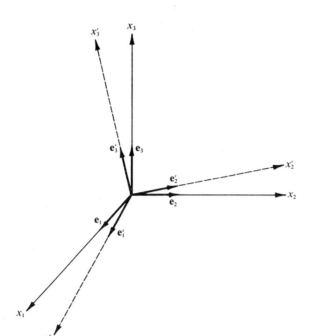

**Figure 3.4**

elements $a_{ij}$. Let $\mathbf{a}_1$, $\mathbf{a}_2$, $\mathbf{a}_3$ be the column vectors of $A$. Then from the Eqs. (3.66) we see that

$$\mathbf{a}_i \cdot \mathbf{a}_j = \mathbf{e}'_i \cdot \mathbf{e}'_j = \delta_{ij},$$

so $A$ is an orthogonal matrix.

If $A$ is a proper orthogonal matrix it can be shown, using concepts to be developed in Chapter 4 (see in particular Exercise 12, Section 4.2), that the transformation $T$ consists of a rotation about a certain line through the origin. If $A$ is improper orthogonal, $T$ consists of a rotation followed by a reflection of one axis in the plane of the remaining two (or a reflection followed by a rotation).[2]

---

[2] The reader familiar with the concepts of "cross product" and "triple scalar product" may observe from Eqs. (3.66) that $\mathbf{e}'_1 \cdot \mathbf{e}'_2 \times \mathbf{e}'_3$ has the value det $A^T =$ det $A$. If $\mathscr{B}$ and $\mathscr{B}'$ are both right-handed (or both left-handed) systems, then det $A$ is positive and $A$ is proper orthogonal. If one of the bases is right-handed and the other left-handed, then det $A$ is negative and $A$ is improper orthogonal.

### Exercises for Section 3.11

1. Which of the following matrices are orthogonal?

   (a) $\begin{bmatrix} 2/\sqrt{5} & 1/\sqrt{5} \\ -1/\sqrt{5} & 2/\sqrt{5} \end{bmatrix}$    (b) $\begin{bmatrix} -3 & 2 \\ 2 & 3 \end{bmatrix}$

   (c) $\begin{bmatrix} 1 & 1 & -1 \\ -1 & 1 & 1 \\ 1 & 0 & 2 \end{bmatrix}$    (d) $\begin{bmatrix} \frac{3}{7} & \frac{6}{7} & \frac{2}{7} \\ \frac{6}{7} & -\frac{2}{7} & -\frac{3}{7} \\ \frac{2}{7} & -\frac{3}{7} & \frac{6}{7} \end{bmatrix}$

2. If $\det A = 1$, is $A$ necessarily orthogonal?

3. Show that $A$ is orthogonal if and only if $A^T$ is orthogonal.

4. If $\mathbf{u}$ and $\mathbf{v}$ are in $R^n$, verify that $2\mathbf{u} \cdot \mathbf{v} = \|\mathbf{u} + \mathbf{v}\|^2 - \|\mathbf{u}\|^2 - \|\mathbf{v}\|^2$.

5. If $A$ and $B$ are orthogonal matrices of the same size, show that $AB$ is orthogonal.

6. Let $A$ be a $2 \times 2$ orthogonal matrix. Show that $A$ has one of the two forms

$$\begin{bmatrix} \cos\theta & -\sin\theta \\ \sin\theta & \cos\theta \end{bmatrix}, \quad \begin{bmatrix} \cos\theta & \sin\theta \\ \sin\theta & -\cos\theta \end{bmatrix}.$$

   Suggestion: if $a^2 + b^2 = 1$ there is an angle $\theta$ such that $a = \cos\theta$ and $b = \sin\theta$.

7. An orthonormal basis $(\mathbf{e}_1, \mathbf{e}_2)$ in $R^2$ is transformed into an orthonormal basis $(\mathbf{e}_1', \mathbf{e}_2')$ by a transformation that consists of a reflection in the line of $\mathbf{e}_2$ followed by a counterclockwise rotation through an angle of $\pi/6$. If a vector $\mathbf{x}$ has the coordinate representations

$$\mathbf{x} = x_1\mathbf{e}_1 + x_2\mathbf{e}_2 = x_1'\mathbf{e}_1' + x_2'\mathbf{e}_2'$$

   express $x_1'$ and $x_2'$ in terms of $x_1$ and $x_2$.

8. An orthonormal base $(\mathbf{e}_1, \mathbf{e}_2, \mathbf{e}_3)$ in $R^3$ is transformed into a base $(\mathbf{e}_1', \mathbf{e}_2', \mathbf{e}_3')$ by a transformation that consists of a rotation of $\mathbf{e}_1$ toward $\mathbf{e}_2$ through an angle of $\pi/4$, followed by a reflection in the plane of $\mathbf{e}_1$ and $\mathbf{e}_2$. If

$$\mathbf{x} = \sum x_i\mathbf{e}_i = \sum x_i'\mathbf{e}_i'$$

   express $x_1'$, $x_2'$, and $x_3'$ in terms of $x_1$, $x_2$, and $x_3$.

9. Let $\mathscr{B}_1$ be an orthonormal basis for $R^n$, let $\mathscr{B}_2$ be any basis for $R^n$, and let $C$ be the matrix of the coordinate transformation from $\mathscr{B}_1$ to $\mathscr{B}_2$. Show that $\mathscr{B}_2$ is an orthonormal basis if and only if $C$ is orthogonal.

**10.** The square matrix $A$ (whose elements may be complex) is called a *unitary matrix* if and only if $A*A = I$, where $A* = \bar{A}^T$. ($A*$ is called the transposed conjugate of $A$.)

(a)  Show that every orthogonal matrix is unitary.

(b)  Show that $A$ is unitary if and only if its column (row) vectors form an orthonormal set.

(c)  If $A$ is unitary show that $|\det A| = 1$.

(d)  A linear transformation $T$ from $C^n$ into $C^n$ is called a *unitary transformation* if $\|T\mathbf{u}\| = \|\mathbf{u}\|$ for every element $\mathbf{u}$. Show that $T$ is unitary if and only if the matrix of $T$ relative to every orthonormal basis is unitary.

# IV

## Characteristic Values

### 4.1 CHARACTERISTIC VALUES

If $A$ and $B$ are square matrices of the same size, we say that $A$ is *similar* to $B$ if there exists a nonsingular matrix $K$ such that

$$B = K^{-1}AK. \tag{4.1}$$

We now investigate the possibility that a given $n \times n$ matrix $A$ is similar to a diagonal matrix.

Let $D = \operatorname{diag}(\lambda_1, \lambda_2, \ldots, \lambda_n)$. Then $A$ is similar to $D$ if and only if there is a nonsingular matrix $K$ such that

$$D = K^{-1}AK$$

or

$$KD = AK. \tag{4.2}$$

Let $K = [\mathbf{k}_1, \mathbf{k}_2, \ldots, \mathbf{k}_n]$. Then the condition (4.2) can be written as

$$[\lambda_1 \mathbf{k}_1, \lambda_2 \mathbf{k}_2, \ldots, \lambda_n \mathbf{k}_n] = [A\mathbf{k}_1, A\mathbf{k}_2, \ldots, A\mathbf{k}_n]. \tag{4.3}$$

This matrix equation corresponds to the system of vector equations

$$A\mathbf{k}_1 = \lambda_1 \mathbf{k}_1, \qquad A\mathbf{k}_2 = \lambda_2 \mathbf{k}_2, \ldots, \qquad A\mathbf{k}_n = \lambda_n \mathbf{k}_n. \tag{4.4}$$

Since $K$ is nonsingular, the vectors $\mathbf{k}_1, \mathbf{k}_2, \ldots, \mathbf{k}_n$ must be linearly independent. Our results thus far may be summarized as follows.

**Theorem 4.1**    The $n \times n$ matrix $A$ is similar to a diagonal matrix if and only if there exist (complex) numbers $\lambda_1, \lambda_2, \ldots, \lambda_n$ and a linearly independent set of vectors $\{\mathbf{k}_1, \mathbf{k}_2, \ldots, \mathbf{k}_n\}$ such that

$$A\mathbf{k}_i = \lambda_i \mathbf{k}_i, \quad 1 \le i \le n.$$

If $K = [\mathbf{k}_1, \mathbf{k}_2, \ldots, \mathbf{k}_n]$, then

$$K^{-1}AK = \operatorname{diag}(\lambda_1, \lambda_2, \ldots, \lambda_n).$$

The number $\lambda_i$ and the matrix $K$ may be complex even when $A$ is real. Each of the vectors $\mathbf{k}_i$ mentioned above must be a nonzero vector. Suppose that we attempt to find a number $\lambda$ and a nonzero vector $\mathbf{k}$ such that

$$A\mathbf{k} = \lambda \mathbf{k}. \tag{4.5}$$

This equation can be written as

$$(\lambda I - A)\mathbf{k} = \mathbf{0}, \tag{4.6}$$

where $I$ is the identity matrix. According to Theorem 2.10, a nonzero vector $\mathbf{k}$ that satisfies the condition (4.6) exists if and only if $\lambda$ is a number such that

$$\det(\lambda I - A) = 0. \tag{4.7}$$

These considerations lead us to formalize certain ideas. If $A$ is an $n \times n$ matrix, a (complex) number $\lambda$ for which the condition (4.7) holds is called a *characteristic value* (or *eigenvalue*) of $A$. If $\lambda$ is such a number, a nonzero (complex) vector $\mathbf{k}$ such that

$$(\lambda I - A)\mathbf{k} = \mathbf{0}$$

is called a *characteristic vector* (or *eigenvector*) of $A$ corresponding to the characteristic value $\lambda$. According to Theorem 4.1, *an $n \times n$ matrix is similar to a diagonal matrix if and only if it possesses a set of $n$ linearly independendent characteristic vectors.* A matrix that is similar to a diagonal matrix is said to be *diagonalizable.*

**Example 1**    Let

$$A = \begin{bmatrix} 1 & -3 \\ -2 & 2 \end{bmatrix}.$$

Then

$$\det(\lambda I - A) = \begin{vmatrix} \lambda - 1 & 3 \\ 2 & \lambda - 2 \end{vmatrix} = (\lambda + 1)(\lambda - 4).$$

The characteristic values of $A$ are therefore $\lambda_1 = -1$ and $\lambda_2 = 4$. When $\lambda = -1$, the equation $(\lambda I - A)\mathbf{k} = 0$ corresponds to the system

$$-2k_1 + 3k_2 = 0,$$
$$2k_1 - 3k_2 = 0$$

for the components $k_1$ and $k_2$ of $\mathbf{k}$. One nontrivial solution is $k_1 = 3$ and $k_2 = 2$. The characteristic vectors of $A$ corresponding to $\lambda_1$ are those vectors of the form $a(3, 2) = (3a, 2a)$, where $a \neq 0$. (The rank and nullity of $\lambda_1 I - A$ are both one.)

When $\lambda = 4$, the equation $(\lambda I - A)\mathbf{k} = 0$ corresponds to the system

$$3k_1 + 3k_2 = 0,$$
$$2k_1 + 2k_2 = 0.$$

Since $k_1 = 1, k_2 = -1$ constitutes a nontrivial solution, the characteristic vectors corresponding to $\lambda_2$ are those vectors of the form $b(1, -1) = (b, -b)$, where $b \neq 0$. The particular characteristic vectors $(3, 2)$ and $(1, -1)$ are linearly independent. Hence $A$ is similar to a diagonal matrix. If

$$K = \begin{bmatrix} 3 & 1 \\ 2 & -1 \end{bmatrix},$$

then

$$K^{-1}AK = \begin{bmatrix} -1 & 0 \\ 0 & 4 \end{bmatrix}.$$

**Example 2**   If

$$A = \begin{bmatrix} 1 & 0 \\ 0 & 1 \end{bmatrix},$$

then

$$\det(\lambda I - A) = \begin{vmatrix} \lambda - 1 & 0 \\ 0 & \lambda - 1 \end{vmatrix} = (\lambda - 1)^2.$$

Thus $A$ has only one characteristic value, $\lambda_1 = 1$. The equation $(\lambda I - A)\mathbf{k} = 0$ becomes

$$0k_1 + 0k_2 = 0,$$
$$0k_1 + 0k_2 = 0.$$

The space of solutions has dimension 2. The vectors $(1, 0)$ and $(0, 1)$ are linearly independent solutions and every nonzero vector is a characteristic vector. Here $A$ is already a diagonal matrix.

**Example 3**   Let

$$A = \begin{bmatrix} 1 & 2 \\ 0 & 1 \end{bmatrix}.$$

Then

$$\det(\lambda I - A) = \begin{vmatrix} \lambda - 1 & -2 \\ 0 & \lambda - 1 \end{vmatrix} = (\lambda - 1)^2,$$

and $\lambda_1 = 1$ is the only characteristic value of $A$. The system for $k_1$ and $k_2$ is

$$0k_1 - 2k_2 = 0,$$
$$0k_1 + 0k_2 = 0.$$

Thus $k_2 = 0$ and the characteristic vectors are those vectors of the form $(a, 0)$, with $a \neq 0$. Since there is only one linearly independent characteristic vector, $A$ is not similar to a diagonal matrix.

Associated with the $n \times n$ matrix $A$ is the function $p$, where

$$p(\lambda) = \det(\lambda I - A). \tag{4.8}$$

This function $p$ is a polynomial of degree $n$, called the *characteristic polynomial* of $A$. To see this, let us write

$$p(\lambda) = \begin{vmatrix} \lambda - a_{11} & -a_{12} & \cdots & -a_{1n} \\ -a_{21} & \lambda - a_{22} & \cdots & -a_{2n} \\ \vdots & \vdots & & \vdots \\ -a_{n1} & -a_{n2} & \cdots & \lambda - a_{nn} \end{vmatrix}.$$

Recalling the definition of a determinant as the sum of products involving one element from each row and column, we observe that

$$(\lambda - a_{11})(\lambda - a_{22}) \cdots (\lambda - a_{nn})$$

is a product in the sum. Also, this is the only product that introduces the $n$th power of $\lambda$. Thus $p$ has the form

$$p(\lambda) = \lambda^n + a_1 \lambda^{n-1} + a_2 \lambda^{n-2} + \cdots + a_n. \tag{4.9}$$

The zeros of $p$ are the characteristic values of $A$. The equation $p(\lambda) = 0$ is called the *characteristic equation* of $A$.

Some of the zeros of $p$ may be complex. For this case we have the following theorem.

**Theorem 4.2**   If $\lambda = \alpha + i\beta$, $\beta \neq 0$, is a complex characteristic value of the real matrix $A$, with corresponding characteristic vector $\mathbf{k}$, then $\bar{\lambda} = \alpha - i\beta$ is also a characteristic value, with corresponding characteristic vector $\bar{\mathbf{k}}$.

PROOF   If $A$ is real then $p$ has real coefficients. Hence if $\lambda = \alpha + i\beta$, $\beta \neq 0$, is a complex zero of $p$, then $\bar{\lambda} = \alpha - i\beta$ is also a zero. Since $A\mathbf{k} = \lambda\mathbf{k}$, we see upon taking complex conjugates that $A\bar{\mathbf{k}} = \bar{\lambda}\bar{\mathbf{k}}$. Hence $\bar{\mathbf{k}}$ is a characteristic vector corresponding to $\bar{\lambda}$.

We conclude this section with the observation that similar matrices have the same characteristic polynomial, and hence the same characteristic values. To see this, let $B = K^{-1}AK$. Then

$$\det(\lambda I - A) = \det(K^{-1}(\lambda I - A)K)$$

$$= \det(\lambda I - K^{-1}AK)$$

$$= \det(\lambda I - B).$$

## Exercises for Section 4.1

In Exercises 1–6, determine the characteristic values of the given matrix $A$. For each characteristic value, determine the corresponding characteristic vectors. Determine whether or not $A$ is similar to a diagonal matrix. If it is, find a nonsinglar matrix $K$ and a diagonal matrix $D$ such that $K^{-1}AK = D$.

1. $\begin{bmatrix} -1 & 1 \\ 4 & 2 \end{bmatrix}$    2. $\begin{bmatrix} 2 & 0 \\ 0 & 2 \end{bmatrix}$

3. $\begin{bmatrix} -1 & 2 \\ -1 & 1 \end{bmatrix}$    4. $\begin{bmatrix} 0 & 4 \\ 1 & 0 \end{bmatrix}$

5. $\begin{bmatrix} 1 & -2 \\ 2 & -3 \end{bmatrix}$    6. $\begin{bmatrix} -2 & 1 \\ -4 & 2 \end{bmatrix}$

In Exercises 7–10, find the characteristic values and characteristic vectors of the given matrix.

**7.**
$$\begin{bmatrix} 1 & 2 & -1 \\ 0 & -2 & 0 \\ 0 & -5 & 2 \end{bmatrix}$$

**8.**
$$\begin{bmatrix} 2 & 0 & 0 \\ 6 & 5 & -3 \\ 9 & 6 & -4 \end{bmatrix}$$

**9.**
$$\begin{bmatrix} 2 & 0 & 0 \\ 0 & 2 & 0 \\ 0 & 0 & 2 \end{bmatrix}$$

**10.**
$$\begin{bmatrix} 2 & 4 & -3 \\ 0 & 2 & 1 \\ 0 & 0 & 2 \end{bmatrix}$$

**11.** If $A$ is a diagonal matrix, show that the characteristic values of $A$ are the diagonal elements.

**12.** If $A$ is a triangular matrix, show that the characteristic values of $A$ are the diagonal elements.

**13.** (a) If $A$ is similar to $B$, show that $B$ is similar to $A$.

(b) If $A$ is similar to $B$ and $B$ is similar to $C$, show that $A$ is similar to $C$.

**14.** Let $p(\lambda) = \lambda^n + a_1\lambda^{n-1} + \cdots + a_n$, where $p$ is the characteristic polynomial of a matrix $A$. Show that

(a) $a_n = (-1)^n \det A$

(b) $a_1 = -(a_{11} + a_{22} + \cdots + a_{nn})$

(c) zero is a characteristic value of $A$ if and only if $A$ is singular.

**15.** Let $T$ be a linear transformation from a complex finite dimensional vector space $V$ into $V$. A number $\lambda$ is called a characteristic value of $T$ if there is a nonzero element $\mathbf{v}$ of $V$ such that $T\mathbf{v} = \lambda\mathbf{v}$. Let $A$ be the matrix of $T$ relative to any basis for $V$. Show that $\lambda$ is a characteristic value of $T$. if and only if $\lambda$ is a characteristic value of the matrix $A$.

## 4.2   LINEAR INDEPENDENCE OF CHARACTERISTIC VECTORS

An $n \times n$ matrix need not have $n$ distinct characteristic values. For instance, the $2 \times 2$ matrices of Examples 2 and 3 of Section 4.1 possessed only one characteristic value. In the general case, if $\lambda_1, \lambda_2, \ldots, \lambda_k$ are the *distinct* characteristic values of an $n \times n$ matrix $A$ (so that $k \leq n$) then the characteristic polynomial of $A$ has the form

$$p(\lambda) = (\lambda - \lambda_1)^{m_1}(\lambda - \lambda_2)^{m_2} \cdots (\lambda - \lambda_k)^{m_k}, \tag{4.10}$$

where $m_1 + m_2 + \cdots + m_k = n$. The integer $m_i$ is called the *multiplicity* of the characteristic value $\lambda_i$.

Let us denote by $V_i$ the set of all characteristic vectors that correspond to $\lambda_i$, together with the zero vector. Then $V_i$ is a vector space, since it is the kernel of the matrix $\lambda_i I - A$ (see Section 3.8). Notice that the zero vector must be specifically included in the description of $V_i$ because a characteristic vector is by definition not the zero vector. We shall denote the dimension of the space $V_i$ by $n_i$. It can be shown,[1] although we shall not prove it here, that

$$1 \le n_i \le m_i, \qquad 1 \le i \le k. \qquad (4.11)$$

That is, the dimension of $V_i$ cannot exceed the multiplicity of $\lambda_i$. It can happen that the dimension is less than the multiplicity. In Example 3 of the last section, $\lambda_1$ has multiplicity two but the dimension of $V_1$ was one. However, in Example 2, the multiplicity of $\lambda_1$ and the dimension of $V_1$ were both two. In order to determine $n_i$ in any specific case, we may apply Theorem 3.15 to the equation $(\lambda_i I - A)\mathbf{k} = \mathbf{0}$. If the rank of the matrix $\lambda_i I - A$ is $r_i$ then $n_i = n - r_i$.

In Example 1 of the last section we saw that the matrix $A$ had two distinct characteristic values, $\lambda_1 = -1$ and $\lambda_2 = 4$. The corresponding characteristic vectors $(3, 2)$ and $(1, -1)$ were linearly independent. This illustrates the following general result.

**Theorem 4.3** Let $\lambda_1, \lambda_2, \ldots, \lambda_k$ be distinct characteristic values of the matrix $A$ and let $\mathbf{u}_1, \mathbf{u}_2, \ldots, \mathbf{u}_k$ be characteristic vectors, with $\mathbf{u}_i$ corresponding to $\lambda_i$. Then these characteristic vectors are linearly independent.

PROOF   Let $m$ be the maximum number of linearly independent vectors in the set $\{\mathbf{u}_1, \mathbf{u}_2, \ldots, \mathbf{u}_k\}$. If $m = k$, the set of $k$ vectors is linearly independent. Suppose that $m < k$. Then there is a subset of $m$ linearly independent vectors, and every subset of $m + 1$ vectors is linearly dependent. Let us assume, without loss of generality, that $\mathbf{u}_1, \mathbf{u}_2, \ldots, \mathbf{u}_m$ are linearly independent. According to Theorem 3.8, there exist numbers $c_1, c_2, \ldots, c_m$ such that

$$\mathbf{u}_{m+1} = c_1 \mathbf{u}_1 + c_2 \mathbf{u}_2 + \cdots + c_m \mathbf{u}_m. \qquad (4.12)$$

Premultiplying both sides of Eq. (4.12) by $A - \lambda_{m+1} I$, we have

$$\mathbf{0} = (\lambda_1 - \lambda_{m+1})c_1 \mathbf{u}_1 + (\lambda_2 - \lambda_{m+1})c_2 \mathbf{u}_2 + \cdots + (\lambda_m - \lambda_{m+1})c_m \mathbf{u}_m.$$

Then $(\lambda_i - \lambda_{m+1})c_i = 0$ and hence $c_i = 0$ for $1 \le i \le m$. But then equation (4.12) implies that $\mathbf{u}_{m+1} = \mathbf{0}$. This is impossible, since $\mathbf{u}_{m+1}$ is a characteristic vector. Our supposition that $m < k$ must be false, so $m = k$.

In case $\lambda_i$ is a multiple characteristic value, it is possible to find $n_i$ linearly

[1] See, for example, Hohn (1964) and Murdock (1957).

independent characteristic vectors associated with $\lambda_i$. (Here $n_i$ is the dimension of the space $V_i$.) That the collection of all these $n_1 + n_2 + \cdots + n_k$ vectors is linearly independent is asserted in the following theorem. Its proof is left as an exercise.

**Theorem 4.4**   Let $\lambda_1, \lambda_2, \ldots, \lambda_k$ be distinct characteristic values of $A$ and let $n_i$ be the dimension of the space $V_i$ associated with $\lambda_i$. Let $\mathbf{u}_{i1}, \mathbf{u}_{i2}, \ldots, \mathbf{u}_{in_i}$ be linearly independent characteristic vectors corresponding to $\lambda_i$. Then the vectors

$$\mathbf{u}_{11}, \mathbf{u}_{12}, \ldots, \mathbf{u}_{1n_1}, \qquad \mathbf{u}_{21}, \mathbf{u}_{22}, \ldots, \mathbf{u}_{2n_2}, \ldots, \qquad \mathbf{u}_{k1}, \mathbf{u}_{k2}, \ldots, \mathbf{u}_{kn_k}$$

are linearly independent.

This result is important in determining whether or not a matrix is similar to a diagonal matrix.

**Theorem 4.5**   An $n \times n$ matrix $A$ has $n$ linearly independent characteristic vectors (and hence is similar to a diagonal matrix) if and only if $n_i = m_i$ for $1 \le i \le k$, where $k$ is the number of distinct characteristic values of $A$; that is, if and only if the maximum number of linearly independent characteristic vectors associated with each characteristic value is equal to the multiplicity of the characteristic value.

PROOF   By Theorem 4.1, $A$ is similar to a diagonal matrix if and only if it has $n$ linearly independent characteristic vectors. According to Theorem 4.4, there is a set of $n_1 + n_1 + \cdots + n_k$ linearly independent characteristic vectors. There can be no larger set of linearly independent characteristic vectors. For if there were, the number of independent vectors in the set associated with some characteristic value, say $\lambda_i$, would have to exceed $n_i$. (If a finite set is linearly independent, so is every subset. See Exercise 4, Section 3.3.) This is impossible by the definition of $n_i$. Since $n_i \le m_i$ for $1 \le i \le k$ and since

$$m_1 + m_2 + \cdots + m_k = n,$$

$A$ has $n$ linearly independent characteristic vectors if and only if $n_i = m_i$ for $1 \le i \le k$.

In particular, an $n \times n$ matrix with $n$ distinct characteristic values is similar to a diagonal matrix. See Exercise 9.

In order to illustrate some of the preceding ideas, we consider the matrix

$$A = \begin{bmatrix} 0 & 2 & 1 \\ -1 & 3 & 1 \\ -1 & 1 & 2 \end{bmatrix}.$$

Some calculation shows that

$$\begin{vmatrix} \lambda & -2 & -1 \\ 1 & \lambda - 3 & -1 \\ 1 & -1 & \lambda - 2 \end{vmatrix} = (\lambda - 1)(\lambda - 2)^2.$$

Setting $\lambda_1 = 1$ and $\lambda_2 = 2$, we see that $m_1 = 1$ and $m_2 = 2$. According to the inequality (4.11) we must have $n_1 = 1$ and $n_2$ is either 1 or 2. The equation $(\lambda_2 I - A)\mathbf{k} = \mathbf{0}$ corresponds to the system

$$2k_1 - 2k_2 - k_3 = 0,$$
$$k_1 - k_2 - k_3 = 0,$$
$$k_1 - k_2 \quad\;\; = 0.$$

Subtracting the third equation from the second and subtracting twice the third equation from the first, we obtain the equivalent system

$$-k_3 = 0,$$
$$-k_3 = 0,$$
$$k_1 - k_2 = 0.$$

The rank of the coefficient matrix of this system is two, so there is only one linearly independent solution. Thus $n_2 = 1$. The given matrix is not diagonalizable.

As we have seen, not every square matrix is similar to a diagonal matrix. Real symmetric matrices, the topic of the next section, can always be diagonalized. We shall state, for the sake of completeness, two additional results about similarity. For proofs we refer the reader to the books on linear algebra listed among the references. Although the proof of Theorem 4.6 is not hard, that of Theorem 4.7 is quite difficult.

**Theorem 4.6** Every square matrix $A$ is similar to an upper triangular matrix and similar to a lower triangular matrix. If $K^{-1}AK = T$, where $T$ is triangular, $K$ and $T$ may be complex even when $A$ is real. However, if all the characteristic values of $A$ are real, there exists a real orthogonal matrix $K$ such that $K^{-1}AK$ is triangular.

In order to describe the next result, we need the idea of a *block form* of a matrix. To illustrate, let

$$A = \begin{bmatrix} 2 & -1 & 3 & 0 & 0 \\ 1 & 2 & 5 & 0 & 0 \\ 1 & 1 & 2 & 3 & 4 \\ 0 & 1 & 3 & 6 & 7 \end{bmatrix}.$$

Setting

$$B = \begin{bmatrix} 2 & -1 & 3 \\ 1 & 2 & 5 \end{bmatrix}, \quad C = \begin{bmatrix} 1 & 1 & 2 \\ 0 & 1 & 3 \end{bmatrix}, \quad D = \begin{bmatrix} 3 & 4 \\ 6 & 7 \end{bmatrix},$$

we write

$$A = \begin{bmatrix} B & 0 \\ C & D \end{bmatrix},$$

where $B$, $C$, $D$, and $0$ are submatrices of $A$.

**Theorem 4.7**   Every square matrix $A$ is similar to a matrix of the form

$$\begin{bmatrix} A_1 & 0 & 0 & \cdots & 0 \\ 0 & A_2 & 0 & \cdots & 0 \\ 0 & 0 & A_3 & \cdots & 0 \\ \vdots & \vdots & \vdots & & \vdots \\ 0 & 0 & 0 & \cdots & A_s \end{bmatrix}, \tag{4.13}$$

where each of the submatrices $A_i$ is a square matrix of the form

$$A_i = \begin{bmatrix} \lambda_i & 1 & 0 & 0 & \cdots & 0 & 0 \\ 0 & \lambda_i & 1 & 0 & \cdots & 0 & 0 \\ 0 & 0 & \lambda_i & 1 & \cdots & 0 & 0 \\ \vdots & \vdots & \vdots & \vdots & & \vdots & \vdots \\ 0 & 0 & 0 & 0 & \cdots & \lambda_i & 1 \\ 0 & 0 & 0 & 0 & \cdots & 0 & \lambda_i \end{bmatrix}. \tag{4.14}$$

The numbers $\lambda_1$, $\lambda_2$, .., $\lambda_s$ are the characteristic values of $A$, but are not necessarily distinct. The matrix (4.13) is called the *Jordan canonical form* of $A$; it is unique except for the order of the blocks $A_i$ in the representation (4.13). The number of blocks corresponding to a particular characteristic value is equal to the dimension of the space of characteristic vectors associated with it.

In practice, it is usually difficult to actually find a matrix $K$ (which may be complex) such that $K^{-1}AK$ has the form (4.13). The use of the canonical form is chiefly for theoretical purposes. For example, it can be shown that every matrix is similar to its transpose by first showing that a matrix of the form (4.13) is similar to its transpose.

## Exercises for Section 4.2

In Exercises 1–6 find the characteristic values of the given matrix. Determine the multiplicity of each characteristic value and the dimension of the space of characteristic vectors associated with it. Is the matrix similar to a diagonal matrix?

1. $\begin{bmatrix} 2 & -2 & 1 \\ 1 & -1 & 1 \\ -3 & 2 & -2 \end{bmatrix}$
2. $\begin{bmatrix} 2 & 1 & 1 \\ 1 & 2 & 1 \\ -2 & -2 & -1 \end{bmatrix}$

3. $\begin{bmatrix} 1 & 2 & 3 \\ -1 & 4 & 3 \\ 1 & -2 & -1 \end{bmatrix}$
4. $\begin{bmatrix} 3 & 0 & 0 \\ 0 & 3 & 0 \\ 0 & 0 & 3 \end{bmatrix}$

5. $\begin{bmatrix} -1 & 1 & 2 \\ -1 & 1 & 1 \\ -2 & 1 & 3 \end{bmatrix}$
6. $\begin{bmatrix} 0 & 1 & 1 \\ -1 & 1 & 1 \\ -1 & 1 & 1 \end{bmatrix}$

7. If $A$ is nonsingular, show that the characteristic values of $A^{-1}$ are the reciprocals of those of $A$.

8. If $A$ is similar to a scalar matrix show that $A$ is a scalar matrix.

9. Show that an $n \times n$ matrix is similar to a diagonal matrix if it has $n$ distinct characteristic values.

10. Prove Theorem 4.4. **Suggestion**: suppose that

$$\sum_{i=1}^{k} \sum_{j=1}^{n_i} c_{ij} \mathbf{u}_{ij} = 0.$$

For each $i$

$$\sum_{j=1}^{n_i} c_{ij} \mathbf{u}_{ij}$$

is either the zero vector or else is a characteristic vector corresponding to $\lambda_i$. Use Theorem 4.3.

11. Let $A$ be an $n \times n$ orthogonal matrix.
   (a) If $\lambda$ is a characteristic value of $A$ show that $|\lambda| = 1$.
   (b) If $n$ is odd and det $A = 1$, show that $\lambda = 1$ is a characteristic value of $A$.

12. Let $T$ be a proper orthogonal transformation from $R^3$ into $R^3$.
   (a) Show that there exists a unit vector $\mathbf{u}_1$ such that $T\mathbf{u}_1 = \mathbf{u}_1$.
   (b) Let $B = (\mathbf{u}_1, \mathbf{u}_2, \mathbf{u}_3)$ be an orthonormal basis for $R^3$, where $\mathbf{u}_1$ is as in part (a). Show that the matrix of $T$, relative to $B$, has the form

$$\begin{bmatrix} 1 & 0 \\ 0 & A \end{bmatrix},$$

where $A$ is a $2 \times 2$ proper orthogonal matrix. Show that $T$ consists of a rotation about the vector $\mathbf{u}_1$.

## 4.3   REAL SYMMETRIC MATRICES

Real symmetric matrices arise in a number of applications. (We recall that $A$ is symmetric if $A^T = A$; that is, if $a_{ji} = a_{ij}$ for all $i$ and $j$.) In this section we shall describe some of the special properties of the characteristic values and vectors of such matrices. We begin with a preliminary result.

**Lemma**   Let $\mathbf{x}$ and $\mathbf{y}$ be $n$-dimensional (real or complex) column vectors and let $A$ be an $n \times n$ symmetric matrix. Then

$$\mathbf{x}^T A \mathbf{y} = \mathbf{y}^T A \mathbf{x}. \tag{4.15}$$

PROOF   Both members of the above equation are $1 \times 1$ matrices, and hence are symmetric. We have

$$\mathbf{x}^T A \mathbf{y} = (\mathbf{x}^T A \mathbf{y})^T = \mathbf{y}^T A^T \mathbf{x} = \mathbf{y}^T A \mathbf{x},$$

where the last equality follows from the symmetry of $A$.

**Theorem 4.8**   Let $A$ be a real symmetric matrix. Then all the characteristic values of $A$ are real, and (real) characteristic vectors of $A$ that correspond to different characteristic values are orthogonal.

PROOF   Suppose that $\lambda = \alpha + i\beta$ is a characteristic value of $A$ and let $\mathbf{x}$ be a corresponding characteristic vector. Then

$$A\mathbf{x} = \lambda\mathbf{x}, \qquad A\bar{\mathbf{x}} = \bar{\lambda}\bar{\mathbf{x}}.$$

Premultiplying in the first equation by $\bar{\mathbf{x}}^T$ and in the second by $\mathbf{x}^T$, we have

$$\bar{\mathbf{x}}^T A \mathbf{x} = \lambda\bar{\mathbf{x}}^T\mathbf{x}, \qquad \mathbf{x}^T A \bar{\mathbf{x}} = \bar{\lambda}\mathbf{x}^T\bar{\mathbf{x}}.$$

Making use of the lemma, we see that

$$\lambda\bar{\mathbf{x}}^T\mathbf{x} = \bar{\lambda}\mathbf{x}^T\bar{\mathbf{x}}.$$

But $\bar{\mathbf{x}}^T\mathbf{x} = \mathbf{x}^T\bar{\mathbf{x}} = [\mathbf{x} \cdot \mathbf{x}]$; therefore

$$(\lambda - \bar{\lambda})\bar{\mathbf{x}}^T\mathbf{x} = 2i\beta\bar{\mathbf{x}}^T\mathbf{x} = [0].$$

Since $\mathbf{x}$ is a characteristic vector, $\|\mathbf{x}\| > 0$, so that we must have $\beta = 0$. Hence $\lambda$ is real.

Next let $\lambda_1$ and $\lambda_2$ be distinct characteristic values of $A$, and let $\mathbf{x}$ and $\mathbf{y}$ be corresponding real characteristic vectors. Then

$$A\mathbf{x} = \lambda_1 \mathbf{x}, \qquad A\mathbf{y} = \lambda_2 \mathbf{y}.$$

Premultiplying in the first of these equations by $\mathbf{y}^T$ and in the second by $\mathbf{x}^T$, we have

$$\mathbf{y}^T A\mathbf{x} = \lambda_1 \mathbf{y}^T \mathbf{x}, \qquad \mathbf{x}^T A\mathbf{y} = \lambda_2 \mathbf{x}^T \mathbf{y}.$$

Again making use of the lemma, we have

$$(\lambda_1 - \lambda_2)\mathbf{x}^T\mathbf{y} = [0].$$

Since $\lambda_1 \neq \lambda_2$, we must have $\mathbf{x}^T\mathbf{y} = [0]$ or $\mathbf{x} \cdot \mathbf{y} = 0$, as we wished to prove.

Our next result is more difficult to prove and we only state it.[2]

**Theorem 4.9** An $n \times n$ real symmetric matrix possesses $n$ real linearly independent characteristic vectors.

If $A$ is a real symmetric matrix, the multiplicity $m_i$ of each characteristic value $\lambda_i$ must be equal to $n_i$, the dimension of the corresponding space $V_i$ of characteristic vectors. Every real symmetric matrix is similar to a diagonal matrix according to Theorem 4.1.

The proof of our next theorem about real symmetric matrices utilizes a lemma that is of importance in its own right.

**Lemma** Let $\mathbf{v}_1, \mathbf{v}_2, \dots, \mathbf{v}_m$, $m > 1$, be linearly independent elements of $R^n$ (or $C^n$). Then the subspace of $R^n$ (or $C^n$) that is spanned by these elements has an orthogonal basis $(\mathbf{u}_1, \mathbf{u}_2, \dots, \mathbf{u}_m)$, where

$$\mathbf{u}_1 = \mathbf{v}_1$$

$$\mathbf{u}_2 = \mathbf{v}_2 - \frac{\mathbf{u}_1 \cdot \mathbf{v}_2}{\|\mathbf{u}_1\|^2} \mathbf{u}_1$$

$$\mathbf{u}_3 = \mathbf{v}_3 - \frac{\mathbf{u}_1 \cdot \mathbf{v}_3}{\|\mathbf{u}_1\|^2} \mathbf{u}_1 - \frac{\mathbf{u}_2 \cdot \mathbf{v}_3}{\|\mathbf{u}_2\|^2} \mathbf{u}_2 \qquad (4.16)$$

$$\dots\dots\dots\dots\dots\dots\dots\dots\dots\dots\dots\dots$$

$$\mathbf{u}_m = \mathbf{v}_m - \frac{\mathbf{u}_1 \cdot \mathbf{v}_m}{\|\mathbf{u}_1\|^2} \mathbf{u}_1 - \cdots - \frac{\mathbf{u}_{m-1} \cdot \mathbf{v}_m}{\|\mathbf{u}_{m-1}\|^2} \mathbf{u}_{m-1}.$$

[2] See, for example, Hohn (1964) and Murdock (1957).

PROOF   Using the formulas (4.16) we first define $\mathbf{u}_1$. If $\mathbf{u}_1$ is not the zero vector then $\|\mathbf{u}_1\| \neq 0$ and we can define $\mathbf{u}_2$. If $\mathbf{u}_2$ is not the zero vector we can define $\mathbf{u}_3$, and so on. We proceed by induction. Certainly $\mathbf{u}_1$ is not the zero vector because $\mathbf{v}_1$ is not the zero vector. Suppose that none of $\mathbf{u}_1, \mathbf{u}_2, \ldots, \mathbf{u}_k$ is the zero vector, where $k$ is any positive integer less than $m$. If $\mathbf{u}_{k+1} = \mathbf{0}$ then $\mathbf{v}_{k+1}$ is a linear combination of $\mathbf{u}_1, \mathbf{u}_2, \ldots, \mathbf{u}_k$ and hence is a linear combination of $\mathbf{v}_1, \mathbf{v}_2, \ldots, \mathbf{v}_k$. By Theorem 3.2, the set $\{\mathbf{v}_1, \mathbf{v}_2, \ldots, \mathbf{v}_{k+1}\}$ is linearly dependent. This is impossible. Hence each of the vectors $\mathbf{u}_1, \mathbf{u}_2, \ldots, \mathbf{u}_m$ is defined and is not the zero vector. It is not hard to show by induction that $\mathbf{u}_i \cdot \mathbf{u}_j = 0$ for $j < i$ and $i = 2, 3, \ldots, m$. As explained in Section 3.5, the mutually orthogonal nonzero vectors are linearly independent.

The method that was used in the proof of the lemma to find an orthogonal basis is known as the *Gram-Schmidt orthogonalization procedure*.

**Theorem 4.10**   An $n \times n$ real symmetric matrix $A$ possesses an orthonormal set $\{\mathbf{k}_1, \mathbf{k}_2, \ldots, \mathbf{k}_n\}$ of real characteristic vectors. The orthogonal matrix $K = [\mathbf{k}_1, \mathbf{k}_2, \ldots, \mathbf{k}_n]$ has the property that

$$K^T A K = \operatorname{diag}(\lambda_1, \lambda_2, \ldots, \lambda_n),$$

where the $\lambda_i$ are the characteristic values of $A$, with $A\mathbf{k}_i = \lambda_i \mathbf{k}_i$.

PROOF   Let $\lambda$ be any characteristic value of $A$, with multiplicity $m$. If $\mathbf{v}_1, \mathbf{v}_2, \ldots, \mathbf{v}_m$ are linearly independent characteristic vectors associated with $\lambda$, they span a space $V_\lambda$ of dimension $m$. According to the lemma, there is an orthogonal basis $\{\mathbf{u}_1, \mathbf{u}_2, \ldots, \mathbf{u}_m\}$ for the space $V_\lambda$. The vectors $\mathbf{w}_i = \mathbf{u}_i/\|\mathbf{u}_i\|$, $1 \leq i \leq m$, constitute an orthonormal basis for the space. Each of the vectors $\mathbf{w}_i$ is a characteristic vector because it is not the zero vector and is a linear combination of $\mathbf{v}_1, \mathbf{v}_2, \ldots, \mathbf{v}_m$. By following this procedure with each characteristic value, we obtain a set of $n$ characteristic vectors for $A$. We have chosen the vectors so that any two vectors corresponding to the same characteristic value are orthogonal, and by Theorem 4.8, vectors corresponding to different characteristic values are orthogonal. The remaining facts stated in the theorem follow from Theorem 4.1 and properties of orthogonal matrices.

**Example**   The matrix

$$A = \begin{bmatrix} -7 & 0 & 6 \\ 0 & 5 & 0 \\ 6 & 0 & 2 \end{bmatrix}$$

is real and symmetric. Its characteristic polynomial is

$$\begin{vmatrix} \lambda + 7 & 0 & -6 \\ 0 & \lambda - 5 & 0 \\ -6 & 0 & \lambda - 2 \end{vmatrix} = (\lambda - 5)^2(\lambda + 10).$$

A characteristic vector corresponding to $\lambda_1 = -10$ is $(2, 0, -1)$. The characteristic vectors corresponding to $\lambda_2 = 5$ are those nonzero vectors of the form $(a, b, 2a)$. The vectors $(1, 0, 2)$ and $(0, 1, 0)$ are mutually orthogonal characteristic vectors corresponding to $\lambda_2$. The matrix

$$K = \begin{bmatrix} 2/\sqrt{5} & 1/\sqrt{5} & 0 \\ 0 & 0 & 1 \\ -1/\sqrt{5} & 2/\sqrt{5} & 0 \end{bmatrix}$$

whose column vectors form an orthonormal set of characteristic vectors of $A$, has the property that

$$K^{-1}AK = K^TAK = \operatorname{diag}(-10, 5, 5).$$

## Exercises for Section 4.3

In Exercises 1–4, verify that the given matrix $A$ is real and symmetric. Then find an orthogonal matrix $K$ and a diagonal matrix $D$ such that $K^TAK = D$.

**1.** $\begin{bmatrix} 7 & 6 \\ 6 & -2 \end{bmatrix}$

**2.** $\begin{bmatrix} 52 & 14 \\ 14 & 148 \end{bmatrix}$

**3.** $\begin{bmatrix} 4 & -12 & 6 \\ -12 & 36 & -18 \\ 6 & -18 & 9 \end{bmatrix}$

**4.** $\begin{bmatrix} -6 & 6 & -6 \\ 6 & 3 & 3 \\ -6 & 3 & 3 \end{bmatrix}$

**5.** Use the Gram–Schmidt procedure to find an orthonormal basis for:
   (a) the subspace of $R^3$ that is spanned by the vectors $(1, 2, -2)$ and $(1, 1, -1)$.
   (b) the subspace of $R^4$ that is spanned by $(1, 0, 1, 0)$, $(0, 1, 1, -1)$, and $(1, 0, 0, 1)$.

**6.** If $A$ is an $n \times n$ real matrix that possesses an orthonormal set of $n$ real characteristic vectors, show that $A$ is symmetric.

**7.** Let $\lambda_1$ be a real characteristic value of $A$ with $\mathbf{x}$ a corresponding characteristic vector. Let $\lambda_2$ be a real characteristic value of $A^T$, with $\mathbf{y}$ a corresponding characteristic vector. If $\lambda_1 \neq \lambda_2$, show that $\mathbf{x}$ and $\mathbf{y}$ are orthogonal.

**8.** (a) If $D$ is a real diagonal matrix with nonnegative elements, show that $D$ can be expressed as $D = D_1^2$, where $D_1$ is real and diagonal.

(b) If $A$ is a real symmetric matrix with nonnegative characteristic values, show that there is a real symmetric matrix $B$ such that $A = B^2$.

9. If $A$ is any real matrix show that $A^T A$ is a real symmetric matrix with nonnegative characteristic values.

10. Let $A$ be real and nonsingular. Show that $A$ can be written as $A = PQ$, where $P$ is orthogonal and $Q$ is real symmetric. Suggestion: $A^T A$ is symmetric and there is an orthogonal matrix $K$ such that $K^T A^T A K = D^2$, where $D$ is diagonal. Let $Q = KDK^T$.

11. A matrix $A$ with complex elements is called an *Hermitian* matrix if $A^* = A$. (Here $A^* = \bar{A}^T$ is the transposed conjugate of $A$.)

(a) Show that every real symmetric matrix is Hermitian.

(b) Show that the diagonal elements of an Hermitian matrix are real.

(c) Show that the characteristic values of an Hermitian matrix are real.

(d) If $\mathbf{x}$ and $\mathbf{y}$ are characteristic vectors of an Hermitian matrix that correspond to different characteristic values, show that $\mathbf{x}$ and $\mathbf{y}$ are orthogonal.

## 4.4   QUADRATIC FORMS

Let $\mathbf{x} = (x_1, x_2, \ldots, x_n)$ be an arbitrary element of $R^n$. A function $f$ from $R^n$ into $R^1$, of the form

$$f(\mathbf{x}) = \sum_{i=1}^{n} \sum_{j=1}^{n} a_{ij} x_i x_j, \qquad (4.17)$$

where the quantities $a_{ij}$ are real numbers, is called a *quadratic form*. For $n = 2$, we have

$$f(\mathbf{x}) = a_{11} x_1^2 + (a_{12} + a_{21}) x_1 x_2 + a_{22} x_2^2, \qquad (4.18)$$

while for $n = 3$, we have

$$f(\mathbf{x}) = a_{11} x_1^2 + a_{22} x_2^2 + a_{33} x_3^2 + (a_{23} + a_{32}) x_2 x_3$$
$$+ (a_{31} + a_{13}) x_3 x_1 + (a_{12} + a_{21}) x_1 x_2. \qquad (4.19)$$

We may always assume that

$$a_{ij} = a_{ji}, \qquad i \neq j \qquad (4.20)$$

because the term

$$(a_{ij} + a_{ji})x_i x_j$$

can be replaced by

$$(a'_{ij} + a'_{ji})x_i x_j,$$

where

$$a'_{ij} = a'_{ji} = \tfrac{1}{2}(a_{ij} + a_{ji}).$$

Associated with the form (4.17) is the $n \times n$ matrix $A$ with elements $a_{ij}$. In view of the assumption (4.20), $A$ is symmetric. If we represent $\mathbf{x}$ by a column vector, then the quantity (4.17) is the single element of the $1 \times 1$ matrix

$$\mathbf{x}^T A \mathbf{x}. \tag{4.21}$$

Verification, which simply involves the computation of the matrix products, is left as an exercise. To consider a specific example, let $n = 3$ and let

$$f(\mathbf{x}) = 5x_1^2 - 2x_3^2 + 6x_1 x_2 - 8x_2 x_3.$$

Writing

$$f(\mathbf{x}) = 5x_1^2 - 2x_3^2 + (3 + 3)x_1 x_2 + (-4 - 4)x_2 x_3,$$

we see that the associated real symmetric matrix is

$$A = \begin{bmatrix} 5 & 3 & 0 \\ 3 & 0 & -4 \\ 0 & -4 & -2 \end{bmatrix}.$$

Going back to the general case, suppose that we make a change of basis, replacing the standard basis $(\mathbf{e}_1, \mathbf{e}_2, \ldots, \mathbf{e}_n)$ by a new orthonormal basis $(\mathbf{e}'_1, \mathbf{e}'_2, \ldots, \mathbf{e}'_n)$. If

$$\mathbf{e}'_j = \sum_{i=1}^{n} c_{ij} \mathbf{e}_i, \qquad 1 \le j \le n,$$

then $\mathbf{x}$ has coordinates $x'_1, x'_2, \ldots, x'_n$, where, according to Theorem 3.20,

$$x_i = \sum_{j=1}^{n} c_{ij} x'_j, \qquad 1 \le i \le n.$$

In terms of column vectors, we have

$$\mathbf{x} = C\mathbf{x}'. \tag{4.22}$$

Then

$$\mathbf{x}^T A \mathbf{x} = (C\mathbf{x}')^T A (C\mathbf{x}') = \mathbf{x}'^T C^T A C \mathbf{x}'.$$

Since $A$ is real and symmetric, we can find an orthogonal matrix $C$ such that

$$C^T A C = C^{-1} A C = \text{diag}(\lambda_1, \lambda_2, \dots, \lambda_n),$$

where the $\lambda_i$ are the characteristic values of $A$. For such a choice of $C$, the single element of the matrix $\mathbf{x}^T A \mathbf{x}$ becomes

$$f(\mathbf{x}) = \lambda_1 x_1'^2 + \lambda_2 x_2'^2 + \cdots + \lambda_n x_n'^2. \tag{4.23}$$

Thus in the new coordinate system, the quadratic form (4.17) has the simpler form (4.23) where there are no terms of the type $x_i' x_j'$ for $i \neq j$.

As an application, let us consider the equation

$$2x_1^2 - 13x_2^2 + 2x_3^2 - 20x_2 x_3 + 40x_3 x_1 - 20x_1 x_2$$
$$+ 48x_1 - 24x_2 - 60x_3 - 90 = 0. \tag{4.24}$$

The real symmetric matrix associated with the quadratic form

$$2x_1^2 - 13x_2^2 + 2x_3^2 - 20x_2 x_3 + 40x_3 x_1 - 20x_1 x_2$$

is

$$A = \begin{bmatrix} 2 & -10 & 20 \\ -10 & -13 & -10 \\ 20 & -10 & 2 \end{bmatrix}. \tag{4.25}$$

The characteristic polynomial of $A$ is found to be

$$p(\lambda) = \lambda^3 + 9\lambda^2 - 648\lambda - 8478$$
$$= (\lambda + 18)^2 (\lambda - 27). \tag{4.26}$$

A unit characteristic vector corresponding to $\lambda = 27$ is $\frac{1}{3}(2, -1, 2)$. Mutually orthogonal unit characteristic vectors corresponding to $\lambda = -18$ are $\frac{1}{3}(2, 2, -1)$

and $\frac{1}{3}(-1, 2, 2)$. If we choose

$$
C = \frac{1}{3} \begin{bmatrix} 2 & -1 & 2 \\ 2 & 2 & -1 \\ -1 & 2 & 2 \end{bmatrix}
$$

and set $\mathbf{x} = C\mathbf{x}'$, then

$$
x^T A x = x'^T C^T A C x' = [-18x_1'^2 - 18x_2'^2 + 27x_3'^2]. \tag{4.27}
$$

The transformation of coordinates is described by the system of equations

$$
\begin{aligned}
x_1 &= \tfrac{1}{3}(2x_1' - x_2' + 2x_3'), \\
x_2 &= \tfrac{1}{3}(2x_1' + 2x_2' - x_3'), \\
x_3 &= \tfrac{1}{3}(-x_1' + 2x_2' + 2x_3').
\end{aligned} \tag{4.28}
$$

This corresponds to the change of basis

$$
\begin{aligned}
\mathbf{e}_1' &= \tfrac{1}{3}(2\mathbf{e}_1 + 2\mathbf{e}_2 - \mathbf{e}_3), \\
\mathbf{e}_2' &= \tfrac{1}{3}(-\mathbf{e}_1 + 2\mathbf{e}_2 - 2\mathbf{e}_3), \\
\mathbf{e}_3' &= \tfrac{1}{3}(2\mathbf{e}_1 - \mathbf{e}_2 + 2\mathbf{e}_3).
\end{aligned} \tag{4.29}
$$

Substituting formulas (4.28) into Eq. (4.24), we obtain the equation

$$
-18x_1'^2 - 18x_2'^2 + 27x_3'^2 + 36x_1' - 72x_2' - 90 = 0.
$$

This reduces to

$$
2x_1'^2 + 2x_2'^2 - 3x_3'^2 - 4x_1' + 8x_2' + 10 = 0
$$

or

$$
2(x_1' - 1)^2 + 2(x_2' + 2)^2 - 3x_3'^2 = 0.
$$

We now recognize that Eq. (4.24) is that of a right circular cone.

By the adoption of a suitable orthonormal basis, any quadratic equation in $n$ variables can be put in the form

$$
a_1 x_1^2 + a_2 x_2^2 + \cdots + a_n x_n^2 + b_1 x_1 + b_2 x_2 + \cdots + b_n x_n + c = 0. \tag{4.30}
$$

We say that a quadratic equation of the form (4.30) is in *standard form.* A function $f$, where

$$f(\mathbf{x}) = F(x_1, x_2, \ldots, x_n),$$

is said to be *positive definite* if $f(\mathbf{0}) = 0$ and $f(\mathbf{x}) > 0$ when $\mathbf{x} \neq \mathbf{0}$. A necessary and sufficient condition that a quadratic form be positive definite is as follows.

**Theorem 4.11**   The quadratic form $f$, where

$$f(\mathbf{x}) = \sum_{i=1}^{n} \sum_{n=1}^{n} a_{ij} x_i x_j,$$

is positive definite if and only if the characteristic values of the real symmetric matrix $A$ are all positive.

PROOF   There is an orthogonal (and hence nonsingular) matrix $C$ such that if $\mathbf{x} = C\mathbf{x}'$, then

$$f(\mathbf{x}) = \lambda_1 x_1'^2 + \lambda_2 x_2'^2 + \cdots + \lambda_n x_n'^2.$$

If $\lambda_1, \lambda_2, \ldots, \lambda_n$ are all positive, then $f(\mathbf{x}) > 0$ except when $\mathbf{x}' = \mathbf{0}$, and hence except when $\mathbf{x} = \mathbf{0}$. On the other hand, if $f$ is positive definite, each characteristic value must be positive. For if $\lambda_i \leq 0$, choose $x_i' = 1$ and $x_j' = 0$ for $j \neq i$. Then $f(\mathbf{x}) = \lambda_i \leq 0$ but $\mathbf{x}' \neq \mathbf{0}$ and $\mathbf{x} \neq \mathbf{0}$.

A real symmetric matrix is said to be *positive definite* if and only if its characteristic values are all positive.

### Exercises for Section 4.4

1.  Find the real symmetric matrix associated with the given quadratic form. (In parts (a) and (b), the size of the matrix is $2 \times 2$, in parts (c) and (d) it is $3 \times 3$.)

    (a)   $-3x_1^2 + 5x_2^2 - 4x_1 x_2$                    (b)   $4x_2^2 + 6x_1 x_2$

    (c)   $-x_1^2 + 5x_3^2 - 4x_2 x_3 + 2x_3 x_1 + 6x_1 x_2$   (d)   $3x_2^2 - 7x_3^2 - 4x_1 x_2$

2.  Find the quadratic form $\mathbf{x}^T A \mathbf{x}$ that is associated with the given real symmetric matrix $A$.

    (a)   $\begin{bmatrix} 2 & -1 \\ -1 & -3 \end{bmatrix}$          (b)   $\begin{bmatrix} 0 & 4 \\ 4 & -1 \end{bmatrix}$

    (c)   $\begin{bmatrix} 2 & -4 & 1 \\ -4 & 1 & 2 \\ 1 & 2 & -1 \end{bmatrix}$          (d)   $\begin{bmatrix} 1 & 1 & -3 \\ 1 & 0 & 2 \\ -3 & 2 & 0 \end{bmatrix}$

In Exercises 3–12, make an orthogonal transformation of coordinates that puts the given equation in standard form. Then identify the graph of the equation. (In Exercise 3–6, the equation is that of a plane curve.)

3. $8x_1^2 + 5x_2^2 - 4x_1x_2 - 36 = 0$

4. $4x_1^2 - 24x_1x_2 + 36x_2^2 - 21\sqrt{10}\,x_1 + 73\sqrt{10}\,x_2 + 360 = 0$

5. $3x_1^2 + 3x_2^2 - 10x_1x_2 - 4x_1 - 4x_2 - 4 = 0$

6. $-5x_1^2 + 5x_2^2 - 24x_1x_2 + 2\sqrt{13}\,(3x_1 + 2x_2) - 65 = 0$

7. $52x_1^2 + 25x_2^2 + 73x_3^2 - 72x_1x_3 + 80x_1 - 50x_2 + 60x_3 + 25 = 0$

8. $2x_1^2 + 2x_2^2 - x_3^2 - 4x_2x_3 + 4x_3x_1 + 2x_1x_2 - 12 = 0$

9. $x_1^2 + x_2^2 + 3x_3^2 + 4x_1x_2 + 4 = 0$

10. $4x_1^2 + 10x_2^2 + 10x_3^2 - 4x_2x_3 + 8x_3x_1 + 8x_1x_2$
    $\quad - \sqrt{6}(2x_1 - x_2 - x_3) - 6 = 0$

11. $x_1^2 + 4x_2^2 - 5x_3^2 + 4x_1x_2 - 4\sqrt{5}(2x_1 - x_2) = 0$

12. $4x_1^2 + 4x_2^2 + x_3^2 - 4x_2x_3 + 4x_3x_1$
    $\quad - 8x_1x_2 - 24x_1 - 12x_2 - 30x_3 + 9 = 0$

13. Verify that the single element of the matrix $\mathbf{x}^T A\mathbf{x}$ has the form (4.17).

14. Prove that the real symmetric matrix

$$\begin{bmatrix} a & b \\ b & c \end{bmatrix}$$

is positive definite if and only if $a > 0$ and $ac - b^2 > 0$.

15. Let $f(x, y)$ be a function of two variables that is continuous along with its first- and second-order partial derivatives in a region. Suppose that $f_x(x_0, y_0) = f_y(x_0, y_0) = 0$ at a point $P_0 : (x_0, y_0)$ in the region. Let

$$A = f_{xx}(x_0, y_0), \qquad B = f_{xy}(x_0, y_0), \qquad C = f_{yy}(x_0, y_0).$$

Show that $f$ has a relative minimum at $P_0$ if

$$A > 0 \quad \text{and} \quad B^2 - AC < 0;$$

a relative maximum if

$$A < 0 \quad \text{and} \quad B^2 - AC < 0;$$

neither a maximum nor a minimum if

$$B^2 - AC > 0.$$

Suggestion: if the directional derivative at $P_0$ in the direction that makes an angle $\theta$ with the positive $x$ axis is

$$\frac{df}{ds} = f_x \cos \theta + f_y \sin \theta,$$

show that

$$\frac{d^2f}{ds^2} = A \cos^2 \theta + 2B \cos \theta \sin \theta + C \sin^2 \theta.$$

Use the results of Exercise 14.

## 4.5   FUNCTIONS OF MATRICES

If $A$ is a square matrix of order $n$ then the product $AA$ is defined and is again a square matrix of order $n$. We define positive integral powers of $A$ in a natural way:

$$A^1 = A, \qquad A^2 = AA, \qquad A^3 = AA^2, \ldots, \qquad A^m = AA^{m-1}, \ldots.$$

We also define $A$ to the zero power as

$$A^0 = I,$$

where $I$ is the identity matrix of order $n$. It can be shown that the law of exponents

$$A^m A^k = A^{m+k} \tag{4.31}$$

holds for all nonnegative integers $m$ and $k$.

If $q$ is a polynomial function, defined by the formula

$$q(x) = c_0 + c_1 x + c_2 x^2 + \cdots + c_m x^m$$

for all real numbers $x$, we define $q(A)$, where $A$ is a square matrix, by the relation

$$q(A) = c_0 I + c_1 A + c_2 A^2 + \cdots + c_m A^m.$$

Note that $q(A)$ is a square matrix of the same size as $A$. If $f$, $g$, $q$, and $r$ are polynomials such that

$$f(x) = q(x) + r(x), \qquad g(x) = q(x)r(x)$$

for all $x$, then it can be shown (by using elementary properties of matrices and formula (4.31)) that

$$f(A) = q(A) + r(A) \tag{4.32}$$

$$g(A) = q(A)r(A) \tag{4.33}$$

for every square matrix $A$. Since $q(x)\, r(x) = r(x)\, q(x)$ for all $x$, it follows that

$$q(A)\, r(A) = r(A)\, q(A). \tag{4.34}$$

In particular, if $p$ is the characteristic polynomial of the square matrix $A$,

$$p(\lambda) = \det(\lambda I - A),$$

then we can consider $p(A)$. A famous theorem, which we shall not prove, is as follows.

**Theorem 4.12** (*Cayley–Hamilton theorem*)   If $A$ is a square matrix with characteristic polynomial $p$, then $p(A) = 0$.
   This result is often described by saying that a square matrix satisfies its characteristic polynomial. We shall verify the theorem for the matrix

$$A = \begin{bmatrix} 1 & -3 \\ -2 & 4 \end{bmatrix}. \tag{4.35}$$

Here

$$p(\lambda) = \begin{vmatrix} \lambda - 1 & 3 \\ 2 & \lambda - 4 \end{vmatrix} = \lambda^2 - 5\lambda - 2.$$

Since

$$A^2 = \begin{bmatrix} 7 & -15 \\ -10 & 22 \end{bmatrix},$$

we have

$$A^2 - 5A - 2I = \begin{bmatrix} 7 & -15 \\ -10 & 22 \end{bmatrix} - 5\begin{bmatrix} 1 & -3 \\ -2 & 4 \end{bmatrix} - 2\begin{bmatrix} 1 & 0 \\ 0 & 1 \end{bmatrix} = \begin{bmatrix} 0 & 0 \\ 0 & 0 \end{bmatrix}$$

as the theorem asserts. The Cayley–Hamilton theorem can be applied to calculate higher powers of $A$. From the relation

$$A^2 - 5A - 2I = 0,\qquad(4.36)$$

we see that

$$A^2 = 5A + 2I.$$

Then, multiplying both sides by $A$,

$$A^3 = 5A^2 + 2A = 5(5A + 2I) + 2A = 27A + 10I,$$
$$A^4 = 27A^2 + 10A = 27(5A + 2I) + 10A = 145A + 54I,$$

and so on. In particular,

$$A^4 = 145\begin{bmatrix} 1 & -3 \\ -2 & 4 \end{bmatrix} + 54\begin{bmatrix} 1 & 0 \\ 0 & 1 \end{bmatrix} = \begin{bmatrix} 199 & -435 \\ -290 & 634 \end{bmatrix}.$$

We can also find the inverse of $A$ from relation (4.36). We have

$$I = \tfrac{1}{2}(A^2 - 5A)$$

and multiplication of both sides of this equation by $A^{-1}$ yields

$$A^{-1} = \tfrac{1}{2}(A - 5I).$$

Calculation shows that

$$A^{-1} = \tfrac{1}{2}\begin{bmatrix} -4 & -3 \\ -2 & -1 \end{bmatrix}.$$

## Exercises for Section 4.5

1. If $f(r) = 2r^2 - 3r + 4$, find $f(A)$ when

(a) $A = \begin{bmatrix} 3 & 1 \\ -2 & -1 \end{bmatrix}$    (b) $A = \begin{bmatrix} 0 & -1 & 2 \\ 1 & 0 & 0 \\ 2 & 1 & -1 \end{bmatrix}$

2. With $A$ as given, find the indicated power of $A$ by using the Cayley–Hamilton theorem.

(a) $A = \begin{bmatrix} 1 & 1 \\ 1 & -2 \end{bmatrix}$; $A^3$    (b) $A = \begin{bmatrix} -4 & 2 \\ -1 & 1 \end{bmatrix}$; $A^4$

(c)   $A = \begin{bmatrix} 2 & 4 \\ 1 & 2 \end{bmatrix}$;   $A^4$      (d)   $A = \begin{bmatrix} 1 & 0 & -2 \\ 0 & 0 & 3 \\ -2 & 1 & 0 \end{bmatrix}$;   $A^4$

3. Use the Cayley–Hamilton theorem to find the inverse of each nonsingular matrix in Exercise 2.

4. Verify the Cayley–Hamilton theorem for the matrices of Exericse 2.

5. Verify properties (4.32) and (4.33). Suggestion: first establish (4.33) for the case where $q(A) = cA^m$.

6. Let $\lambda$ be a characteristic value of $A$, with $\mathbf{k}$ as a corresponding characteristic vector. If $f$ is a polynomial, show that $f(\lambda)$ is a characteristic value of $f(A)$ with $\mathbf{k}$ as a corresponding characteristic vector. Suggestion: if $A\mathbf{k} = \lambda\mathbf{k}$ then $A^2\mathbf{k} = \lambda A\mathbf{k} = \lambda^2\mathbf{k}$.

7. Let $A$ and $B$ be square matrices such that $K^{-1}AK = B$. Show that
   (a)   $K^{-1}A^2K = B^2$.
   (b)   $K^{-1}A^mK = B^m$, where $m$ is any positive integer.
   (c)   $K^{-1}f(A)K = f(B)$, where $f$ is any polynomial.

8. (a)   Show that the Cayley–Hamilton theorem holds for any diagonal matrix.
   (b)   Show that the Cayley–Hamilton theorem holds for any matrix that is similar to a diagonal matrix. (Use the results of Exercise 7.)

## 4.6   SERIES OF MATRICES

Having defined *polynomial* functions with square matrix arguments we next try to define quantities such as $e^A$ and $\sin A$ by means of the formulas

$$e^A = I + A + \frac{1}{2!}A^2 + \cdots + \frac{1}{k!}A^k + \cdots,$$

$$\sin A = A - \frac{1}{3!}A^3 + \frac{1}{5!}A^5 - \cdots + (-1)^k \frac{1}{(2k-1)!}A^{2k-1} + \cdots.$$

In this approach we attempt to replace the number $x$ in the relations

$$e^x = \sum_{k=0}^{\infty} \frac{1}{k!}x^k, \qquad \sin x = \sum_{k=1}^{\infty} \frac{(-1)^k}{(2k-1)!}x^{2k-1}$$

by a square matrix $A$. First, however, we must define what we mean by an infinite series of matrices.

We define the *norm* of an $m \times n$ matrix $A$, written $|A|$, as

$$|A| = \max_{\substack{1 \le i \le m \\ 1 \le j \le n}} |a_{ij}|. \tag{4.37}$$

Thus the norm of $A$ is simply the largest of the absolute values of all the elements of $A$. It follows from this definition that

$$|cA| = |c| \, |A| \tag{4.38}$$

for every number $c$. If $A$ and $B$ are matrices of the same size it can be shown (Exercise 1) that

$$|A + B| \le |A| + |B|. \tag{4.39}$$

Also, if $A$ is an $m \times n$ matrix and $B$ is an $n \times q$ matrix, then

$$|AB| \le n|A| \, |B|. \tag{4.40}$$

To see this, let $C = AB$. Then

$$c_{ij} = a_{i1}b_{1j} + a_{i2}b_{2j} + \cdots + a_{in}b_{nj}$$

and

$$
\begin{aligned}
|c_{ij}| &\le |a_{i1}| \, |b_{1j}| + |a_{i2}| \, |b_{2j}| + \cdots + |a_{in}| \, |b_{nj}| \\
&\le |A| \, |B| + |A| \, |B| + \cdots + |A| \, |B| \\
&\le n|A| \, |B|
\end{aligned}
$$

for $1 \le i \le m$ and $1 \le j \le q$. Hence $|C| \le n|A| \, |B|$.

Let

$$(A_1, A_2, A_3, \ldots) \tag{4.41}$$

be an infinite sequence of $m \times n$ matrices. This sequence is said to converge to the $m \times n$ matrix $A$ if

$$\lim_{k \to \infty} |A_k - A| = 0. \tag{4.42}$$

If the sequence (4.41) converges to $A$ we write

$$\lim_{k \to \infty} A_k = A.$$

Notice that, for each positive integer $k$, $|A_k - A|$ is a real number. Let the element in the $i$th row and $j$th column of $A_k$ be denoted by $a_{ij}^{(k)}$. Since

$$|A_k - A| = \max_{i, j} |a_{ij}^{(k)} - a_{ij}|,$$

it can be seen that the sequence (4.41) converges to $A$ if and only if the sequence of numbers

$$(a_{ij}^{(1)}, a_{ij}^{(2)}, a_{ij}^{(3)}, \ldots)$$

converges to $a_{ij}$ for $1 \le i \le m$ and $1 \le j \le n$. For example, the sequence of matrices whose first few terms are

$$\begin{bmatrix} 2 & 1 \\ 3 & 4 \end{bmatrix}, \quad \begin{bmatrix} 0 & 1 \\ 2 & -2 \end{bmatrix}, \quad \begin{bmatrix} -1 & 3 \\ 2 & 5 \end{bmatrix}, \ldots$$

converges to the matrix

$$\begin{bmatrix} a & b \\ c & d \end{bmatrix}$$

if and only if the sequences

$$(2, 0, -1, \ldots), (1, 1, 3, \ldots), (3, 2, 2, \ldots), (4, -2, 5, \ldots)$$

converge to $a$, $b$, $c$, and $d$, respectively.

In order to define convergence for an infinite series of $m \times n$ matrices.

$$\sum_{k=1}^{\infty} B_k, \tag{4.43}$$

we form the sequence

$$(S_1, S_2, S_3, \ldots), \tag{4.44}$$

where

$$S_1 = B_1, \qquad S_2 = B_1 + B_2, \qquad S_3 = B_1 + B_2 + B_3,$$

and so on. We say that the series (4.43) converges to the $m \times n$ matrix $B$ if the sequence (4.44) converges to $B$. We see that

$$s_{ij}^{(k)} = b_{ij}^{(1)} + b_{ij}^{(2)} + \cdots + b_{ij}^{(k)},$$

where $s_{ji}^{(k)}$ and $b_{ji}^{(p)}$ are the elements in the $i$th row and $j$th column of $S_k$ and $B_p$, respectively. Consequently the series (4.43) converges to $B$ if and only if the series of numbers

$$\sum_{k=1}^{\infty} b_{ij}^{(k)}$$

converges to $b_{ij}$ for $1 \le i \le m$ and $1 \le j \le n$. Thus the series

$$\begin{bmatrix} 3 & 5 \\ -1 & 1 \end{bmatrix} + \begin{bmatrix} 2 & 0 \\ 4 & 6 \end{bmatrix} + \begin{bmatrix} -2 & 3 \\ 2 & 4 \end{bmatrix} + \cdots$$

converges to

$$\begin{bmatrix} a & b \\ c & d \end{bmatrix}$$

if and only if

$$3 + 2 - 2 + \cdots = a, \qquad 5 + 0 + 3 + \cdots = b,$$
$$-1 + 4 + 2 + \cdots = c, \qquad 1 + 6 + 4 + \cdots = d.$$

Using the notion of the norm of a matrix, we can formulate the following test for the convergence of an infinite series of matrices.

**Theorem 4.13** Let $B_1$, $B_2$, $B_3$,... be $m \times n$ matrices. If the series of numbers

$$\sum_{k=1}^{\infty} |B_k| \tag{4.45}$$

converges, then the series of matrices

$$\sum_{k=1}^{\infty} B_k \tag{4.46}$$

converges.

PROOF   Fixing $i$ and $j$, we compare the series

$$\sum_{k=1}^{\infty} b_{ij}^{(k)} \tag{4.47}$$

with the series (4.45). Since

$$|b_{ij}^{(k)}| \leq |B_k|, \quad k = 1, 2, 3, \ldots$$

the series (4.47) converges absolutely for each $i$ and $j$. Hence the series (4.46) converges.

We now define $e^A$ for a square $n \times n$ matrix $A$. We first observe that

$$|A^2| \leq n|A|\,|A| = n|A|^2,$$

$$|A^3| \leq n|A|\,|A^2| \leq n^2|A|^3,$$

$$\ldots\ldots\ldots\ldots\ldots\ldots\ldots\ldots\ldots\ldots$$

$$|A^k| \leq n|A|\,|A^{k-1}| \leq n^{k-1}|A|^k.$$

Since the series of numbers

$$1 + \sum_{k=1}^{\infty} \frac{1}{k!} n^{k-1} |A|^k$$

converges (as can be shown by the ratio test), the series of matrices

$$I + \sum_{k=1}^{\infty} \frac{1}{k!} A^k$$

converges, by Theorem 4.13. We may now define

$$e^A = I + \sum_{k=1}^{\infty} \frac{1}{k!} A^k \tag{4.48}$$

for any square matrix $A$. Note that $e^A$ is a square matrix of the same size as $A$. Matrices of the form (4.48) are important in the study of certain systems of differential equations. We shall return to them in Chapter 6.

## Exercises for Section 4.6

**1.** Prove the inequality $|A + B| \leq |A| + |B|$.

**2.** Let

$$\sum_{k=1}^{\infty} B_k$$

be a series of matrices that converges to the matrix $B$. If $A$ and $C$ are matrices such that $AB$ and $BC$ are defined, and if $b$ is a number, prove that the series

$$\sum_{k=1}^{\infty} AB_k, \qquad \sum_{k=1}^{\infty} B_k C, \qquad \sum_{k=1}^{\infty} bB_k$$

converge to $AB$, $BC$, and $bB$, respectively.

**3.** If

$$\sum_{k=1}^{\infty} A_k = A, \qquad \sum_{k=1}^{\infty} B_k = B,$$

where $A$ and $B$ are of the same size, show that

$$\sum_{k=1}^{\infty} (A_k + B_k) = A + B.$$

**4.** Show that the series

$$\sum_{k=1}^{\infty} \frac{(-1)^{k-1}}{(2k-1)!} A^{2k-1}$$

converges for every square matrix $A$.

**5.** Let $A$ be an $n \times n$ matrix. Show that

$$(I - A)^{-1} = I + A + A^2 + \cdots + A^k + \cdots$$

provided that $|A| < 1/n$. (Use the results of Exercises 2 and 3.)

**6.** If $D = \operatorname{diag}(d_1, d_2, \ldots, d_n)$ show that

$$e^D = \operatorname{diag}(e^{d_1}, e^{d_2}, \ldots, e^{d_n}).$$

**7.** Given

$$A = \begin{bmatrix} 0 & 1 & 0 \\ 0 & 0 & 4 \\ 0 & 0 & 0 \end{bmatrix},$$

verify that $A^3 = 0$ and find $e^A$.

**8.** If $A$ is similar to a diagonal matrix $D$ and $K^{-1}AK = D$, show that

$$e^A = Ke^D K^{-1}.$$

**9.** Use the result of Exercise 8 to find $e^A$ if

$$A = \begin{bmatrix} 2 & -1 \\ 3 & -2 \end{bmatrix}.$$

**10.** Let

$$f(x) = \sum_{k=0}^{\infty} c_k x^k$$

for all $x$. (The power series converges, and hence converges absolutely, for all $x$.) If $A$ is any square matrix, show that the series of matrices in the definition

$$f(A) = \sum_{k=0}^{\infty} c_k A^k$$

is convergent.

# V

Linear
Differential Equations

## 5.1  INTRODUCTION

A linear differential equation of order $n$ is of the form

$$a_0 y^{(n)} + a_1 y^{(n-1)} + \cdots + a_{n-1} y' + a_n y = F, \tag{5.1}$$

where the functions $a_i$ and $F$ are specified on some interval $\mathscr{I}$. The functions $a_i$ are called the *coefficients* of the differential equation. If $F$ is the zero function, the equation is said to be *homogeneous*; otherwise it is said to be *non-homogeneous*. A homogeneous equation always has the zero function as one of its solutions. This solution is sometimes called the *trivial solution*.

Associated with Eq. (5.1) is the linear differential operator

$$L = a_0 D^n + a_1 D^{n-1} + \cdots + a_{n-1} D + a_n. \tag{5.2}$$

Such operators were discussed in Section 3.7. If $\mathscr{F}^n$ is the space of all functions that possess at least $n$ derivatives on $\mathscr{I}$, then $L$ maps $\mathscr{F}^n$ into $\mathscr{F}^0$, or $\mathscr{F}$. A function $f$ in $\mathscr{F}^n$ for which $Lf = 0$ is a solution of the homogeneous equation

$$Ly = 0 \tag{5.3}$$

on $\mathscr{I}$. The set of all solutions of Eq. (5.3) on $\mathscr{I}^1$ constitutes a vector space,

---

[1] Suppose that $L$ satisfies the conditions in Theorem 5.1. Then if $g$ is a solution on an interval $\mathscr{J}$ that is contained in $\mathscr{I}$, it can be shown (see Chapter 8) that there is a solution $f$ on $\mathscr{I}$ such that $g(x) = f(x)$ for all $x$ in $\mathscr{J}$. For this reason we consider only solutions that exist throughout $\mathscr{I}$.

namely the kernel of the operator $L$. Consequently, if $f$ is a solution and $c$ is a number, then $cf$ is also a solution. If $f$ and $g$ are both solutions, then so is $f + g$. Finally, if $f_1, f_2, \ldots, f_m$ are solutions and if $c_1, c_2, \ldots, c_m$ are numbers, then $c_1 f_1 + c_2 f_2 + \cdots + c_m f_m$ is again a solution.

These properties of solutions do not occur for nonlinear differential equations. For example, let us consider the equation

$$y' = -y^2, \tag{5.4}$$

which is nonlinear and separable. The solutions of this equation consist of the zero function and those functions of the form

$$y = \frac{1}{x + c},$$

where $c$ is an arbitrary constant. Taking $c = 0$ and $c = 1$, we see that the functions

$$f_1(x) = \frac{1}{x}, \qquad f_2(x) = \frac{1}{x + 1}$$

are both solutions on the interval $(0, \infty)$. Setting $f = f_1 + f_2$, we see that

$$f'(x) = -\frac{1}{x^2} - \frac{1}{(x + 1)^2}$$

and

$$-[f(x)]^2 = -\frac{1}{x^2} - \frac{1}{(x + 1)^2} - \frac{2}{x(x + 1)}.$$

Since $f' \neq -f^2$, $f_1 + f_2$ is not a solution of Eq. (5.4).

The space $\mathscr{F}^n$ is not finite-dimensional, as was shown in Section 3.5. However, the *kernel* of the operator $L$ is a finite-dimensional subspace of $\mathscr{F}^n$. To be precise, the following theorem can be proved.

**Theorem 5.1** Let the functions $a_0, a_1, \ldots, a_n$ be continuous, with $a_0$ never zero on an interval $\mathscr{I}$. Then the kernel of the linear operator (5.2) is a vector space of dimension $n$.

To describe the set of all solutions of the homogeneous equation $Ly = 0$, we need to find a basis for the kernel of $L$. If $\{f_1, f_2, \ldots, f_n\}$ is such a basis, then the set of all solutions (the general solution) consists of all functions of the form

$$c_1 f_1 + c_2 f_2 + \cdots + c_n f_n.$$

In order to find the specific solution that satisfies the initial conditions

$$y(x_0) = k_0, \quad y'(x_0) = k_1, \ldots, \quad y^{(n-1)}(x_0) = k_{n-1}$$

we must determine the constants $c_i$ from the requirements

$$
\begin{aligned}
&c_1 f_1(x_0) + c_2 f_2(x_0) + \cdots + c_n f_n(x_0) = k_0, \\
&c_1 f_1'(x_0) + c_2 f_2'(x_0) + \cdots + c_n f_n'(x_0) = k_1, \\
&\cdots\cdots\cdots\cdots\cdots\cdots\cdots\cdots\cdots\cdots\cdots\cdots\cdots\cdots\cdots\cdots \\
&c_1 f_1^{(n-1)}(x_0) + c_2 f_2^{(n-1)}(x_0) + \cdots + c_n f_n^{(n-1)}(x_0) = k_{n-1}.
\end{aligned}
\tag{5.5}
$$

To illustrate, we consider the second-order differential equation

$$y'' - 3y' + 2y = 0 \tag{5.6}$$

on the interval $(-\infty, \infty)$. It is easy to verify that the functions $f_1$ and $f_2$, where

$$f_1(x) = e^x, \qquad f_2(x) = e^{2x}$$

are solutions and hence belong to the kernel of the operator

$$L = D^2 - 3D + 2.$$

Furthermore, these functions are linearly independent, since their Wronskian

$$W(x) = \begin{vmatrix} e^x & e^{2x} \\ e^x & 2e^{2x} \end{vmatrix} = e^{3x}$$

does not vanish for all $x$. Hence the general solution of Eq. (5.6) is

$$y = c_1 e^x + c_2 e^{2x}.$$

If we wish to find a specific solution that satisfies a set of initial conditions, such as

$$y(0) = 4, \qquad y'(0) = 1, \tag{5.7}$$

we need only determine the constants $c_1$ and $c_2$ in such a way that the conditions are met. Since

$$y' = c_1 e^x + 2c_2 e^{2x}$$

the conditions (5.7) become

$$c_1 + c_2 = 4, \qquad c_1 + 2c_2 = 1.$$

We find that $c_1 = 7$ and $c_2 = -3$. Hence the desired solution is

$$y = 7e^x - 3e^{2x}.$$

If a set of functions is linearly independent on an interval $\mathscr{I}$ the Wronskian of the functions may still vanish at every point of $\mathscr{I}$. An example was given in Section 3.4. However, if the functions $f_1, f_2, \ldots, f_n$ are linearly independent solutions of the $n$th-order linear differential equation $Ly = 0$, this cannot happen. It can be shown that if $L$ is subject to the conditions of Theorem 5.1, the Wronskian is *never* zero on $\mathscr{I}$. Thus the system of equations (5.5) for $c_1, c_2, \ldots, c_n$ always possesses a unique solution. A fuller discussion of these matters is postponed until Section 5.11.

The main problem, of course, is how to find the linearly independent solutions of the differential equation. In the next few sections we shall investigate certain classes of linear equations for which this can readily be accomplished.

A theory for the nonhomogeneous equation $Ly = F$ can be based on the theory for the associated homogeneous equation $Ly = 0$. Nonhomogeneous equations will be discussed in Sections 5.6–5.8.

### Exercises for Section 5.1

1. Find at least one solution of the equation $y'' + e^x y = 0$.

2. Verify that the given functions form a basis for the space of solutions of the given differential equation.

   (a) $y'' + y = 0$, $f_1(x) = \cos x$, $f_2(x) = \sin x$
   (b) $y' - 2xy = 0$, $f_1(x) = \exp(x^2)$
   (c) $x^2 y'' - 2xy' + 2y = 0$, $f_1(x) = x$, $f_2(x) = x^2$, $x > 0$
   (d) $y''' - y' = 0$, $f_1(x) = 1$, $f_2(x) = e^x$, $f_2(x) = e^{-x}$

3. Find the solution of the corresponding differential equation in Exercise 2 that satisfies the given initial conditions.

   (a) $y(\pi) = 0$, $y'(\pi) = -2$      (b) $y(0) = 5$
   (c) $y(1) = 0$, $y'(1) = 0$      (d) $y(0) = 3$, $y'(0) = 1$, $y''(0) = 1$

4. If $a_0, a_1, \ldots, a_n$ possess derivatives of all orders on $\mathscr{I}$ and if $a_0$ is never zero on $\mathscr{I}$, show that every solution of Eq. (5.3) possesses derivatives of all orders on $\mathscr{I}$, Suggestion: write

$$y^{(n)} = -\frac{a_1}{a_0} y^{(n-1)} - \cdots - \frac{a_n}{a_0} y.$$

**5.** What can be said about the solution of the differential equation

$$y''' + xy' - (\sin x)y = 0, \quad \text{for all } x,$$

if $y(1) = y'(1) = y''(1) = 0$?

**6.** If $u$ and $v$ are both solutions of the equation $Ly = F$ show that $u - v$ is a solution of the equation $Ly = 0$.

## 5.2  POLYNOMIAL OPERATORS

Linear differential equations whose coefficients are constant functions are of particular interest. One reason for this is that they are easy to solve. Another is that important problems in mechanics and electric circuits can be described by such equations. Examples of such problems are discussed in Sections 5.9 and 5.10.

An $n$th-order linear equation with constant coefficients has the form

$$a_0 y^{(n)} + a_1 y^{(n-1)} + \cdots + a_{n-1}y' + a_n y = F, \tag{5.8}$$

where the $a_i$ are constants, with $a_0 \neq 0$. This equation can be written as $Ly = F$, where

$$L = a_0 D^n + a_1 D^{n-1} + \cdots + a_{n-1}D + a_n. \tag{5.9}$$

We say that $L$ is a linear differential operator with constant coefficients. Associated with this operator in an obvious way is the polynomial $P$, where

$$P(r) = a_0 r^n + a_1 r^{n-1} + \cdots + a_{n-1}r + a_n. \tag{5.10}$$

We therefore write

$$P(D) = a_0 D^n + a_1 D^{n-1} + \cdots + a_{n-1}D + a_n \tag{5.11}$$

and call $P(D)$ a *polynomial operator*.

The sum, $P(D) + Q(D)$, of two polynomial operators is evidently a polynomial operator. Its associated polynomial is $P + Q$. For example, if

$$P(D) = 2D^2 - 3D + 1, \qquad Q(D) = D - 4,$$

then

$$P(D) + Q(D) = 2D^2 - 2D - 3.$$

For the $P$ and $Q$ in this example, let us consider the operator $Q(D) P(D)$, which maps $\mathscr{F}^3$ into $\mathscr{F}$. For any function $f$ in $\mathscr{F}^3$ we have

$$
\begin{aligned}
Q(D) P(D) f &= Q(D)[P(D)f] \\
&= (D - 4)(2f'' - 3f' + f) \\
&= 2f''' - 3f'' + f' - 8f'' + 12f' - 4f \\
&= (2D^2 - 11D + 13D - 4)f.
\end{aligned}
$$

Thus $Q(D) P(D)$ is the polynomial operator whose associated polynomial is $PQ$ because

$$
Q(r) P(r) = (2r^2 - 3r + 1)(r - 4) = 2r^3 - 11 r^2 + 13r - 4.
$$

More generally, if

$$
P(D) = a_0 D^m + a_1 D^{m-1} + \cdots + a_m,
$$
$$
Q(D) = b_0 D^n + b_1 D^{n-1} + \cdots + b_n,
$$

it can be shown (Exercise 12) that $Q(D) P(D)$ is a polynomial operator whose associated polynomial is $QP$. Since ordinary polynomials commute, that is, $P(r) Q(r) = Q(r) P(r)$ for all $r$, we have the important result that polynomial operators commute. Thus

$$
P(D) Q(D) = Q(D) P(D). \tag{5.12}
$$

If $P$ is any ploynomial that can be written in the factored form

$$
P(r) = P_1(r) P_2(r) \cdots P_k(r),
$$

it follows from the previously described properties of polynomial operators that

$$
P(D) = P_1(D) P_2(D) \cdots P_k(D). \tag{5.13}
$$

The order and manner of grouping of the polynomial operators in this expression do not matter.

It is well known that a polynomial of degree $n$ can be written as the product of $n$ linear factors. If

$$
P(r) = a_0 r^n + a_1 r^{n-1} + \cdots + a_{n-1}r + a_n,
$$

then

$$
P(r) = a_0(r - r_1)(r - r_2) \cdots (r - r_n), \tag{5.14}
$$

where the numbers $r_i$ need not be distinct. (The numbers $r_i$ are called the *zeros* of $P$ and the *roots* of the equation $P(r) = 0$.) Some or all of the $r_i$ may be complex. However, if $P$ has real coefficients, whenever $r_1 = a + ib$ is a complex zero, then $r_2 = a - ib$ must also be a zero. The second-degree polynomial

$$(r - r_1)(r - r_2) = [r - (a + ib)][r - (a - ib)] = (r - a)^2 + b^2$$

has real coefficients. Therefore any polynomial operator $P(D)$ with real co-efficients can be written as the product of first- and second-order polynomial operators with real coefficients. For example, let

$$P(D) = D^3 - 5D^2 + 9D - 5.$$

Since

$$P(r) = (r - 1)(r - 2 - i)(r - 2 + i) = (r - 1)(r^2 - 4r + 5),$$

we may write

$$P(D) = (D - 1)(D^2 - 4D + 5) = (D^2 - 4D + 5)(D - 1).$$

If a function is to be a solution of the equation

$$P(D)y = 0, \tag{5.15}$$

where

$$P(D) = a_0 D^n + a_1 D^{n-1} + \cdots + a_n, \tag{5.16}$$

a certain linear combination of that function and its first $n$ derivatives must vanish. If $f$ is an exponential function, of the form $f(x) = e^{rx}$, the derivatives of $f$ are multiples of $f$ itself. We have

$$D^m e^{rx} = r^m e^{rx} \tag{5.17}$$

for every nonnegative integer $m$. We therefore attempt to find solutions of Eq. (5.15) that are of this form. In view of formula (5.17) we have

$$(a_0 D^n + a_1 D^{n-1} + \cdots + a_n) e^{rx} = (a_0 r^n + a_1 r^{n-1} + \cdots + a_n) e^{rx}$$

or

$$P(D) e^{rx} = P(r) e^{rx}. \tag{5.18}$$

The polynomial $P$ is called the *auxiliary polynomial* associated with the differential equation (5.15). Using formula (5.14) we may write

$$P(D) e^{rx} = a_0(r - r_1)(r - r_2) \cdots (r - r_n) e^{rx}. \tag{5.19}$$

If $r_i$ is a real zero of $P$, the function $e^{r_i x}$ is a solution of Eq. (5.15). If $r_1, r_2,, \ldots r_n$ are real and distinct, each of the functions

$$e^{r_1 x}, e^{r_2 x}, \ldots, e^{r_n x}$$

is a solution. These solutions are linearly independent, as was shown in Section 3.4. In this case the general solution of Eq. (5.15) is

$$y = c_1 e^{r_1 x} + c_2 e^{r_2 x} + \cdots + c_n e^{r_n x}.$$

**Example 1**  Consider the equation

$$y'' - y' - 2y = 0,$$

which may be written as

$$(D^2 - D - 2)y = 0.$$

The auxiliary polynomial is

$$r^2 - r - 2 = (r + 1)(r - 2).$$

The zeros are $r_1 = -1$ and $r_2 = 2$. Hence $e^{-x}$ and $e^{2x}$ are linearly independent solutions and the general solution is

$$y = c_1 e^{-x} + c_2 e^{2x}.$$

In the general case of Eq. (5.15), the polynomial $P$ may not have $n$ distinct zeros. Even if it does, some of the zeros may be complex. The next two examples illustrate these possibilities.

**Example 2**

$$y'' - 4y' + 4y = 0.$$

Here the auxiliary polynomial

$$r^2 - 4r + 4 = (r - 2)^2$$

has only one distinct zero, $r_1 = 2$. Thus $e^{2x}$ is a solution, but we need still another solution to be able to describe the general solution.

**Example 3**

$$y'' - 2y' + 5y = 0.$$

The auxiliary polynomial equation

$$r^2 - 2r + 5 = 0$$

has the complex roots $r_1 = 1 + 2i$ and $r_2 = 1 - 2i$. Thus we are unable (as yet) to write down any nontrivial solution.

The difficulties caused by the appearance of complex zeros of the auxiliary polynomial will be removed in the next section.

### Exercises for Section 5.2

1.  Write the differential equation in the form $P(D) y = 0$.

    (a)  $y'' - 3y' + 2y = 0$          (b)  $y' + 4y = 0$

    (c)  $y''' - 3y'' - y' + y = 0$          (d)  $y^{(4)} - y'' + y = 0$

2.  Write the differential equation in factored form, in terms of real polynomial operators of first- and second-order.

    (a)  $(D^2 + D - 2)y = 0$          (b)  $(D^4 - 1)y = 0$

    (c)  $(D^3 - 3D + 2)y = 0$          (d)  $(D^3 + D^2 - 4D + 6)y = 0$

3.  Find a differential equation of the form $P(D) y = 0$, with real coefficients, whose associated polynomial $P$ has the given numbers among its zeros. The order of $P$ should be as low as possible.

    (a)  $r_1 = 2$,  $r_2 = -1$          (b)  $r_1 = 3 - 2i$

    (c)  $r_1 = r_2 = 2$,  $r_3 = 0$          (d)  $r_1 = 2i$,  $r_2 = -1$,  $r_3 = 2$

4.  If the function $u$ is a solution of the equation $P(D) y = 0$ and $v$ is a solution of the equation $Q(D) y = 0$, show that $u$ and $v$ are both solutions of the equation $P(D) Q(D) y = 0$.

In Exercises 5–10, express the general solution of the given differential equation in terms of exponential functions, if possible.

5.  $y'' - 5y' + 6y = 0$          6.  $2y'' + 5y' - 3y = 0$

7.  $y''' + 5y'' - y' - 5y = 0$          8.  $y'' + 4y = 0$

**9.**   $y'' - 6y' + 9y = 0$                          **10.**   $y''' - 4y' = 0$

**11.**   Show that no differential equation of the form $P(D) y = 0$ has the function $f$, where $f(x) = 1/x$, $x > 0$, as a solution.

**12.**   (a)   If $P(D)$ is any polynomial operator and $Q(D) = cD^k$, show that $Q(D) P(D)$ is a polynomial operator whose associated polynomial is $QP$.
(b)   Use the result of part (a) to show that if $P$ and $Q$ are any two polynomials, then $Q(D) P(D)$ is a polynomial operator whose corresponding polynomial is $QP$.

## 5.3   COMPLEX SOLUTIONS

In applications that give rise to differential equations, we are almost always concerned with real solutions of the equations. However, it is sometimes convenient to extract a desired real solution from a complex solution, as will be seen later on.

A complex function $w$ of a real variable can be regarded as an ordered pair of real functions $(u, v)$. We write

$$w(x) = u(x) + i \, v(x), \tag{5.20}$$

where $i$ is the imaginary unit with the property that $i^2 = -1$. We call $u$ the real part and $v$ the imaginary part of $w$. The derivative of the complex function $w$ is defined by the relation

$$w'(x) = u'(x) + i \, v'(x), \tag{5.21}$$

provided that $u'(x)$ and $v'(x)$ both exist.

In this chapter we are concerned with linear differential equations with *real* coefficients. Let

$$L = a_0 D^n + a_1 D^{n-1} + \cdots + a_{n-1} D + a_n,$$

where the functions $a_i$ are real valued. Since

$$a_k D^{n-k} w = a_k D^{n-k} u + i a_k D^{n-k} v$$

for $k = 0, 1, \ldots, n$, it follows that

$$Lw = Lu + i \, Lv,$$

where $Lu$ and $Lv$ are both real functions. If $w$ is a complex solution of the homogeneous equation $Ly = 0$, then

$$Lu + i\,Lv = 0.$$

Since a complex number can be zero only if its real and imaginary parts are both zero, it follows that $Lu = 0$ and $Lv = 0$. Consequently the real and imaginary parts, $u$ and $v$, of a complex solution $w$ are real solutions of the differential equation.

We shall also be interested in nonhomogeneous linear equations of the form

$$Ly = F, \tag{5.22}$$

where $F$ is a complex function,

$$F(x) = f(x) + i\,g(x).$$

If $w = u + iv$ is a solution of Eq. (5.22), then

$$Lw = F$$

or

$$Lu + i\,Lv = f + i\,g.$$

But this means that

$$Lu = f, \qquad Lv = g.$$

Thus by finding a complex solution of Eq. (5.22) we obtain real solutions for the two equations

$$Ly = f, \qquad Ly = g.$$

We shall be particularly concerned with a class of complex functions known as complex exponential functions. In order to define these functions we first define the complex number $e^{a+ib}$, where $e$ is the base of natural logarithms, as

$$e^{a+ib} = e^a \cos b + i e^a \sin b. \tag{5.23}$$

When $a + ib$ happens to be real ($b = 0$) we see that $e^a$ has its usual real value. Other special cases of interest are

$$e^{ib} = \cos b + i \sin b, \qquad e^{-ib} = \cos b - i \sin b. \tag{5.24}$$

From these relations, we obtain the formulas

$$\cos b = \frac{1}{2}(e^{ib} + e^{-ib}), \qquad \sin b = \frac{1}{2i}(e^{ib} - e^{-ib}). \tag{5.25}$$

The laws of exponents

$$e^{z_1} \cdot e^{z_2} = e^{z_1 + z_2}, \tag{5.26a}$$

$$e^{z_1}/e^{z_2} = e^{z_1 - z_2}, \tag{5.26b}$$

where $z_1$ and $z_2$ are arbitrary complex numbers, can be derived from the definition (5.23). We shall consider the rule (5.26a) here, leaving the second to the exercises. Writing $z_1 = a_1 + ib_1$ and $z_2 = a_2 + ib_2$, we have

$$
\begin{aligned}
e^{z_1} \cdot e^{z_2} &= e^{a_1}(\cos b_1 + i \sin b_1) \cdot e^{a_2}(\cos b_2 + i \sin b_2) \\
&= e^{a_1 + a_2}[(\cos b_1 \cos b_2 - \sin b_1 \sin b_2) \\
&\quad + i(\cos b_1 \sin b_2 + \sin b_1 \cos b_2)] \\
&= e^{a_1 + a_2}[\cos(b_1 + b_2) + i \sin(b_1 + b_2)] \\
&= e^{(a_1 + a_2) + i(b_1 + b_2)} \\
&= e^{z_1 + z_2}.
\end{aligned}
$$

We now consider a complex function $F$ of the form

$$F(x) = e^{h(x)},$$

where $h(x) = u(x) + i\,v(x)$. We seek a formula for the derivative of this function. Writing

$$F(x) = e^{u(x)} \cos v(x) + i e^{u(x)} \sin v(x)$$

and using the definition (5.21), we have

$$
\begin{aligned}
F' &= u'(e^u \cos v + i e^u \sin v) + v'(-e^u \sin v + i e^u \cos v) \\
&= (u' + iv')(e^u \cos v + i e^u \sin v).
\end{aligned}
$$

Hence

$$\frac{de^{h(x)}}{dx} = h'(x)e^{h(x)}. \tag{5.27}$$

A function $w$ of the form

$$w(x) = e^{cx}, \tag{5.28}$$

where $c = a + ib$ is a complex constant, is called a *complex exponential function*. Using the rule (5.27), we see that

$$\frac{de^{cx}}{dx} = ce^{cx} \tag{5.29}$$

for all $x$. From Eqs. (5.24) and (5.25) we obtain the formulas

$$e^{ibx} = \cos bx + i \sin bx,$$
$$e^{-ibx} = \cos bx - i \sin bx, \tag{5.30}$$

and

$$\cos bx = \frac{1}{2}(e^{ibx} + e^{-ibx}),$$
$$\sin bx = \frac{1}{2i}(e^{ibx} - e^{-ibx}), \tag{5.31}$$

which relate the real trigonometric functions and the complex exponential functions. As an example of their use, let us consider the homogeneous differential equation $(D^2 + 4)y = 0$. Seeking a solution of the form $y = e^{cx}$, we find that

$$(D^2 + 4)e^{cx} = (c^2 + 4)e^{cx} = (c + 2i)(c - 2i)e^{cx}.$$

Since this is identically zero when $c = \pm 2i$ the functions $e^{2ix}$ and $e^{-2ix}$ are complex solutions. But then the real and imaginary parts, $\cos 2x$ and $\sin 2x$, are real solutions.

Next we consider the nonhomogeneous differential equation

$$Ly = Ae^{ibx}, \tag{5.32}$$

where $L$ has real coefficients and $A$ is a real constant. According to the formulas (5.30),

$$Ae^{ibx} = A \cos bx + i A \sin bx.$$

If we can find a complex solution of Eq. (5.32), then the real and imaginary parts of this solution will be real solutions of the equations

$$Ly = A \cos bx, \qquad Ly = A \sin bx, \qquad (5.33)$$

respectively. The reason for preferring to deal with complex exponential functions rather than real trigonometric functions is that the differentiation formulas for the former are simpler. A number of examples are presented in Sections 5.4 and 5.7.

If $c = a + ib$ is any complex number, and if $\alpha$ is any positive real number, we define

$$\alpha^c = e^{c \ln \alpha}. \qquad (5.34)$$

Notice that when $c$ is real ($b = 0$) this formula agrees with the definition of $\alpha^c$ given in calculus. The laws of exponents

$$\alpha^{z_1} \cdot \alpha^{z_2} = \alpha^{z_1 + z_2}, \qquad (5.35a)$$

$$\alpha^{z_1} / \alpha^{z_2} = \alpha^{z_1 - z_2}, \qquad (5.35b)$$

follow easily from the laws (5.26). Their derivations are left to the exercises.

If $x$ is a positive real number we have

$$x^c = e^{c \ln x} \qquad (5.36)$$

according to the definition (5.34). Application of the differentiation formula (5.27) yields

$$\frac{dx^c}{dx} = \frac{c}{x} e^{c \ln x} = \frac{c}{x} x^c$$

or

$$\frac{dx^c}{dx} = cx^{c-1}, \quad x > 0. \qquad (5.37)$$

Notice that we have not defined $\alpha^\beta$ when $\alpha$ and $\beta$ are both complex, or even when $\alpha$ is real and negative. To give proper definitions would take us further into the theory of complex variables than is necessary for our study of linear differential equations.

## Exercises for Section 5.3

1. (a)  If $w = u + iv$ is a complex function and $c = a + ib$ is a complex constant, show that

$$\frac{d[cw(x)]}{dx} = c\,\frac{dw(x)}{dx}.$$

(b)  If $w_1 = u_1 + iv_1$ and $w_2 = u_2 + iv_2$ are complex functions, show that

$$(w_1 + w_2)' = w_1' + w_2'$$

2. Let $w_1 = u_1 + iv_1$ and $w_2 = u_2 + iv_2$ be complex functions. Show that

(a)  $(w_1 w_2)' = w_1' w_2 + w_1 w_2'$

(b)  $(w_1/w_2)' = (w_1' w_2 - w_1 w_2')/w_2^2$

3. Express each complex exponential function in terms of trigonometric functions and express each trigonometric function in terms of complex exponential functions.

(a)  $e^{3ix}$      (b)  $e^{-2ix}$      (c)  $e^{(2-3)ix}$      (d)  $e^{(-2+i)x}$

(e)  $\cos 2x$      (f)  $\sin 5x$      (g)  $\sin x$      (h)  $\cos 5x$

4. Show that $e^{z_1}/e^{z_2} = e^{z_1 - z_2}$ when $z_1$ and $z_2$ are arbitrary complex numbers.

5. The *modulus* of a complex number $c = a + ib$, written $|c|$, is defined to be $|c| = (a^2 + b^2)^{1/2}$. Show that $|e^{a+ib}| = e^a$.

6. For every real number $\theta$ and every integer $m$ show that

$$(e^{i\theta})^m = e^{im\theta}$$

and hence that

$$(\cos \theta + i \sin \theta)^m = \cos m\theta + i \sin m\theta.$$

This is known as DeMoivre's formula. Suggestion: first let $m$ be non-negative and use induction.

7. Show that the complex function $e^{(-1+2i)x}$ is a solution of the differential equation

$$(D^2 + 2D + 5)y = 0.$$

Use this fact to find two linearly independent real solutions.

8. For each of the following differential equations find all numbers $r$, real and complex, such that $e^{rx}$ is a solution. Find two linearly independent real solutions.

   (a) $(D^2 + 9)y = 0$          (b) $(D^2 - 3D + 2)y = 0$
   (c) $(D^2 - 4D + 5)y = 0$     (d) $(D^2 + 4D + 5)y = 0$

9. Let $P(D) = D^2 - D + 5$. Find a complex solution of the equation

$$P(D)y = 10e^{2ix}$$

   of the form $y = Ae^{2ix}$, where $A$ is a complex constant. Use your answer to find a real solution of each of the equations

$$P(D)\, y = 10 \cos 2x, \qquad P(D)\, y = 10 \sin 2x.$$

10. Derive the laws of exponents (5.35) from the laws (5.26).

11. Find all solutions of the differential equation

$$x^2 y'' + xy' + 4y = 0$$

   on the interval $(0, \infty)$ that are of the form $y = x^c$, where $c$ may be complex. Use your answer to find two linearly independent real solutions.

12. Show that the set of all complex functions defined on an interval forms a vector space over the complex numbers.

## 5.4  EQUATIONS WITH CONSTANT COEFFICIENTS

By making use of complex exponential functions we can now solve any differential equation of the type

$$P(D)y = 0, \tag{5.38}$$

provided that the auxiliary polynomial $P$ does not have multiple zeros. If $a + ib$ and $a - ib$ are complex zeros of $P$, then

$$e^{(a+ib)x} = e^{ax}(\cos bx + i \sin bx),$$
$$e^{(a-ib)x} = e^{ax}(\cos bx - i \sin bx) \tag{5.39}$$

are complex solutions of Eq. (5.38). The real and imaginary parts

$$e^{ax} \cos bx, \qquad e^{ax} \sin bx \tag{5.40}$$

are real solutions. Thus, to every pair of complex conjugate zeros of $P$ there corresponds a pair of real solutions of Eq. (5.38), and to each real zero $c$ of $P$ there corresponds a real solution $e^{cx}$. It will be shown in this section later on that the set of all real solutions so obtained is linearly independent. First we consider some examples.

**Example 1**

$$y'' + a^2 y = 0. \tag{5.41}$$

The polynomial equation, $r^2 + a^2 = 0$, has the roots $ai$ and $-ai$. Hence

$$e^{iax} = \cos ax + i \sin ax$$

is a complex solution. The real and imaginary parts, $\cos ax$ and $\sin ax$, are real solutions. The general solution is

$$y = c_1 \cos ax + c_2 \sin ax. \tag{5.42}$$

Equation (5.41) is important for many applications. We rewrite formula (5.42) as

$$y = (c_1^2 + c_2^2)^{1/2}[p \cos ax + q \sin ax],$$

where

$$p = \frac{c_1}{(c_1^2 + c_2^2)^{1/2}}, \qquad q = \frac{c_2}{(c_1^2 + c_2^2)^{1/2}}.$$

Since $p^2 + q^2 = 1$, there exist angles $\alpha$ and $\beta$ such that

$$\cos \alpha = p, \qquad \sin \alpha = -q,$$
$$\sin \beta = p, \qquad \cos \beta = q.$$

Consequently the general solution is also described by either of the formulas

$$y = A \cos (ax + \alpha), \qquad y = B \sin(ax + \beta), \tag{5.43}$$

where $A$, $B$, $\alpha$, and $\beta$ are arbitrary constants. (See Exercise 28.)

**Example 2**

$$(D^3 - 5D^2 + 9D - 5)y = 0.$$

The polynomial

$$P(r) = r^3 - 5r^2 + 9r - 5 = (r - 1)(r^2 - 4r + 5)$$

has the zeros $r_1 = 1$, $r_2 = 2 + i$, and $r_2 = 2 - i$. Thus

$$e^{(2+i)x} = e^{2x}(\cos x + i \sin x)$$

is a complex solution and $e^{2x} \cos x$, $e^{2x} \sin x$ are real solutions. The general solution is

$$y = c_1 e^x + c_2 e^{2x} \cos x + c_3 e^{2x} \sin x.$$

Let us now consider the general case of an $n$th-order equation (5.38) where the polynomial $P$,

$$P(r) = a_0(r - r_1)(r - r_2) \cdots (r - r_n),$$

has $n$ distinct zeros. The functions

$$e^{r_1 x}, e^{r_2 x}, \dots, e^{r_n x}, \tag{5.44}$$

some of which may be complex, belong to the vector space of complex functions that are defined on $(-\infty, \infty)$. The linear independence of the functions (5.44) follows from a consideration of their Wronskian. We shall now show that the corresponding set of $n$ real solutions is a linearly independent set of elements of the space of real functions that are defined on $(-\infty, \infty)$. Suppose that $c_1, c_2, \dots, c_n$ are real numbers such that

$$c_1 e^{ax} \cos bx + c_2 e^{ax} \sin bx + \cdots = 0$$

for all $x$. Then

$$c_1' e^{(a+ib)x} + c_2' e^{(a-ib)x} + \cdots = 0$$

for all $x$, where

$$c_1 = c_1' + c_2', \qquad c_2 = i(c_1' - c_2').$$

But then $c_1' = c_2' = \cdots = 0$ so $c_1 = c_2 = \cdots = 0$. Hence the $n$ real solutions are linearly independent.

In order to treat the case where the polynomial $P$ has multiple zeros, we need the following result.

**Lemma**   If $r$ is any number, real or complex, and if the function $w$, which may be complex, has at least $m$ derivatives, then

$$(D - r)^m [e^{rx} w(x)] = e^{rx} D^m w(x). \tag{5.45}$$

This formula evidently holds when $m = 0$. When $m = 1$, we have

$$(D - r) [e^{rx} w(x)] = e^{rx} w'(x) + r e^{rx} w(x) - r e^{rx} w(x)$$
$$= e^{rx} D w(x).$$

The reader who understands mathematical induction should be able to establish that formula (5.45) holds for every nonnegative integer $m$.

Now suppose that $r_1$ is a zero of $P$ of multiplicity $k$, so that

$$P(r) = a_0 (r - r_1)^k (r - r_2) \cdots (r - r_{n-k+1}).$$

Then

$$P(D) = Q(D)(D - r_1)^k,$$

where $Q(D)$ is a polynomial operator of order $n - k$. We shall show that each of the $k$ functions

$$x^j e^{r_1 x}, \qquad 0 \le j \le k - 1, \tag{5.46}$$

is a solution of the equation $P(D) y = 0$. Using the lemma, we have

$$P(D)(x^j e^{r_1 x}) = Q(D)(D - r_1)^k x_j e^{r_1 x}$$
$$= Q(D) e^{r_1 x} D^k x^j$$
$$= 0$$

since $D^k x^j = 0$ when $j < k$. This proves our assertion. Of course if $r_1$ is complex, the functions (5.46) will be complex.

If $r_1 = a + ib$ is a complex zero of $P$ of multiplicity $k$, then $\bar{r}_1 = a - ib$ is also a zero of multiplicity $k$. Then each of the $2k$ functions

$$x^j e^{(a+ib)x}, \qquad x^j e^{(a-ib)x}, \qquad 0 \le j \le k - 1$$

is a complex solution of the differential equation $P(D) y = 0$. But since the real and imaginary parts of a complex solution are both real solutions, we know that each of the $2k$ real functions

$$x^j e^{ax} \cos bx, \qquad x^j e^{ax} \sin bx, \qquad 0 \le j \le k - 1$$

is a real solution.

Thus even when $P$ has complex and multiple zeros, we can still find $n$ seemingly distinct real solutions of the $n$th-order equation $P(D) y = 0$. Proofs that these $n$ solutions are linearly independent are given by Coddington (1961), Kaplan (1958), Kreider *et al.* (1968), and Rabenstein (1966). We summarize our results as follows.

**Theorem 5.2** If $r_1$ is a real root of multiplicity $k$ of the $n$th-degree polynomial equation $P(r) = 0$, then each of the $k$ functions

$$x^j e^{r_1 x}, \qquad 0 \le j \le k - 1 \qquad\qquad (5.47)$$

is a solution of the $n$th-order differential equation $P(D) y = 0$. If $a + ib$, $b \ne 0$, is a complex root of multiplicity $k$ (in which case $a - ib$ is also a root of multiplicity $k$) then each of the $2k$ real functions

$$x^j e^{ax} \cos bx, \qquad x^j e^{ax} \sin bx, \qquad 0 \le j \le k - 1 \qquad (5.48)$$

is a solution. The $n$ real solutions of the differential equation (corresponding to the $n$ zeros of the polynomial) that are described here are linearly independent on every interval.

**Example 3**

$$y'' + 4y' + 4y = 0.$$

Since $r^2 + 4r + 4 = (r + 2)^2$, we have $r_1 = r_2 = -2$. According to formula (5.47), $e^{-2x}$ and $xe^{-2x}$ are solutions. The general solution is

$$y = c_1 e^{-2x} + c_2 x e^{-2x}.$$

**Example 4**

$$D^2(D + 2)^3(D - 3)y = 0.$$

The polynomial equation $r^2(r + 2)^3(r - 3) = 0$ has roots $0, 0, -2, -2, -2$, and 3. Each of the functions

$$1, \quad x, \quad e^{-2x}, \quad xe^{-2x}, \quad x^2 e^{-2x}, \quad e^{3x}$$

is a solution of the differential equation. (Remember that $e^{0x} = 1$.) The general solution is

$$y = c_1 + c_2 x + (c_3 + c_4 x + c_5 x^2)e^{-2x} + c_6 e^{3x}.$$

**Example 5**

$$(D^4 + 8D^2 + 16)y = 0.$$

The auxiliary polynomial equation is

$$r^4 + 8r^2 + 16 = (r^2 + 4)^2 = 0,$$

with roots $2i$, $2i$, $-2i$, and $-2i$. Thus

$$e^{2ix}, \quad xe^{2ix}, \quad e^{-2ix}, \quad xe^{-2ix}$$

are complex solutions. The functions

$$\cos 2x, \quad x \cos 2x, \quad \sin 2x, \quad x \sin 2x$$

are real solutions and the general solution is

$$y = (c_1 + c_2 x) \cos 2x + (c_3 + c_4 x) \sin 2x.$$

Differential equations of certain classes can be transformed into equations with constant coefficients by means of a change of variable. One such class is considered in the next section.

## Exercises for Section 5.4

In Exercises 1–20, find the general solution of the differential equation.

1. $y'' - y' - 6y = 0$
2. $2y'' - 5y' + 2y = 0$
3. $y'' + 2y' = 0$
4. $y''' + 2y'' - y' - 2y = 0$
5. $y''' + 3y'' - 4y' = 0$
6. $y^{(4)} - 10y'' + 9y = 0$
7. $y'' + 2y' + y = 0$
8. $y'' - 6y' + 9y = 0$
9. $y''' - 6y'' + 12y' - 8y = 0$
10. $y''' + 5y'' + 3y' - 9y = 0$
11. $y''' + y'' = 0$
12. $(D - 1)^3(D + 2)^2 y = 0$
13. $y'' + 9y = 0$
14. $y'' + 2y' + 10y = 0$
15. $y'' - 6y' + 13y = 0$
16. $y''' + 2y'' + y' + 2y = 0$
17. $y^{(4)} + 2y'' + y = 0$
18. $(D^2 - 2D + 5)^2 y = 0$
19. $(D - 2)^2(D^2 + 2)y = 0$
20. $y^{(5)} + 4y''' = 0$

In Exercises 21–25, find the solution of the initial value problem.

**21.** $y'' - 4y' + 3y = 0$, $y(0) = -1$, $y'(0) = 3$

**22.** $y'' - 4y' + 4y = 0$, $y(0) = 2$, $y'(0) = 1$

**23.** $y'' + 4y = 0$, $y(\pi) = 1$, $y'(\pi) = -4$

**24.** $y'' + 2y' + 2y = 0$, $y(0) = 2$, $y'(0) = -3$

**25.** $y''' + y'' = 0$, $y(0) = 2$, $y'(0) = 1$, $y''(0) = -1$

**26.** Show that every solution of the differential equation $P(D)\, y = 0$ tends to zero as $x$ becomes infinite if and only if all the roots of the polynomial equation $P(r) = 0$ have negative real parts.

**27.** Show that the general solution of the equation $y'' - a^2 y = 0$, where $a$ is a constant, can be written either as

$$y = c_1 e^{ax} + c_2 e^{-ax}$$

or as

$$y = C_1 \cosh ax + C_2 \sinh ax.$$

**28.** Show that the general solution of the equation $y'' + a^2 y = 0$, where $a$ is a constant, can be written in the forms (5.43).

**29.** Find a linear homogeneous differential equation with real constant coefficients, whose order is as low as possible, that has the given function as a solution.

(a) $xe^{-2x}$                       (b) $x - e^{3x}$

(c) $\cos 2x$                   (d) $e^x \sin 2x$

(e) $x \sin 3x$                (f) $\cos 2x + 3e^{-x}$

## 5.5 CAUCHY–EULER EQUATIONS

A linear differential equation of the form

$$(b_0 x^n D^n + b_1 x^{n-1} D^{n-1} + \cdots + b_{n-1} x D + b_n)y = 0, \qquad (5.49)$$

where $b_0, b_1, \ldots, b_n$ are constants, is known as a *Cauchy–Euler equation*, or as an *equidimensional equation*. Examples of Cauchy–Euler equations are

$$x^2 y'' - 3xy' + 4y = 0, \qquad x^3 y''' + 2xy' = 0.$$

Equations of this type can be transformed into equations with constant coefficients by means of a change of independent variable.

Let

$$x = e^t, \qquad t = \ln x, \tag{5.50}$$

where as $t$ varies over the set of all real numbers, $x$ varies over the interval $(0, \infty)$. In what follows we assume that $x > 0$. The case $x < 0$ is treated in Exercise 19. It will be convenient to use the operator notation

$$\mathscr{D} = \frac{d}{dt}.$$

Since $dt/dx = 1/x$, we have

$$Du = \frac{du}{dx} = \frac{du}{dt}\frac{dt}{dx} = \frac{1}{x}\mathscr{D}u$$

for every differentiable function $u$. That is,

$$D = \frac{1}{x}\mathscr{D}.$$

Then

$$xDy = x\left(\frac{1}{x}\mathscr{D}\right)y = \mathscr{D}y$$

and

$$x^2 D^2 y = x^2 D\left(\frac{1}{x}\mathscr{D}y\right)$$

$$= x^2\left[D\left(\frac{1}{x}\right)\mathscr{D}y + \frac{1}{x}D(\mathscr{D}y)\right].$$

According to the chain rule,

$$D(\mathscr{D}y) = \mathscr{D}(\mathscr{D}y)\frac{1}{x}$$

so that

$$x^2 D^2 y = x^2\left[-\frac{1}{x^2}\mathscr{D}y + \frac{1}{x^2}\mathscr{D}^2 y\right]$$

$$= (\mathscr{D}^2 - \mathscr{D})y$$

$$= \mathscr{D}(\mathscr{D} - 1)y.$$

By the use of mathematical induction, it can be shown that

$$x^m D^m y = \mathscr{D}(\mathscr{D} - 1)(\mathscr{D} - 2) \cdots (\mathscr{D} - m + 1)y \tag{5.51}$$

for every positive integer $m$.

Using formula (5.51) we see that the original equation (5.49) becomes

$$[b_0 \mathscr{D}(\mathscr{D} - 1) \cdots (\mathscr{D} - n + 1) + b_1 \mathscr{D}(\mathscr{D} - 1) \cdots (\mathscr{D} - n + 2)$$
$$+ \cdots + b_{n-1}\mathscr{D} + b_n]y = 0. \tag{5.52}$$

This equation with constant coefficients has the auxiliary polynomial $Q$, where

$$Q(r) = b_0 r(r - 1) \cdots (r - n + 1) + b_1 r(r - 1) \cdots (r - n + 2)$$
$$+ \cdots + b_{n-1}r + b_n. \tag{5.53}$$

This polynomial is of degree $n$. If $r_1$ is a real zero of multiplicity $k$, then each of the $k$ functions

$$t^j e^{r_1 t}, \qquad 0 \le j \le k - 1 \tag{5.54}$$

is a solution of Eq. (5.52) and each of the functions

$$(\ln x)^j x^{r_1}, \qquad 0 \le j \le k - 1 \tag{5.55}$$

are solutions of the original equation (5.49). If $a + ib$ and $a - ib$ are zeros of $Q$ of multiplicity $k$, the functions

$$t^j e^{at} \cos bt, \qquad t^j e^{at} \sin bt, \qquad 0 \le j \le k - 1 \tag{5.56}$$

are solutions of Eq. (5.52). The corresponding solutions of the original equation are

$$x^a(\ln x)^j \cos(b \ln x), \qquad x^a(\ln x)^j \sin(b \ln x), \qquad 0 \le j \le k - 1. \tag{5.57}$$

**Example 1**
$$2x^2 y'' - 5xy' + 3y = 0.$$

The change of variable $x = e^t$ leads to the equation

$$[2\mathscr{D}(\mathscr{D} - 1) - 5\mathscr{D} + 3]y = 0$$

or

$$(2\mathscr{D} - 1)(\mathscr{D} - 3)y = 0.$$

Hence

$$y = c_1 e^{t/2} + c_2 e^{3t}$$

or

$$y = c_1 x^{1/2} + c_2 x^3, \quad x > 0.$$

**Example 2**

$$x^2 y'' - xy' + 5y = 0.$$

This equation becomes

$$[\mathscr{D}(\mathscr{D} - 1) - \mathscr{D} + 5]y = 0$$

or

$$(\mathscr{D}^2 - 2\mathscr{D} + 5)y = 0.$$

The roots of the auxiliary equation are $1 + 2i$ and $1 - 2i$. Thus

$$y = c_1 e^t \cos 2t + c_2 e^t \sin 2t$$

or

$$y = c_1 x \cos(2 \ln x) + c_2 x \sin (2 \ln x).$$

The equation

$$Ly \equiv (b_0 x^n D^n + b_1 x^{n-1} D^{n-1} + \cdots + b_{n-1} xD + b_n)y = 0$$

can be solved more directly by attempting to find solutions of the form $y = x^r$, without any change of variable. Observing that

$$D^k x^r = r(r - 1)(r - 2) \cdots (r - k + 1) x^{r-k}$$

and

$$x^k D^k x^r = r(r - 1)(r - 2) \cdots (r - k + 1) x^r$$

we have

$$L(x^r) = [b_0(r-1)\cdots(r-n+1) + b_1r(r-1)\cdots(r-n+2)$$
$$+ \cdots + b_{n-1}r + b_n]\, x^r$$

or

$$L(x^r) = Q(r)\, x^r,$$

where $Q$ is the polynomial (5.53). If $r_1$ is a real zero of $Q$ of multiplicity $k$, the functions

$$x^{r_1}(\ln x)^j, \quad 0 \le j \le k-1$$

are solutions. If $a + ib$ and $a - ib$ are zeros of multiplicity $k$, then the functions

$$x^a(\ln x)^j \cos(b \ln x), \qquad x^a(\ln x)^j \sin(b \ln x), \quad 0 \le j \le k-1$$

are solutions.

**Example 3**

$$x^3 y''' - x^2 y'' + xy' = 0.$$

Setting $y = x^r$, we must have

$$r(r-1)(r-2) - r(r-1) + r = 0$$

or

$$r(r-2)^2 = 0.$$

The general solution is

$$y = c_1 + (c_2 + c_3 \ln x)x^2.$$

### Exercises for Section 5.5

In Exercises 1–12, find the general solution of the differential equation if $x$ is restricted to the interval $(0, \infty)$.

1. $x^2 y'' - 2y = 0$
2. $x^2 y'' + 3xy' - 3y = 0$

3.  $3xy'' + 2y' = 0$

4.  $x^3 y''' + x^2 y'' - 2xy' + 2y = 0$

5.  $4x^2 y'' + y = 0$

6.  $x^2 y'' - 3xy' + 4y = 0$

7.  $xy''' + 2y'' = 0$

8.  $x^3 y''' + 6x^2 y'' + 7xy' + y = 0$

9.  $x^2 y'' + xy' + 4y = 0$

10. $x^2 y'' - 5xy' + 13y = 0$

11. $x^3 y''' + 2x^2 y'' + xy' - y = 0$

12. $x^4 y^{(4)} + 6x^3 y''' + 15x^2 y'' + 9xy' + 16y = 0$

In Exercises 13–15, find the solution of the initial value problem on the interval $(0, \infty)$.

13. $x^2 y'' + 4xy' + 2y = 0,$    $y(1) = 1,$   $y'(1) = 2$

14. $x^2 y'' - 3xy' + 4y = 0,$    $y(1) = 2,$   $y'(1) = 1$

15. $x^2 y'' + xy' + 4y = 0,$    $y(1) = 1,$   $y'(1) = 4$

In Exercises 16–18, find all solutions on the interval $(0, \infty)$ that have a finite limit as $x$ tends to zero.

16. $4x^2 y'' + 4xy' - y = 0$    17. $x^2 y'' + 2xy' - 2y = 0$

18. $x^2 y'' + 6xy' + 6y = 0$

19. If the function $f$ is a solution of Eq. (5.49) on the interval $(0, \infty)$ show that the function $g$, where $g(x) = f(-x)$, is a solution on $(-\infty, 0)$.

20. Show that the change of variable $t = ax + b$ transforms the equation

$$b_0(ax + b)^2 y'' + b_1(ax + b)y' + b_2 y = 0$$

into a Cauchy–Euler equation.

21. Use the result of Exercise 20 to find the general solution of the given equation.

    (a)  $(x - 3)^2 y'' + 3(x - 3)y' + y = 0,$   $x > 3$

    (b)  $(2x + 1)^2 y'' + 4(2x + 1)y' - 24y = 0,$   $x > -1/2$

22. What conditions must the zeros of the polynomial $Q$ satisfy in order that every solution of Eq. (5.49) tend to zero as

    (a)  $x$ approaches zero through positive values?

    (b)  $x$ becomes positively infinite?

23. If the functions $f_1, f_2, \dots, f_n$ are linearly independent on the set of all real numbers and if $g_i(x) = f_i(\ln x)$ for $x > 0$, show that the functions $g_i$ are linearly independent on $(0, \infty)$.

## 5.6 NONHOMOGENEOUS EQUATIONS

The linear equation

$$Ly = F,$$ (5.58)

where

$$L = a_0 D^n + a_1 D^{n-1} + \cdots + a_{n-1} D + a_n$$ (5.59)

(the functions $a_i$ need not be constants) is called *nonhomogeneous* if $F$ is not the zero function. Associated with the nonhomogeneous equation is the homogeneous equation

$$Ly = 0.$$ (5.60)

It turns out that we can solve the nonhomogeneous equation if we can solve the homogeneous equation and if we can also find just one particular solution of the nonhomogeneous equation.

**Theorem 5.3**  Let $u_1, u_2, \ldots, u_n$ be linearly independent solutions of the homogeneous equation (5.60) on an interval $\mathscr{I}$ [2] and let $u_p$ be any particular solution of the nonhomogeneous equation (5.58) on $\mathscr{D}$. Then the set of all solutions of equation (5.58) on $\mathscr{I}$ consists of all functions of the form

$$c_1 u_1 + c_2 u_2 + \cdots + c_n u_n + u_p,$$ (5.61)

where $c_1, c_2, \ldots, c_n$ are constants.

PROOF  First let us verify that every function of the form (5.61) is a solution of Eq. (5.58). Since $Lu_i = 0$, $1 \le i \le n$, and $Lu_p = F$, we have

$$L(c_1 u_1 + c_2 u_2 + \cdots + c_n u_n + u_p) = c_1 Lu_1 + c_2 Lu_2 + \cdots + c_n Lu_n + Lu_p$$
$$= 0 + 0 + \cdots + 0 + F$$
$$= F$$

which we wished to show. Next we must show that every solution of Eq. (5.58) is of the form (5.61). Let $u$ be any solution. Then $Lu = F$. Since also $Lu_p = F$ we have

$$L(u - u_p) = Lu - Lu_p = F - F = 0.$$

---

[2] The functions $a_i$ and $F$ are assumed to be continuous on $\mathscr{I}$ with $a_0$ never zero on $\mathscr{I}$.

Thus the function $u - u_p$ is a solution of the homogeneous equation $Ly = 0$. By Theorem 5.1, $u - u_p$ must be of the form

$$u - u_p = c_1 u_1 + c_2 u_2 + \cdots + c_n u_n .$$

Then

$$u = c_1 u_1 + c_2 u_2 + \cdots + c_n u_n + u_p ,$$

which we wished to show.

To illustrate the use of this theorem, let us consider the equation

$$y'' - 4y' + 4y = 9e^{-x} .$$

It is easy to verify that a particular solution is $u_p(x) = e^{-x}$. The associated homogeneous equation has $e^{2x}$ and $xe^{2x}$ as linearly independent solutions. Hence the general solution of the nonhomogeneous equation is

$$y = (c_1 + c_2 x)e^{2x} + e^{-x} .$$

In case the associated homogeneous equation has constant coefficients, or is of the Cauchy–Euler type, we can solve it. There remains the problem of finding one solution $u_p$ of the nonhomogeneous equation. A method that applies in certain cases is described in the next section. A more general method is discussed in Section 5.8. In finding particular solutions, the following result is often useful.

**Theorem 5.4**   If $u_p$ and $v_p$ are solutions of the equations $Ly = f$ and $Ly = g$, respectively, then $u_p + v_p$ is a solution of the equation $Ly = f + g$.

PROOF   Since $Lu_p = f$ and $Lv_p = g$, we have

$$L(u_p + v_p) = Lu_p + Lv_p = f + g .$$

For example, suppose that $u_p$ and $v_p$ are solutions of the equations

$$Ly = 3 e^x, \qquad Ly = -2 \sin x ,$$

respectively. Then $u_p + v_p$ is a solution of the equation

$$Ly = 3e^x - 2 \sin x .$$

### Exercises for Section 5.6

1.  Verify that $u_p$, where $u_p(x) = \sin 2x$, is a solution of the differential equation

$$y'' - y = -5 \sin 2x.$$

Use this fact to find the general solution of the equation.

2. Show that the nonhomogeneous equation $P(D) y = ce^{ax}$ has a solution of the form $y = Ae^{ax}$ if and only if $P(a) \neq 0$. Show that in this case the solution is $y = ce^{ax}/P(a)$.

3. Use the result of Exercise 2 to find the general solution of the given differential equation.

   (a) $y' - 2y = 6e^{5x}$       (b) $y'' - 2y' + y = -9e^{-2x}$

   (c) $y'' + y = 4e^{x}$          (d) $(D + 1)(D + 2)(D + 3)y = 6e^{-4x}$

4. If $a_n$ is a nonzero constant and $c$ is a constant, show that a solution of the equation
$$a_0 y^{(n)} + a_1 y^{(n-1)} + \cdots + a_n y = c$$
is $y = c/a_n$.

5. Show that the equation
$$b_0 x^2 y'' + b_1 xy' + b_2 y = c\, x^a$$
has a solution of the form $y = Ax^a$ provided that
$$b_0 a(a - 1) + b_1 a + b_2 \neq 0.$$

6. If $F$ belongs to the kernel of the operator $Q(D)$, show that every solution of the nonhomogeneous equation $P(D) y = F$ is a solution of the homogeneous equation $Q(D) P(D) y = 0$.

7. Use the result of Exercise 2 and Theorem 5.4 to solve the given differential equation.

   (a) $y'' - 4y' + 3y = -2e^{2x} + 8e^{-x}$

   (b) $y'' + 3y' = 4e^{-x} - 2e^{-2x}$

8. Let $T$ be a linear transformation from a vector space $V$ into a vector space $W$. For a given element $\mathbf{w}$ of $W$ consider the equation $T\mathbf{v} = \mathbf{w}$. If $\mathbf{v}_p$ is a solution, show that the general solution consists of all elements of $V$ of the form $\mathbf{z} + \mathbf{v}_p$, where $\mathbf{z}$ belongs to the kernel of $T$. Explain how Theorem 5.3 can be regarded as a special case of this result.

## 5.7 THE METHOD OF UNDETERMINED COEFFICIENTS

In this section we describe a method that yields a particular solution of the nonhomogeneous equation $Ly = F$ when the following two conditions are both met.

   (a)  The operator $L$ has constant coefficients.

   (b)  The function $F$ is itself a solution of some linear homogeneous differential equation with constant coefficients.

Here $F$ must consist of a linear combination of functions of the types

$$x^j, \quad x^j e^{cx}, \quad x^j e^{ax} \cos bx, \quad x^j e^{ax} \sin bx. \tag{5.62}$$

Actually, according to Theorem 5.4, we can concentrate on the case where $F$ is a constant multiple of just *one* of these functions. For instance, to find a solution of the equation

$$Ly = 5e^x \sin 2x - 4x^2 e^{-x} \tag{5.63}$$

we first find solutions $u$ and $v$ of the equations

$$Ly = 5e^x \sin 2x, \quad Ly = -4x^2 e^{-x},$$

respectively. Then $u + v$ will be a solution of Eq. (5.63).

   Also, if $w = u + iv$ is a complex solution of the equation

$$Ly = Ax^j e^{(a+ib)x},$$

then $u$ and $v$ will be solutions of the equations

$$Ly = Ax^j e^{ax} \cos bx, \quad Ly = Ax^j e^{ax} \sin bx,$$

respectively.

   We therefore consider an $n$th-order equation

$$P(D)y = F, \tag{5.64}$$

where $F$ is of the form

$$F(x) = Ax^j e^{cx},$$

and where $c$ may be real or complex. (In particular, $c$ may be zero.) Then there exists a polynomial operator $Q(D)$, with real coefficients, such that

$$Q(D) F = 0.$$

We say that the operator $Q(D)$ *annihilates* $F$. If $c$ is real, we may take

$$Q(D) = (D - c)^{j+1}.$$

If $c = a + ib$, $b \neq 0$, we may take

$$Q(D) = [(D - c)(D - \bar{c})]^{j+1} = [(D - a)^2 + b^2]^{j+1}.$$

With these choices, the order of $Q(D)$ is as low as possible.

Suppose that the order of $Q(D)$ is $m$. If we operate on both members of Eq. (5.64) with $Q(D)^3$ we see that every solution of Eq. (5.64) is also a solution of the homogeneous equation

$$Q(D) P(D) y = 0. \tag{5.65}$$

(However, not every solution of Eq. (5.65) need be a solution of Eq. (5.64).) The order of this equation is $m + n$. Every solution of the equation

$$P(D) y = 0 \tag{5.66}$$

is a solution of Eq. (5.65), but the latter equation also possesses additional solutions. Let the general solution of Eq. (5.65) be

$$A_1 u_1 + \cdots + A_m u_m + B_1 v_1 + \cdots + B_n v_n,$$

where the functions $u_i$ and $v_i$ are of the types (5.62) and the functions $v_i$ are linearly independent solutions of Eq. (5.66).

Since every solution of the nonhomogeneous equation (5.64) is also of the above form, it must be possible to choose the constants $A_i$ and $B_i$ in such a way that

$$P(D)(A_1 u_1 + \cdots + A_m u_m + B_1 v_1 + \cdots + B_n v_n) = F.$$

Since

$$P(D)(B_1 v_1 + \cdots + B_n v_n) = 0$$

for every choice of the $B_i$, it must be possible to find constants $A_i$ such that

$$A_1 u_1 + \cdots + A_m u_m \tag{5.67}$$

is a solution of Eq. (5.64). If the functions $u_i$ are known, the constants $A_i$ can be determined by substituting the expression (5.67) into the differential equation (5.64) and requiring that the latter be satisfied identically. The expression (5.67) is called a *trial solution* for Eq. (5.64).

---

[3] Every solution of Eq. (5.64) possesses derivatives of all orders, so the derivatives of $y$ in $Q(D) P(D) y$ all exist.

Let us pause to consider some specific cases.

**Example 1**

$$y'' - y' - 2y = 20e^{4x}.$$

This equation may be written as

$$(D + 1)(D - 2)y = 20e^{4x}. \tag{5.68}$$

The operator $D - 4$ annihilates $e^{4x}$. Operating on both sides of equation (5.68) with $D - 4$, we see that every solution of Eq. (5.68) is also a solution of the homogeneous equation

$$(D - 4)(D + 1)(D - 2)y = 0.$$

The solutions of this equation are of the form

$$y = Ae^{4x} + B_1e^{-x} + B_2e^{2x},$$

and hence every solution of Eq. (5.68) is of this form. Thus there exist constants $A$, $B_1$, and $B_2$ such that

$$(D + 1)(D - 2)(Ae^{4x} + B_1e^{-x} + B_2e^{2x}) = 20e^{4x}.$$

Since $e^{-x}$ and $e^{2x}$ are solutions of the homogeneous equation

$$(D + 1)(D - 2)y = 0,$$

it must be possible to choose $A$ so that

$$(D + 1)(D - 2)(Ae^{4x}) = 20e^{4x}.$$

This yields the requirement

$$(D^2 - D - 2)(Ae^{4x}) = 20e^{4x},$$

$$10Ae^{4x} = 20e^{4x},$$

or

$$A = 2.$$

Hence a particular solution of Eq. (5.68) is $y_p(x) = 2e^{4x}$. The general solution is

$$y = c_1 e^{-x} + c_2 e^{2x} + 2 e^{4x},$$

where $c_1$ and $c_2$ are arbitrary constants.

**Example 2**

$$y'' - 2y' + y = 6 \sin x. \tag{5.69}$$

We consider instead the equation

$$(D - 1)^2 y = 6e^{ix}. \tag{5.70}$$

The operator $(D - i)(D + i) = D^2 + 1$ annihilates $e^{ix}$. Operating on both sides of Eq. (5.70) with $D^2 + 1$, we see that every solution of Eq. (5.70) is also a solution of the equation

$$(D^2 + 1)(D - 1)^2 y = 0.$$

The solutions of this equation that are not solutions of the homogeneous equation $(D-1)^2 y = 0$ are $e^{ix}$ and $e^{-ix}$. Hence equation (5.70) has a solution of the form $Ae^{ix} + Be^{-ix}$. But $B$ is destined to be zero (since $e^{ix}$ and $e^{-ix}$ are linearly independent complex functions) so we take as our trial solution

$$y_p(x) = Ae^{ix}.$$

Substituting in Eq. (5.70), we obtain the requirement

$$(D^2 - 2D + 1)(Ae^{ix}) = 6e^{ix},$$
$$-2Aie^{ix} = 6e^{ix},$$

or

$$A = 3i.$$

Since the imaginary part of

$$y_p(x) = 3ie^{ix} = 3i(\cos x + i \sin x)$$

is $3 \cos x$, a particular solution of the original equation (5.69) is $3 \cos x$. The general solution is

$$y = (c_1 + c_2 x)e^x + 3 \cos x.$$

We now return to the general case and seek to determine the nature of the functions $u_i$ in the trial solution (5.67). These functions are the solutions of the equation $Q(D) P(D) y = 0$ that are not solutions of the equation $P(D) y = 0$. If $P$ and $Q$ have no common zeros, the functions $u_i$ are simply the solutions of the equation $Q(D) y = 0$. Hence our trial solution for the equation

$$P(D) y = Ax^j e^{cx} \tag{5.71}$$

is of the form[4]

$$y_p(x) = (A_0 + A_1 x + \cdots + A_j x^j)e^{cx}. \tag{5.72}$$

Next, suppose that $c$ is a zero of $P$, of multiplicity $m$. Then

$$e^{cx}, xe^{cx}, \ldots, x^{m-1}e^{cx} \tag{5.73}$$

are solutions of the equation $P(D) y = 0$. Since $c$ is a zero of $QP$ of multiplicity $m + j + 1$, the functions

$$e^{cx}, xe^{cx}, \ldots, x^{m+j}e^{cx} \tag{5.74}$$

are solutions of the equation $Q(D) P(D) y = 0$. Selecting those functions in the set (5.74) that are not in the set (5.73), we set that our trial solution takes the form

$$y_p(x) = x^m(A_0 + A_1 x + \cdots + A_j x^j)e^{cx}. \tag{5.75}$$

We may formulate the following rule for finding a particular solution of Eq. (5.71). Our tentative trial solution is given by formula (5.72). But if $x^{m-1}e^{cx}$ is a solution of the homogeneous equation $P(D) y = 0$, and $m$ is the largest integer for which this is true, then this tentative trial solution must be modified by multiplying each term by $x^m$, as in formula (5.75). In any case, the constants in the expression (5.72) or (5.75) are determined by substituting in Eq. (5.71) and requiring that it be satisfied identically.

We now consider several more examples.

**Example 3**

$$y'' - 3y' + 2y = 6e^{3x}.$$

---

[4] If $c$ is not real, it can be shown that no function of the form $x^m e^{\bar{c}x}$ is a solution of Eq. (5.71).

The general solution of the associated homogeneous equation

$$y'' - 3y' + 2y = 0$$

is

$$c_1 e^x + c_2 e^{2x}.$$

The tentative trial solution is

$$y_p = Ae^{3x}.$$

Since no function of the form $x^m e^{3x}$ is a solution of the homogeneous equation, the trial solution need not be modified. The constant $A$ is determined by the requirement that

$$(D^2 - 3D + 2)(Ae^{3x}) = 6e^{3x}$$

or

$$2Ae^{3x} = 6e^{3x}.$$

Thus $A = 3$ and a particular solution of the nonhomogeneous equation is

$$y_p = 3e^{3x}.$$

The general solution of the nonhomogeneous equation is

$$y = c_1 e^x + c_2 e^{2x} + 3e^{3x}.$$

**Example 4**

$$y'' - 4y' + 4y = 12xe^{2x}.$$

The general solution of the associated homogeneous equation is

$$c_1 e^{2x} + c_2 xe^{2x}.$$

The tentative trial solution is

$$y_p = (A + Bx)e^{2x}.$$

But since $xe^{2x}$ is a solution of the homogeneous equation, we must take

$$y_p = (Ax^2 + Bx^3)e^{2x}.$$

Differentiation shows that

$$y_p' = [2Ax + (2A + 3B)x^2 + 2Bx^3]e^{2x}$$
$$y_p'' = [2A + (8A + 6B)x + (4A + 12B)x^2 + 4Bx^3]e^{2x}.$$

Calculation shows that

$$(D^2 - 4D + 4)y_p = (2A + 6Bx)\, e^{2x}$$

and this must be equal to $12xe^{2x}$ if $y_p$ is to be a solution of the original equation. Hence we take $A = 0$ and $B = 2$. Then

$$y_p = 2x^3 e^{2x}$$

and the general solution is

$$y = (c_1 + c_2 x)e^{2x} + 2x^3 e^{2x}.$$

**Example 5**

$$y'' + y' - 2y = 4 \sin 2x.$$

We consider instead the equation

$$y'' + y' - 2y = 4e^{2ix}.$$

By taking the imaginary part of a solution of this equation, we obtain a real solution of the original equation. The general solution of the homogeneous equation is $c_1 e^x + c_2 e^{-2x}$. Our tentative trial solution is

$$y_p = Ae^{2x}.$$

Since the homogeneous equation has no solution of the form $x^m e^{2ix}$, this is correct as is. We have

$$y_p' = 2iAe^{2ix}, \qquad y_p'' = -4Ae^{2ix},$$

and

$$(D^2 + D - 2)y_p = (-6 + 2i)Ae^{2ix}.$$

We require that

$$(-6 + 2i)Ae^{2ix} = 4e^{2ix}$$

or

$$A = \frac{-2}{3 - i} = -\frac{3 + i}{5}.$$

Then

$$y_p = -\frac{3 + i}{5} e^{2ix} = -\frac{3 + i}{5}(\cos 2x + i \sin 2x).$$

The real and imaginary parts of $y_p$ are

$$u_p = -\tfrac{3}{5}\cos 2x + \tfrac{1}{5}\sin 2x, \qquad v_p = -\tfrac{1}{5}\cos 2x - \tfrac{3}{5}\sin 2x,$$

respectively. The general solution of the equation

$$(D^2 + D - 2)y = 4 \sin 2x$$

is

$$y = c_1 e^x + c_2 e^{-2x} - \tfrac{1}{5}\cos 2x - \tfrac{3}{5}\sin 2x.$$

As a by-product, we have also found that the general solution of the equation

$$(D^2 + D - 2)y = 4 \cos 2x$$

is

$$y = c_1 e^x + c_2 e^{-2x} - \tfrac{3}{5}\cos 2x + \tfrac{1}{5}\sin 2x.$$

**Example 6**

$$y''' + y' = 4 \cos x.$$

We consider the equation

$$y''' + y' = 4e^{ix}$$

and take the real part of a particular solution. The general solution of the homogeneous equation is

$$c_1 + c_2 \cos x + c_3 \sin x.$$

Since $e^{ix}$ is a solution of the homogeneous equation, we take

$$y_p = Axe^{ix}$$

as our trial solution. We find that

$$y'_p = A(ix + 1)e^{ix}, \qquad y''_p = A(-x + 2i)e^{ix}, \qquad y'''_p = A(-ix - 3)e^{ix}$$

and

$$(D^3 + D)y_p = -2Ae^{ix}.$$

Taking $A = -2$, we have

$$y_p = -2xe^{ix} = -2x(\cos x + i \sin x).$$

Then $-2x \cos x$ is a real solution of our original equation. The general solution is

$$y = c_1 + c_2 \cos x + c_3 \sin x - 2x \cos x.$$

**Example 7**

$$x^2 y'' - 2xy' + 2y = 6 \ln x, \quad x > 0.$$

The change of variable $x = e^t$ leads to the equation

$$(\mathscr{D} - 1)(\mathscr{D} - 2)y = 6t,$$

where $\mathscr{D} = d/dt$. The general solution of the homogeneous equation with constant coefficients is

$$y = c_1 e^t + c_2 e^{2t}.$$

The tentative trial solution is

$$y_p = A + Bt.$$

Since the homogeneous equation has no solution of the form $t^m$, this is correct as it stands. Calculation shows that $A = \frac{9}{2}$, $B = 3$. The general solution of the equation with constant coefficients is

$$y = c_1 e^t + c_2 e^{2t} + 3t + \tfrac{9}{2}.$$

The general solution of the original equation, obtained by setting $t = \ln x$, is

$$y = c_1 x + c_2 x^2 + 3 \ln x + \tfrac{9}{2}.$$

## Exercises for Section 5.7

In Exercises 1–20, find the general solution of the differential equation. If initial conditions are given, also find the solution that satisfies those conditions.

1. $y'' + 2y' - 3y = 5e^{2x}$,   $y(0) = 5$,   $y'(0) = 2$
2. $y'' + 4y' + 4y = 3e^{-x}$,   $y(0) = 3$,   $y(0) = 1$
3. $y'' + 3y' + 2y = 36xe^x$
4. $y'' + y' - 2y = 6e^{-x} + 4e^{-3x}$,   $y(0) = -1$,   $y'(0) = 1$
5. $y'' + 3y' + 2y = 20 \cos 2x$,   $y(0) = -1$,   $y'(0) = 6$
6. $y'' + y = 5e^x \sin x$
7. $y'' - y' - 6y = 2$
8. $y'' + 3y' + 2y = 4x^2$
9. $(D-1)^2(D+1)y = 10 \cos 2x$
10. $y'' + 2y' + y = -4e^{-3x} \sin 2x$
11. $y'' + 3y' + 2y = 5e^{-2x}$
12. $y'' - y = 4e^x - 3e^{2x}$
13. $y'' - y' - 2y = -6xe^{-x}$
14. $y'' + 4y' + 4y = 4e^{-2x}$
15. $y'' + 4y' + 3y = 6x^2 e^{-x}$
16. $y'' + 2y' = -4$
17. $y'' + y' = 3x^2$
18. $y'' + y = 4 \sin x$
19. $y'' + 2y' + 5y = 4e^{-x} \cos 2x$
20. $y'' + 4y = 16x \sin 2x$
21. Show that a particular solution of the nonhomogeneous equation

$$P(D) y = Ae^{cx}$$

is

$$y_p = \frac{A}{P(c)} e^{cx}$$

provided that $c$ is not a root of the equation $P(r) = 0$.

**22.** Use the result of Exercise 21 to find the general solution of the differential equation.

(a)  $y'' - 3y' + 2y = 6e^{-x}$    (b)  $y'' - 2y' + y = -3e^{2x}$

(c)  $y'' - y' - 2y = 10 \cos x$    (d)  $y'' - 3y' + 2y = 4e^{3x} + 6e^{-x}$

(e)  $(D - 1)^2(D + 1)y = 9e^{2x}$    (f)  $y'' + y' = 6 \sin 2x$

**23.** If the number $c$ is an $m$-fold root of the polynomial equation $P(r) = 0$ then $P(r) = Q(r)(r - c)^m$, where $Q(c) \neq 0$. Show that in this case the differential equation

$$P(D) y = Ae^{cx}$$

possesses the solution

$$y = \frac{A}{m! \, Q(c)} \, x^m e^{cx}.$$

**24.** Use the result of Exercise 23 to find the general solution of the differential equation.

(a)  $y'' - y' - 2y = 6e^{2x}$    (b)  $y'' - 4y' + 4y = 4e^{2x}$

(c)  $y'' + y = 4 \cos x$    (d)  $(D + 1)(D - 2)^3 y = 6e^{2x}$

In Exercises 25–30, find the general solution of the differential equation. Assume that the independent variable is restricted to the interval $(0, \infty)$.

**25.**  $x^2 y'' - 6y = 6x^4$    **26.**  $x^2 y'' + xy' - y = 9x^2 \ln x$

**27.**  $x^2 y'' - 3xy' + 3y = -6$    **28.**  $x^2 y'' - xy' = -4$

**29.**  $x^2 y'' + 2xy' - 2y = 6x$    **30.**  $x^2 y'' - xy' + y = 6x \ln x$

## 5.8   VARIATION OF PARAMETERS

The method of undetermined coefficients discussed in the last section allows us to find a particular solution of the nonhomogeneous equation $Ly = F$ only in special cases. The operator $L$ must have constant coefficients, and $F$ must belong to a certain class of functions. The method of variation of parameters is more general. It gives a formula for a particular solution provided only that the general solution of the homogeneous equation $Ly = 0$ is known. The operator $L$ need not have constant coefficients and there is no restriction (other than continuity) on $F$. However, the method of undetermined coefficients is usually easier to use when that method applies. The method of variation of parameters is valuable for theoretical purposes, as well as for actually finding solutions of equations. Some examples of its use are given in the exercises at the end of this section, and also in those at the end of Section 5.11.

In the method of variation of parameters, we assume that $n$ linearly independent solutions $u_1, u_2, \ldots, u_n$ of the homogeneous equation

$$a_0(x)\, y^{(n)} + a_1(x)\, y^{(n-1)} + \cdots + a_{n-1}(x)\, y' + a_n(x)\, y = 0 \qquad (5.76)$$

are known. We attempt to find a solution of the nonhomogeneous equation

$$a_0(x)\, y^{(n)} + a_1(x)\, y^{(n-1)} + \cdots + a_{n-1}(x)\, y' + a_n(x)\, y = F(x) \qquad (5.77)$$

that is of the form

$$y = C_1(x)\, u_1(x) + C_2(x)\, u_2(x) + \cdots + C_n(x)\, u_n(x), \qquad (5.78)$$

where the functions $C_1, C_2, \ldots, C_n$ are to be determined. If we simply calculate the first $n$ derivatives of the expression (5.78) and substitute them into the differential equation (5.77), we shall obtain one condition to be satisfied by the $n$ functions. We shall impose $n - 1$ other conditions *en route*. Differentiating once, we have

$$y' = (C_1 u_1' + \cdots + C_n u_n') + (C_1' u_1 + \cdots + C_n' u_n). \qquad (5.79)$$

We now impose the requirement

$$C_1' u_1 + \cdots + C_n' u_n = 0. \qquad (5.80)$$

This simplifies expression (5.79) for the first derivative since it now becomes

$$y' = C_1 u_1' + \cdots + C_n u_n'. \qquad (5.81)$$

We have also obtained one condition to be satisfied by the functions $C_i$. Differentiating again, we have

$$y'' = (C_1 u_1'' + \cdots + C_n u_n'') + (C_1' u_1' + \cdots + C_n' u_n'). \qquad (5.82)$$

This time we require that

$$C_1' u_1' + \cdots + C_n' u_n' = 0. \qquad (5.83)$$

The second derivative simplifies to

$$y'' = C_1 u_1'' + \cdots + C_n u_n''. \qquad (5.84)$$

We have now imposed two conditions, (5.80) and (5.83), on the functions $C_i$. Continuing in this way through $n - 1$ differentiations, we impose the $n - 1$ conditions

$$C_1' u_1^{(k)} + C_2' u_2^{(k)} + \cdots + C_n' u_n^{(k)} = 0, \qquad 0 \le k \le n - 2, \qquad (5.85)$$

on the functions $C_i$, and the derivatives of $y$ are given by the formula

$$y^{(k)} = C_1 u_1^{(k)} + C_2 u_2^{(k)} + \cdots + C_n u_n^{(k)}, \qquad 0 \le k \le n - 1. \qquad (5.86)$$

Then

$$y^{(n)} = C_1 u_1^{(n)} + \cdots + C_n u_n^{(n)} + C_1' u_1^{(n-1)} + \cdots + C_n' u_n^{(n-1)}. \qquad (5.87)$$

To obtain a final $n$th condition on the functions $C_i$, we substitute the expressions (5.86) and (5.87) into the differential equation (5.77). We find that

$$a_0[C_1 u_1^{(n)} + \cdots + C_n u_n^{(n)} + C_1' u_1^{(n-1)} + \cdots + C_n' u_n^{(n-1)}]$$
$$+ a_1[C_1 u_1^{(n-1)} + \cdots + C_n u_n^{(n-1)}] + \cdots + a_n[C_1 u_1 + \cdots + C_n u_n] = F,$$

or, upon regrouping terms,

$$a_0[C_1' u_1^{(n-1)} + \cdots + C_n' u_n^{(n-1)}] + C_1[a_0 u_1^{(n)} + \cdots + a_n u_1]$$
$$+ \cdots + C_n[a_0 u_n^{(n)} + \cdots + a_n u_n] = F.$$

This reduces to

$$C_1' u_1^{(n-1)} + \cdots + C_n' u_n^{(n-1)} = F/a_0 \qquad (5.88)$$

since the functions $u_i$ are solutions of the homogeneous equation (5.76). We have now obtained the $n$ conditions

$$\begin{aligned}
C_1' u_1 + \cdots + C_n' u_n &= 0, \\
C_1' u_1' + \cdots + C_n' u_n' &= 0, \\
C_1' u_1'' + \cdots + C_n' u_n'' &= 0, \\
\cdots\cdots\cdots\cdots\cdots\cdots\cdots\cdots\cdots \\
C_1' u_1^{(n-2)} + \cdots + C_n' u_n^{(n-2)} &= 0, \\
C_1' u_1^{(n-1)} + \cdots + C_n' u_n^{(n-1)} &= F/a_0
\end{aligned} \qquad (5.89)$$

for $C_1', C_2', \ldots, C_n'$. The determinant of this system is the Wronskian of the functions $u_1, u_2, \ldots, u_n$. As mentioned in Section 5.1, this Wronskian is never zero on an interval where $a_0$ does not vanish. Consequently the system (5.89) possesses a unique solution. The solution functions are continuous, for reasons explained in Section 2.7. If we solve for the quantities $C_i'$, the functions $C_i$ can be found by integration.

In arriving at the conditions (5.89), we proceeded under the assumption that the Eq. (5.77) had a solution of the form (5.78). To make our argument rigorous, let the functions $C_i$ be functions which satisfy the conditions (5.89). Then the corresponding function (5.78) has derivatives given by the formulas (5.86) and (5.87). It can now be verified that the function (5.78) is indeed a solution of the differential equation. We sum up our results as follows.

**Theorem 5.5**   Let the functions $a_i$ and $F$ be continuous on an interval $I$, with $a_0$ never zero on $I$. Let $u_1, u_2, \ldots, u_n$ be linearly independent solutions of the homogeneous equation (5.76) on $I$. If the functions $C_i$ are such that their derivatives $C_i'$ satisfy the system of equations (5.89), then the function

$$y_p = C_1 u_1 + C_2 u_2 + \cdots + C_n u_n \tag{5.90}$$

is a solution of the nonhomogeneous equation (5.77).

**Example 1**

$$y'' - 3y' + 2y = -\frac{e^{2x}}{e^x + 1}. \tag{5.91}$$

The functions $u_1$ and $u_2$, where

$$u_1(x) = e^x, \qquad u_2(x) = e^{2x},$$

are linearly independent solutions of the associated homogeneous equation

$$y'' - 3y' + 2y = 0.$$

We seek a solution of the nonhomogeneous equation (5.91) of the form

$$y_p = C_1 e^x + C_2 e^{2x}.$$

The conditions (5.89) in this case are

$$C_1' e^x + C_2' e^{2x} = 0,$$

$$C_1' e^x + 2C_2' e^{2x} = -\frac{e^{2x}}{e^x + 1}$$

or, upon dividing through in each equation by $e^x$,

$$C_1' + C_2' e^x = 0,$$

$$C_1' + 2C_2' e^x = -\frac{e^x}{e^x + 1}.$$

Solving, we find that

$$C_1' = \frac{e^x}{e^x + 1}, \qquad C_2' = \frac{-1}{e^x + 1} = -\frac{e^{-x}}{1 + e^{-x}}.$$

Then we may take

$$C_1 = \ln(e^x + 1), \qquad C_2 = \ln(1 + e^{-x}).$$

A particular solution of Eq. (5.91) is

$$y_p = e^x \ln(e^x + 1) + e^{2x} \ln(1 + e^{-x}).$$

The general solution is

$$y = c_1 e^x + c_2 e^{2x} + e^x \ln(e^x + 1) + e^{2x} \ln(1 + e^{-x}).$$

**Example 2**

$$x^2 y'' + xy' - y = -2x^2 e^x, \quad x > 0. \tag{5.92}$$

Notice that if we make the change of variable $x = e^t$, the equation becomes

$$[\mathscr{D}(\mathscr{D} - 1) + \mathscr{D} - 1]y = -2e^{2t} \exp(e^t).$$

The nonhomogeneous term is one for which the method of undetermined coefficients does not apply; therefore, we must use the method of variation of parameters. The homogeneous equation

$$x^2 y'' + xy' - y = 0$$

possesses the solutions $u_1(x) = x$ and $u_2(x) = x^{-1}$. We seek a solution of Eq. (5.92) that is of the form

$$y_p = C_1(x)x + C_2(x) x^{-1}.$$

Conditions (5.89) become

$$C_1' x + C_2' x^{-1} = 0,$$
$$C_1' - C_2' x^{-2} = -2e^x.$$

(Here $a_0(x) = x^2$, therefore $F(x)/a_0(x) = -2e^x$.) We find that

$$C_1' = -e^x, \qquad C_2' = x^2 e^x,$$

so we may take

$$C_1 = -e^x, \qquad C_2 = (x^2 e^x - 2x + 2).$$

Our general solution is

$$y = c_1 x + c_2 x^{-1} - x e^x + x^{-1} e^x (x^2 - 2x + 2)$$

or

$$y = c_1 x + c_2 x^{-1} + 2(x^{-1} - 1)e^x.$$

The method of variation of parameters requires that the associated homogeneous equation can be solved. Thus far we are able to do this when the homogeneous equation

(a) has constant coefficients
(b) is of the Cauchy–Euler type
(c) is a second-order equation with dependent or independent variable absent (Section 1.8).

One other situation might be mentioned here. If one nontrivial solution of the homogeneous equation is known, the problem of solving the equation can be reduced to that of solving an equation whose order is one less. If the original equation is of second order, we arrive at a first-order linear equation that can be treated by the method of Section 1.4.

Suppose that the function $u_1$ is a nontrivial solution of the equation

$$a_0(x) y'' + a_1(x) y' + a_2(x) y = 0,$$

and that we wish to solve the equation

$$a_0(x) y'' + a_1(x) y' + a_2(x) y = F(x), \tag{5.93}$$

where $F$ may or may not be the zero function. We introduce a new dependent variable $v$, where $y = u_1(x) \, v$. Then

$$y' = u_1 v' + u_1' v, \qquad y'' = u_1 v'' + 2u_1' v' + u_1'' v.$$

Substituting into Eq. (5.93), we have

$$a_0(u_1 v'' + 2u_1' v' + u_1'' v) + a_1(u_1 v' + u_1' v) + a_2 u_1 v = F$$

or

$$a_0 u_1 v'' + (2a_0 u_1' + a_1 u_1)v' + (a_0 u_1'' + a_1 u_1' + a_2 u_1)v = F.$$

Here the coefficient of $V$ is zero since $u_1$ is a solution of the homogeneous equation. The equation

$$a_0 u_1 v'' + (2a_0 u_1' + a_1 u_1)v' = F$$

has the dependent variable missing. Setting $w = v'$, we obtain the first-order equation

$$a_0 u_1 w' + (2a_0 u_1' + a_1 u_1)w = F.$$

We solve this equaton for $w$, find $v$ by integration, and then multiply $v$ by $u_1$ to obtain the solutions of the original equation (5.93).

**Example 3**

$$x^2(x + 1)y'' - 2xy' + 2y = 0.$$

We observe that a solution is $u_1(x) = x$. Setting $y = vx$, we have

$$x^2(x + 1)(xv'' + 2v') - 2x(xv' + v) + 2xv = 0$$

or

$$(x + 1)v'' + 2v' = 0.$$

This first-order equation for $v'$ has the solutions

$$v' = -\frac{c_1}{(x + 1)^2};$$

therefore

$$v = \frac{c_1}{x+1} + c_2 .$$

The general solution of the original equation is

$$y = vx = c_1 \frac{x}{x+1} + c_2 x .$$

**Example 4**

$$xy'' + 2(1-x)y' + (x-2)y = 2e^x .$$

It is easy to verify that $u_1(x) = e^x$ is a solution of the associated homogeneous equation. Setting $y = ve^x$, we find that

$$[x(v'' + 2v' + v) + (2-2x)(v' + v) + (x-2)v]e^x = 2e^x$$

or

$$xv'' + 2v' = 2 .$$

Then

$$v' = -\frac{c_1}{x^2} + 1$$

and

$$v = \frac{c_1}{x} + c_2 + x .$$

The general solution of the original equation is

$$y = [(c_1/x) + c_2 + x]e^x .$$

## Exercises for Section 5.8

In Exercises 1–10, find the general solution of the differential equation.

1. $y'' - y = \dfrac{2}{e^x + 1}$

2. $y'' - 2y' + y = \dfrac{e^x}{x}$

3. $y'' + 2y' + y = 4e^{-x} \ln x$

4. $y'' + 2y' + y = e^{-x} \sec^2 x$

5. $y'' + y = \csc x$

6. $y'' + y = \tan x \sec x$

7. $y'' + 2y' + 2y = 2e^{-x} \tan^2 x$

8. $x^2 y'' - 2xy' + 2y = x^3 e^x$

9. $(D-1)^3 y = 2\dfrac{e^x}{x^2}$

10. $(D-1)(D+1)(D+2)y = \dfrac{6}{e^x + 1}$

**11.** Let $F$ be defined and continuous on the interval $[0, \infty)$.

(a)  Show that the general solution of the equation

$$y'' + k^2 y = F(x)$$

may be written as

$$y = A \sin(kx + \alpha) - \frac{1}{k} \int_{x_0}^{x} \sin k(t - x)\, F(t)\, dt\,,$$

where $A$ and $\alpha$ are arbitrary constants and $x_0$ is any nonnegative number.

(b)  Suppose that there exist numbers $M$, $x_1$, and $a$, $a > 1$, such that

$$|F(x)| \le M\, x^{-a}, \quad x \ge x_1\,.$$

Show that every solution of the equation in part (a) is bounded on the interval $[0, \infty)$.

(c)  If $\int_0^{\infty} |F(x)|\, dx$ converges, show that every solution of the equation in part (a) is bounded on $[0, \infty)$.

**12.** Let $a$ and $b$ be positive real numbers with $a \ne b$. Let the function $F$ be defined and continuous on $[0, \infty)$.

(a)  Show that the general solution of the equation

$$(D + a)(D + b)y = F(x)$$

may be written as

$$y = c_1 e^{-ax} + c_2 e^{-bx} + \frac{1}{b - a} \int_{x_0}^{x} \left[ e^{-a(x-t)} - e^{-b(x-t)} \right] F(t)\, dt\,,$$

where $x_0$ is any nonnegative number.

(b)  If $F$ is bounded (that is, $|F(x)| \le M$ for some number $M$ and $x \ge 0$), show that every solution of the equation in part (a) is bounded on $[0, \infty)$.

(c)  if $\int_0^{\infty} |F(x)|\, dx$ converges, show that every solution of the equation in part (a) is bounded.

(d)  If $\lim_{x \to \infty} F(x) = L$, show that every solution of the equation in part (a) tends to the limit $L/(ab)$ as $x$ becomes infinite.

In Exercises 13–16, find the general solution of the differential equation, given one solution of the homogeneous equation.

13. $x^3 y'' + xy' - y = 0, \quad y = x$

14. $xy'' + (1 - 2x)y' + (x - 1)y = 0, \quad y = e^x$

15. $2xy'' + (1 - 4x)y' + (2x - 1)y = e^x, \quad y = e^x$

16. $x^2(x + 2)y'' + 2xy' - 2y = (x + 2)^2, \quad y = x$

17. Suppose that $u_1$ and $u_2$ are linearly independent solutions of the third-order equation

$$a_0 y''' + a_1 y'' + a_2 y' + a_3 y = 0.$$

Show that the change of variable $y = u_1(x) v$ leads to a second-order equation for $v' = w$. Find a solution of this equation, in terms of $u_2$, and use it to reduce the equation to one of first order.

## 5.9  SIMPLE HARMONIC MOTION

Suppose that a spring hangs vertically from a support, as in Fig. 5.1a. Let $L$ denote the length of the spring when it is at rest. When the spring is stretched or compressed a distance $s$ by a force $F$ applied at the ends, it is found by experiment that the magnitude of the force is approximately proportional to the distance $s$, at least when $s$ is not too large. Thus

$$F = ks,$$

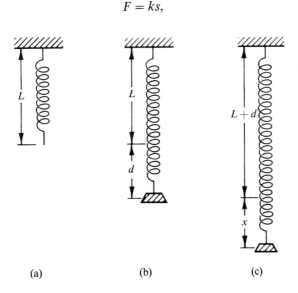

(a)  (b)  (c)

**Figure 5.1**

where $k$ is a positive constant of proportionality known as the *spring constant*. For example, if a force of 30 lb is required to stretch a spring 2 in., then

$$30 = 2k;$$

therefore $k = 15$ lb/in.

When a body of mass $m$, and weight $mg$, is attached to the free end of the spring, the body will remain at rest in a position such that the spring has length $L + d$ (Fig. 5.1b), where $mg = kd$. The downward force $mg$ due to gravity is balanced by the restoring force $kd$ of the spring. Let us denote by $x$ the directed distance downward of the center of mass of the body from its position of rest, or equilibrium (Fig. 5.1c). Then the equation of motion $m\ddot{x} = F$ becomes

$$m\ddot{x} = mg - k(x + d)$$

or

$$m\ddot{x} + kx = 0. \tag{5.94}$$

The general solution of this equation is

$$x = c_1 \cos \omega t + c_2 \sin \omega t, \tag{5.95}$$

where

$$\omega = (k/m)^{1/2}. \tag{5.96}$$

If the body is held in the position $x = x_0$ and released from rest at time $t = 0$, the initial conditions are

$$x(0) = x_0, \qquad \dot{x}(0) = 0. \tag{5.97}$$

Using these conditions to determine the constants $c_1$ and $c_2$ in the formula (5.95), we find that

$$x = x_0 \cos \omega t. \tag{5.98}$$

Thus the body oscillates about the equilibrium position $x = 0$ between the points $x = \pm|x_0|$ without ever coming to rest.

If the body is struck sharply when it is in the equilibrium position, giving it a velocity $v_0$, the initial conditions become

$$x(0) = 0, \qquad \dot{x}(0) = v_0. \tag{5.99}$$

This time we find that

$$x = \frac{v_0}{\omega} \sin \omega t. \qquad (5.100)$$

In the more general case where

$$x(0) = x_0, \qquad \dot{x}(0) = v_0,$$

the solution is the sum of the solutions (5.98) and (5.100). We have

$$x = x_0 \cos \omega t + \frac{v_0}{\omega} \sin \omega t. \qquad (5.101)$$

Let us write

$$A = [x_0^2 + (v_0/\omega)^2]^{1/2}.$$

Then there is an angle $\theta_1$ such that

$$\cos \theta_1 = x_0/A, \qquad \sin \theta_1 = -v_0/(\omega A),$$

and formula (5.101) can be written

$$x = A \cos(\omega t + \theta_1). \qquad (5.102)$$

If we put

$$\sin \theta_2 = x_0/A, \qquad \cos \theta_2 = v_0/(\omega A),$$

it becomes

$$x = A \sin(\omega t + \theta_2). \qquad (5.103)$$

Straight line motion that can be described by a function of the form

$$x = A \cos(\omega t + \theta) \qquad \text{or} \qquad x = A \sin(\omega t + \theta) \qquad (5.104)$$

is called *simple harmonic motion*. The number $|A|$ is called the *amplitude* of the motion. Notice that $x$ fluctuates between $-|A|$ and $|A|$ periodically. The period $P$ of the motion is given by the formula $P = 2\pi/\omega$. This is the time required for the body to move through one cycle. The frequency $f$ is the number of cycles per unit time. Thus $f = 1/P = \omega/(2\pi)$.

The presence of a damping force equal to $c$ times the velocity may be indicated by means of a *dashpot*, as shown in Figure 5.2. In this case, the body is said to exhibit *damped harmonic motion*. The equation of motion becomes

$$m\ddot{x} + c\dot{x} + kx = 0.\qquad(5.105)$$

The auxiliary polynomial equation is

$$mr^2 + cr + k = 0,$$

with roots

$$r = \frac{1}{2m}[-c \pm (c^2 - 4mk)^{1/2}].$$

We consider separately the following cases:
(1)   $c^2 < 4mk$ (two complex roots)
(2)   $c^2 > 4mk$ (two distinct real roots)
(3)   $c^2 = 4mk$ (equal real roots).
These cases are called the *underdamped*, *overdamped*, and *critically damped* cases, respectively.

For case 1, we may write the general solution of the equation of motion (5.105) as

$$x = c_1 e^{-\alpha t} \cos \omega t + c_2 e^{-\alpha t} \sin \omega t,\qquad(5.106)$$

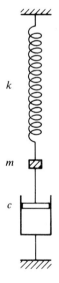

**Figure 5.2**

where

$$\alpha = \frac{c}{2m}, \qquad \omega = \frac{1}{2m}(4mk - c^2)^{1/2}.$$

The solution that satisfies the initial conditions

$$x(0) = x_0, \qquad x(0) = 0$$

is

$$x = x_0 e^{-\alpha t}\left(\cos \omega t + \frac{\alpha}{\omega}\sin \omega t\right). \tag{5.107}$$

This may be written

$$x = x_0\left[1 + \left(\frac{\alpha}{\omega}\right)^2\right]^{1/2} e^{-\alpha t}\sin(\omega t + \theta_1), \tag{5.108}$$

where

$$\theta_1 = \tan^{-1}\frac{\omega}{\alpha}$$

or

$$x = x_0\left[1 + \left(\frac{\alpha}{\omega}\right)^2\right]^{1/2} e^{-\alpha t}\cos(\omega t + \theta_2), \tag{5.109}$$

where

$$\theta_2 = -\tan^{-1}\frac{\alpha}{\omega}.$$

In this case the body still oscillates back and forth across the equilibrium position, but the amplitude decreases exponentially with time. The situation is illustrated in Fig. 5.3. In the overdamped and critically damped cases, the body no longer exhibits this oscillatory behavior. The derivations of the exact forms of the solutions in these cases are left as exercises.

If an external force $F(t)$ is applied to the body on the spring, the equation becomes

$$m\ddot{x} + c\dot{x} + kx = F(t). \tag{5.110}$$

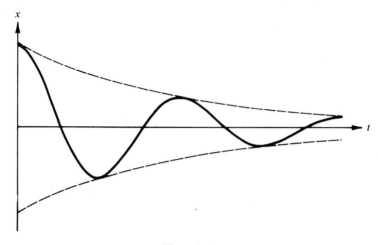

**Figure 5.3**

We shall consider here the case where $F$ is a periodic force of the form

$$F(t) = A \sin \omega_1 t$$

and where damping is absent ($c = 0$). Then Eq. (5.110) becomes

$$m\ddot{x} + kx = A \sin \omega_1 t. \qquad (5.111)$$

In order to obtain a particular solution, we consider the equation

$$m\ddot{x} + kx = Ae^{i\omega_1 t}. \qquad (5.112)$$

If $x = Ce^{i\omega_1 t}$, then $\ddot{x} = -C\omega_1^2 e^{i\omega_1 t}$ and we require that

$$(-Cm\omega_1^2 + kC)e^{i\omega_1 t} = Ae^{i\omega_1 t}$$

or

$$C = \frac{A}{k - m\omega_1^2}.$$

Using the notation $\omega = (k/m)^{1/2}$, we have

$$C = \frac{A}{m(\omega^2 - \omega_1^2)}$$

provided that $\omega_1 \neq \omega$. The general solution of Eq. (5.111) is

$$x = c_1 \cos \omega t + c_2 \sin \omega t + \frac{A}{m(\omega^2 - \omega_1^2)} \sin \omega_1 t. \qquad (5.113)$$

However, if $\omega_1 = \omega$, we must seek a particular solution of Eq. (5.112) that is of the form

$$x = Cte^{i\omega t}.$$

A particular solution turns out to be

$$x = -\frac{Ai}{2\omega m} te^{i\omega t}.$$

We take the imaginary part to obtain a real solution of Eq. (5.111). The general solution of this equation is

$$x = c_1 \cos \omega t + c_2 \sin \omega t - \frac{A}{2\omega m} t \cos \omega t. \qquad (5.114)$$

In this case, $\omega_1 = \omega$, the magnitude of the oscillations increases indefinitely because of the presence of the term $t \cos \omega t$. This phenomena is called *resonance*. Actually, when the oscillations become sufficiently large the law $|F| = ks$ does not hold and our mathematical model no longer applies.

### Exercises for Section 5.9

1. (a) Find the general solution of the equation of motion (5.105) in the overdamped case $c^2 > 4mk$. Show that every solution tends to zero as $t$ becomes infinite. For convenience, let

$$\alpha = -\frac{1}{2m}[-c + (c^2 - 4mk)^{1/2}], \qquad \beta = -\frac{1}{2m}[-c - (c^2 - 4mk)^{1/2}].$$

   (b) Find the solution for which $x(0) = x_0$, $\dot{x}(0) = 0$. Draw a graph of the solution for $t \geq 0$.

   (c) Find the solution for which $x(0) = 0$, $\dot{x}(0) = v_0$. Draw a graph of the solution for $t \geq 0$.

2. (a) Find the general solution of the equation of motion (5.105) in the critically damped case $c^2 = 4mk$. Show that every solution tends to zero as $t$ becomes infinite. For convenience let $\gamma = c/(2m)$.

   (b) Find the solution for which $x(0) = x_0$, $\dot{x}(0) = 0$ and draw its graph.

   (c) Find the solution for which $x(0) = 0$, $\dot{x}(0) = v_0$ and draw its graph.

**3.** If an external force $F(t)$ is applied to the body on the spring, its equation of motion becomes

$$m\ddot{x} + c\dot{x} + kx = F(t).$$

Assume that $c \neq 0$ and that $F$ is a periodic function of the form

$$F(t) = A \sin \omega_1 t.$$

(a) When $t$ is large, show that every solution is approximately equal to

$$x_p(t) = B \sin(\omega_1 t - \theta),$$

where

$$B = \frac{A}{D}, \quad D = [(k - m\omega_1^2)^2 + c^2\omega_1^2]^{1/2}, \quad \theta = \cos^{-1}\frac{k - m\omega_1^2}{D}.$$

This is called the *steady-state solution* of the equation.

(b) If $c^2 \geq 4mk$ (overdamping or critical damping), show that the amplitude $|B|$ is a strictly decreasing function of $\omega_1$.

(c) If $c^2 < 4mk$ (underdamping), the homogeneous equation has damped oscillatory solutions of the form

$$x = e^{-\alpha t}(c_1 \cos \omega t + c_2 \sin \omega t)$$

where

$$\alpha = \frac{c}{2m}, \quad \omega = \frac{1}{2m}(4mk - c^2)^{1/2}.$$

Show that the amplitude $|B|$ of the steady-state solution, considered as a function of $k$, is largest when $k = m\omega_1^2$, in which case $\omega = (\omega_1^2 - \alpha^2)^{1/2}$.

**4.** If the applied force $F$ is constant, $F(t) = F_0$, find the limiting position of the body on the spring.

## 5.10   ELECTRIC CIRCUITS

Let us consider an electric circuit in which a resistance, capacitance, and inductance are connected in series with a voltage source, as in Fig. 5.4. When the switch is closed at $t = 0$, a current will flow in the loop. We denote the value of the current at time $t$ by $I(t)$. The arrow in the figure gives the loop a direction. We understand that $I$ is positive when the flow is in the direction of the arrow and negative when in the opposite direction. In Fig. 5.5 we

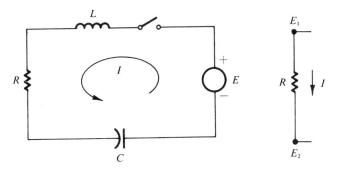

Figure 5.4                    Figure 5.5

have isolated one circuit element, the resistance. The voltages at the terminals are denoted by $E_1$ and $E_2$. When the current is positive, in the direction of the arrow, $E_1 < E_2$ and the *voltage drop* across the element is $E_1 - E_2$, a positive quantity. Thus a positive current flows in the direction of decreasing voltage. This is a matter of convention, and the term "current" as used here should not be identified with a flow of electrons. Actually it may help the reader to think of the current as a flow of positively charged particles.

We shall use the following system of units: *amperes* for the current $I$, *volts* for the voltage, *ohms* for the resistance $R$, *henrys* for the inductance $L$, *farads* for the capacitance $C$, *coulombs* for the charge on the capacitance, and *seconds* for the time $t$. Using this system of units, the voltage drop across the resistance is $RI$ and that across the inductance is $L\,dI/dt$. The voltage drop across the capacitance is $Q/C$, where $Q$ is the charge on the capacitance. The charge and current are related by the equations

$$I = \frac{dQ}{dt}, \qquad Q(t) = \int_0^t I(s)\,ds + Q_0, \tag{5.115}$$

where $Q_0$ is the charge on the capacitance at $t = 0$.

According to one of Kirchoff's two laws (the other will be considered presently) the sum of the voltage drops around the loop must be equal to the applied voltage. Therefore the equality

$$L\frac{dI}{dt} + RI + \frac{1}{C}Q = E(t) \tag{5.116}$$

must hold for $t \geq 0$, provided that the sign of $E(t)$ is chosen in accordance with the $+$ and $-$ signs in Fig. 5.4. Upon differentiating with respect to $t$, and using relations (5.115), we arrive at the second-order differential equation

$$L\frac{d^2 I}{dt^2} + R\frac{dI}{dt} + \frac{1}{C}I = E'(t) \tag{5.117}$$

for $I$. We notice the resemblance of this equation to that for a harmonic oscillator,

$$m \frac{d^2x}{dt^2} + c \frac{dx}{dt} + kx = F(t). \tag{5.118}$$

In particular, the term $Ld^2I/dt^2$ in the circuit equation corresponds to the inertia term $m\, d^2x/dt^2$ in Eq. (5.118). This means that the current passing through the inductance must be the same immediately before and after a sudden change[5] or jump in the voltage drop across it. Since the current was zero before the switch was closed we must have

$$I(0) = 0. \tag{5.119}$$

Using this fact we can find the initial value of $dI/dt$ from Eq. (5.116). Assuming that the initial charge on the capacitance is zero, we have

$$LI'(0) + R\,I(0) = E(0)$$

or

$$I'(0) = \frac{E(0)}{L}. \tag{5.120}$$

In many applications, the applied voltage is approximately constant, as in the case of a battery, or sinusoidal, as in the case of an alternating current generator. Let us consider the cases where the applied voltage has the constant value $E_0$. Then $E'(t) = 0$ and Eq. (5.117) becomes

$$L \frac{d^2I}{dt^2} + R \frac{dI}{dt} + \frac{1}{C} I = 0. \tag{5.121}$$

The initial conditions are

$$I(0) = 0, \qquad I'(0) = \frac{E_0}{L}. \tag{5.122}$$

The form of the solution depends on whether the quantity $R^2 - 4L/C$ is positive, negative, or zero. We consider here only the case where $R^2 - 4L/C > 0$.

---

[5] Unless the change is infinite, as can happen in some idealized situations.

The solution of the initial value problem, as found by routine methods, is

$$I(t) = \frac{E_0}{(R^2 - 4L/C)^{1/2}} (e^{-\alpha t} - e^{-\beta t}), \qquad (5.123)$$

where

$$\alpha = \frac{1}{2L} [R - (R^2 - 4L/C)^{1/2}], \qquad \beta = \frac{1}{2L} [R + (R^2 - 4L/C)^{1/2}].$$

Since $\alpha$ and $\beta$ are both positive, $I(t)$ tends to zero as $t$ becomes infinite.

We next consider the circuit of Fig. 5.6, in which the applied voltage is sinusoidal, of the form $E(t) = E_0 \cos \omega t$. No inductance is present. Kirchhoff's law leads to the equation

$$RI + \frac{1}{C}Q = E_0 \cos \omega t \qquad (5.124)$$

or

$$R \frac{dI}{dt} + \frac{1}{C} I = \omega E_0 \sin \omega t. \qquad (5.125)$$

From Eq. (5.124) we obtain the initial condition

$$I(0) = \frac{E_0}{R}, \qquad (5.126)$$

again assuming that the initial charge on the capacitance is zero. The solution

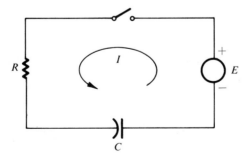

**Figure 5.6**

of the first-order equation (5.125) that satisfies the condition (5.126) is found to be

$$I(t) = I_1(t) + I_2(t),$$    (5.127)

where

$$I_1(t) = \frac{E_0}{R}\left[1 + \frac{\omega^2}{(RC)^{-2} + \omega^2}\right]e^{-t/(RC)}$$    (5.128)

and

$$I_2(t) = \frac{E_0\,\omega}{R[(RC)^{-2} + \omega^2]}\left(\frac{1}{RC}\sin \omega t - \omega\cos \omega t\right).$$    (5.129)

The function $I_1$ dies out as $t$ becomes infinite. It is called the *transient solution* of the initial-value problem. When $t$ is large the solution $I$ is very nearly equal to $I_2$; this is called the *steady-state* solution.

We next examine a circuit that consists of a resistance, inductance, and capacitance connected in parallel with a current source. Such a circuit is shown in Fig. 5.7. The switch is opened at $t = 0$. The circuit has three loops (once the switch is opened) and two junctures, or nodes. Here it is advantageous to use Kirchhoff's other law, which says that the current entering a node must be equal to the current leaving it. If $E(t)$ is the increase in voltage from node 2 to node 1 (or the voltage drop from node 1 to node 2), then the current flowing from node 1 to node 2 is

$$\frac{1}{R}E \quad \text{through the resistance,}$$

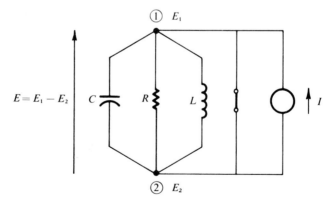

**Figure 5.7**

$$C\frac{dE}{dt} \quad \text{through the capacitance,}$$

and

$$\frac{1}{L}\int_0^t E(s)\,ds \quad \text{through the inductance.}$$

If $I(t)$ is the applied current, we must then have

$$C\frac{dE}{dt} + \frac{1}{R}E + \frac{1}{L}\int_0^t E(s)\,ds = I(t). \tag{5.130}$$

Differentiation with respect to $t$ yields the second-order equation

$$C\frac{d^2E}{dt^2} + \frac{1}{R}\frac{dE}{dt} + \frac{1}{L}E = I'(t) \tag{5.131}$$

for the voltage $E$.

It can be shown that the voltage drop across a capacitance is the same immediately before and after a sudden change in the current through it. Consequently $E(0) = 0$. From Eq. (5.130), we find that

$$C\,E'(0) + 0 + 0 = I(0)$$

or

$$E'(0) = \frac{1}{C}I(0).$$

Thus our initial conditions are

$$E(0) = 0, \qquad E'(0) = \frac{1}{C}I(0). \tag{5.132}$$

The mathematics of this problem is similar to that for problems considered earlier in this section, and we shall proceed no further.

Electric circuits that involve more than one loop and more than one pair of nodes will be considered in Chapter 6, which deals with systems of differential equations.

## Exercises for Section 5.10

1. A resistance and an inductance are connected in series with a battery of constant voltage $E_0$, as shown in Fig. 5.8. The switch is closed at $t = 0$.

   (a) Find a formula for the current as a function of $t$.

   (b) Find the voltage drop across the resistance and that across the inductance.

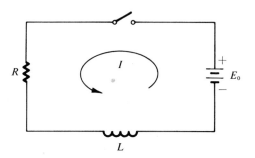

**Figure 5.8**

2. In the circuit of Fig. 5.4, suppose that $E(t) = E_0$ and that $Q_0 = 0$.

   (a) Find the current in the loop for the case $R^2 = 4L/C$.

   (b) Find the current if $R^2 < 4L/C$.

   (c) Show that the current tends to zero as $t$ becomes infinite regardless of whether $R^2 - 4L/C$ is positive, negative, or zero so long as $R \neq 0$.

   (d) Show that the charge $Q(t)$ on the capacitance tends to the value $E_0 C$ regardless of whether $R^2 - 4L/C$ is positive, negative, or zero. Suggestion: look at Eq. (5.116).

3. In the circuit of Fig. 5.4 suppose that the applied-voltage is sinusoidal of the form

$$E(t) = A \, \sin(\omega_1 t + \alpha),$$

   where $A$, $\omega_1$, and $\alpha$ are constants. Find the steady-state solution. Suggestion: look for a particular solution of the form

$$a e^{i(\omega_1 t + \alpha)}.$$

**4.** In the circuit of Fig. 5.9 suppose that

$$E(t) = A \sin \omega_1 t.$$

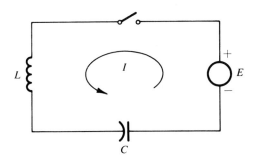

**Figure 5.9**

Let $\omega = (LC)^{-1/2}$. Find a formula for the current if

(a) $\omega_1 \neq \omega$        (b) $\omega_1 = \omega$

**5.** In the circuit of Fig. 5.10, as charge $Q_0$ has been placed on the capacitance. If the switch is closed at $t = 0$, find a formula for the current.

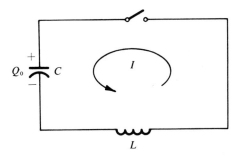

**Figure 5.10**

**6.** A current source is connected in parallel with an inductance and a resistance, as shown in Fig. 5.11. The switch is opened at $t = 0$. If $I(t)$ has the constant value $I_0$, find

(a) the voltage $E(t)$
(b) the current through the resistance
(c) the current through the inductance.

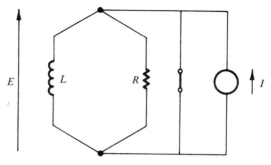

**Figure 5.11**

7. In the circuit of Fig. 5.12 the switch is opened at $t = 0$. If $I(t) = A \sin \omega t$, find

(a) the voltage $E(t)$
(b) the steady-state current through the resistance
(c) the steady-state current through the capacitance.

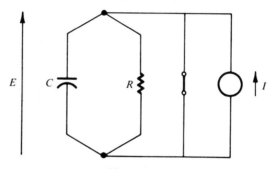

**Figure 5.12**

## 5.11   THEORY OF LINEAR DIFFERENTIAL EQUATIONS

Let us consider the initial value problem

$$a_0(x)\, y^{(n)} + a_1(x)\, y^{(n-1)} + \cdots + a_{n-1}(x)\, y' + a_n(x)\, y = F(x),$$
$$y^{(i)}(x_0) = k_i, \quad 0 \le i \le n-1,$$

(5.133)

where the $k_i$ are constants and the functions $a_i$ and $F$ are continuous on an interval $\mathscr{I}$ that contains the point $x_0$. We assume that $a_0$ is never zero on $\mathscr{I}$. The most basic result in the theory of linear differential equations is given by the following theorem.

**Theorem 5.6** The initial value problem (5.133) possesses a solution that exists throughout the interval $\mathcal{I}$. Furthermore, any two solutions that exist on an interval that contains $x_0$ are identical on that interval.

This theorem is an *existence* theorem because it says that the initial value problem has a solution. It is also a *uniqueness* theorem because it says that the problem has essentially only one solution.

In the case of an initial value problem that involves an equation with constant coefficients, such as

$$y'' - y' - 2y = 0 ,$$
$$y(0) = k_0 , \quad y'(0) = k_1 , \tag{5.134}$$

we can actually exhibit the solution. Any function of the form

$$y = c_1 e^{-x} + c_2 e^{2x}$$

is a solution of the equation for all $x$, and if we require that

$$y(0) = c_1 + c_2 = k_0 ,$$
$$y'(0) = -c_1 + 2c_2 = k_1 ,$$

then

$$c_1 = \tfrac{1}{3}(2k_0 - k_1), \qquad c_2 = \tfrac{1}{3}(k_0 + k_1).$$

Hence the solution of the problem is

$$y = \tfrac{1}{3}[(2k_0 - k_1)e^{-x} + (k_0 + k_1 e)^{2x}].$$

However, for an initial value problem such as

$$y'' + (\sin x)y' - e^x y = x^2 ,$$
$$y(1) = k_0 , \quad y'(1) = k_1 , \tag{5.135}$$

we can no longer find a simple formula for the solution. Nevertheless, Theorem 5.6 assures us that a solution does exist for all $x$, and that there is only one solution.

One important result of Theorem 5.6 is that the space of solutions of the $n$th-order linear equation $Ly = 0$ contains a set of $n$ linearly independent functions. To see this, let $u_1, u_2, \ldots, u_n$ be the solutions for which

$$u_i^{(j-1)}(x_0) = \delta_{ij} ,$$

*i* and *j* varying from 1 to *n*. (Theorem 5.6 guarantees that these solutions exist.) The Wronskian of these functions at $x_0$ is the determinant of the identity matrix $I_n$, and hence is not equal to zero. Thus the solutions $u_i$ are linearly independent.

A proof of the existence of a solution to the problem (5.133) is deferred until Chapter 8. We only remark here that the proof involves the construction of a sequence of functions that can be shown to converge to a function that is a solution of the problem. The functions of the sequence are expressed in terms of intetrals and it is seldom possible to find a simple formula for the limit of the sequence.

The restriction that $a_0$, the coefficient of the highest derivative in the differential equation of problem (5.133), is never zero should be emphasized. Let us consider the example

$$xy' + y = 2x, \quad y(0) = 1, \tag{5.136}$$

where $a_0(x) = x$. Obviously $a_0$ does vanish at $x = 0$. If the problem has a solution it must satisfy the relation

$$(xy)' = 2x$$

or

$$xy = x^2 + c.$$

Setting $x = 0$ and $y = 1$, we see that $c = 0$. Hence $xy = x^2$ or $y = x$. But then the initial condition $y(0) = 1$ is not satisfied. We conclude that the problem can have no solution.

The uniqueness part of Theorem 5.6 is not so difficult to prove. The proof that we shall present here is based on the following lemma

**Lemma**   On an interval $\mathcal{I}$ containing the point $x_0$ let the function *w* be continuous and nonnegative, and satisfy the inequality

$$w(x) \le M \left| \int_{x_0}^x w(s)\, ds \right|, \tag{5.137}$$

whese M is a positive constant. Then *w* is identically zero on $\mathcal{I}$.

PROOF   We define the function *W* by means of the formulas

$$W(x) = \int_{x_0}^x w(s)\, ds, \quad x \ge x_0,$$

$$W(x) = -\int_{x_0}^x w(s)\, ds, \quad x < x_0.$$

Notice that $W$ is nonnegative and that $W(x_0) = 0$.

We consider first the case where $x \geq x_0$. Then $W'(x) = w(x)$ and the inequality (5.137) can be written as

$$W'(x) - M W(x) \leq 0, \quad x \geq x_0.$$

If we multiply through in this inequality by $\exp[-M(x - x_0)]$, which is always positive, we find that

$$\frac{d}{dx} [W(x)e^{-M(x-x_0)}] \leq 0, \quad x \geq x_0.$$

Integrating from $x_0$ to $x$, we have

$$W(x)e^{-M(x-x_0)} - W(x_0) \leq 0$$

or

$$W(x) \leq W(x_0)e^{M(x-x_0)}, \quad x \geq x_0.$$

Since $W(x_0) = 0$ we must have $W(x) = 0$ for $x \geq x_0$. From the inequality (5.137) we see that $w(x) = 0$ for $x \geq x_0$. The case where $x < x_0$ can be treated similarly. The details are left as an exercise.

We can now prove that a solution of the initial value problem (5.133) is unique. Since $a_0$ is never zero, we shall first divide through by $a_0$ in the differential equation and consider the problem

$$Ly = y^{(n)} + b_1(x)y^{(n-1)} + \cdots + b_{n-1}(x)y' + b_n(x)y = G(x),$$
$$y^{(i)}(x_0) = k_i, \quad 0 \leq i \leq n - 1, \tag{5.138}$$

where $b_i = a_i/a_0$ and $G = F/a_0$. The functions $b_i$ and $G$ are continuous on an interval $\mathscr{I}$ that contains $x_0$.

**Theorem 5.7** Suppose that the functions $u$ and $v$ are both solutions of the initial value problem (5.138) on an interval $\mathscr{J}$ that is contained in $\mathscr{I}$ and which contains $x_0$. Then $u(x) = v(x)$ for all $x$ in $\mathscr{J}$.

PROOF Let $w = u - v$. Since $Lu = G$ and $Lv = G$ we have $Lw = Lu - Lv = 0$. Thus $w$ is a solution of the homogeneous equation $Ly = 0$, so that

$$w^{(n)} = -[b_1 w^{(n-1)} + \cdots + b_{n-1}w' + b_n w] \tag{5.139}$$

on $\mathscr{J}$. Also,

$$w^{(i)}(x_0) = 0, \quad 0 \leq i \leq n - 1. \tag{5.140}$$

Let us write

$$w_1 = w, \quad w_2 = w', \quad w_3 = w'', \ldots, \quad w_n = w^{(n-1)}. \tag{5.141}$$

Then

$$w_1(x) = \int_{x_0}^{x} w_2(s)\, ds,$$

$$w_2(x) = \int_{x_0}^{x} w_3(s)\, ds, \tag{5.142}$$

$$\cdots\cdots\cdots\cdots\cdots\cdots\cdots$$

$$w_{n-1}(x) = \int_{x_0}^{x} w_n(s)\, ds,$$

and, from Eq. (5.139),

$$w_n(x) = -\int_{x_0}^{x} [b_n(s)w_1(s) + \cdots + b_1(s)w_n(s)]\, ds. \tag{5.143}$$

Let $\mathcal{K}$ be a bounded closed interval that contains $x_0$ and is contained in $\mathcal{J}$. Each point of $\mathcal{J}$ belongs to such an interval. Since the functions $b_i$ are continuous on $\mathcal{K}$ there is a constant $M$ such that $|b_i(x)| \leq M$ for $x$ in $\mathcal{K}$, $1 \leq i \leq n$. From Eq. (5.143), we have

$$|w_n(x)| \leq M \left| \int_{x_0}^{x} [|w_1(s)| + \cdots + |w_n(s)|]\, ds \right|. \tag{5.144}$$

From Eqs. (5.142) we have

$$|w_i(x)| \leq \left| \int_{x_0}^{x} |w_{i+1}(s)|\, ds \right|, \qquad 1 \leq i \leq n-1. \tag{5.145}$$

Adding the $n$ inequalities (5.144) and (5.145), and considering the cases $x \geq x_0$ and $x < x_0$ separately, we find that

$$|w_1(x)| + \cdots + |w_n(x)| \leq (M+1) \left| \int_{x_0}^{x} [|w_1(s)| + \cdots + |w_n(s)|]\, ds \right|.$$

By the lemma, $|w_1| + |w_2| + \cdots + |w_n| = 0$. In particular $w_1 = w$ is zero on $\mathcal{K}$. But this means that $u = v$ on $\mathcal{K}$. Hence $u = v$ on $\mathcal{J}$, which we wished to show.

Theorem 5.7 has several important consequences. We recall that in Section 3.4 we showed that the Wronskian of a set of linearly dependent functions was identically zero. However, it was possible for a set of linearly independent functions to have a Wronskian that vanished identically also. In the case of $n$ functions that are all solutions of the same $n$th order linear homogeneous differential equation this is no longer true, as we shall now show.

**Theorem 5.8**  Let the functions $u_1, u_2, \ldots, u_n$ be solutions of the differential equation

$$y^{(n)} + b_1(x)\, y^{(n-1)} + \cdots + b_{n-1}(x)\, y' + b_n(x)\, y = 0 \qquad (5.146)$$

on an interval $\mathscr{I}$ where the functions $b_i$ are continuous. If the functions are linearly dependent their Wronskian is identically zero on $\mathscr{I}$, but if they are linearly independent their Wronskian never vanishes on $\mathscr{I}$. Thus the functions are linearly dependent if and only if their Wronskian is identically zero.

PROOF   When the functions $u_i$ are linearly dependent we know that their Wronskian vanishes identically by the results of Section 3.4. We shall now show that if the functions are linearly independent their Wronskian can never vanish.

Suppose that the Wronskian does vanish at a point $x_0$. Then there exist constants $c_1, c_2, \ldots, c_n$, not all zero, such that

$$
\begin{aligned}
c_1\, u_1(x_0) + \cdots + c_n\, u_n(x_0) &= 0, \\
c_1\, u_1'(x_0) + \cdots + c_n\, u_n'(x_0) &= 0, \\
&\ \vdots \\
c_1\, u_1^{(n-1)}(x_0) + \cdots + c_n\, u_n^{(n-1)}(x_0) &= 0.
\end{aligned}
\qquad (5.147)
$$

This is because the determinant of the system of equations for the $c_i$ is zero. Let us define the function $w$ by means of the formula

$$w(x) = c_1 u_1(x) + \cdots + c_n u_n(x).$$

Then $w$ is a solution of the homogeneous equation (5.146) on $\mathscr{I}$ and $w^{(i)}(x_0) = 0$ for $0 \leq i \leq n-1$, according to Eqs. (5.147). By Theorem 5.7, $w$ must be identically zero on $\mathscr{I}$. Hence

$$c_1\, u_1(x) + \cdots + c_n\, u_n(x) = 0$$

for $x$ in $\mathscr{I}$ and the constants $c_i$ are not all zero. But this is impossible since the functions $u_i$ are linearly independent. Consequently the Wronskian cannot vanish at any point of $\mathscr{I}$.

The result of Theorem 5.8 was used in Section 5.8 in finding a particular solution of a nonhomogeneous equation by the method of variation of parameters. The next result was stated without proof in Section 5.1.

**Theorem 5.9**   Let $u_1, u_2, \ldots, u_n$ be linearly independent solutions of the equation

$$y^{(n)} + b_1(x)\, y^{(n-1)} + \cdots + b_{n-1}(x)\, y' + b_n(x)\, y = 0 \qquad (5.148)$$

on an interval $\mathscr{I}$ where the $b_i$ are continuous. Then every solution that exists on $\mathscr{I}$ is of the form $c_1 u_1 + c_2 u_2 + \cdots + c_n u_n$, where the $c_i$ are constants.

PROOF   Let $v$ be a solution of Eq. (5.148) on $\mathscr{I}$. Let $x_0$ be any fixed point of $\mathscr{I}$. Since the Wronskian of the functions $u_i$ is not zero at $x_0$ (or at any other point) there exist numbers $c_1, c_2, \ldots, c_n$ such that

$$c_1\, u_1(x_0) + \cdots + c_n\, u_n(x_0) = v(x_0),$$
$$c_1\, u_1'(x_0) + \cdots + c_n\, u_n'(x_0) = v'(x_0),$$
$$\cdots\cdots\cdots\cdots\cdots\cdots\cdots\cdots\cdots\cdots\cdots\cdots\cdots\cdots$$
$$c_1\, u_1^{(n-1)}(x_0) + \cdots + c_n\, u_n^{(n-1)}(x_0) = v^{(n-1)}(x_0).$$

Let $w(x) = c_1\, u_1(x) + c_2\, u_2(x) + \cdots + c_n\, u_n(x)$. Then $w$ is also a solution of Eq. (5.148) and $w^{(i)}(x_0) = v^{(i)}(x_0)$ for $0 \le i \le n - 1$. By Theorem 5.7, $v(x) = w(x)$ for $x$ in $I$. That is,

$$v(x) = c_1\, u_1(x) + c_2\, u_2(x) + \cdots + c_n\, u_n(x),$$

which we wished to show.

### Exercises for Section 5.11

1. This problem furnishes a proof of the uniqueness of solutions for the initial value problem (5.133) when the differential equation has constant coefficients. Let the function $f$ be a solution of the equation

$$P(D)\, y = 0, \qquad (1)$$

where

$$P(D) = D^n + a_1 D^{n-1} + \cdots + a_{n-1} D + a_n$$

on a closed bounded interval $I$.

(a) Show that $f$ possesses derivatives of all orders at every point of $I$.

(b) Since $f$ and its derivatives are continuous on $I$, there exist constants $K_i$ such that

$$|f^{(i)}(x)| \le K_i, \qquad 0 \le i \le n-1,$$

for $x$ in $I$. If

$$K = \max(K_0, K_1, \ldots, K_{n-1})$$

and

$$M = \max(1, |a_1|, |a_2|, \ldots, |a_n|),$$

show that

$$|f^{(m)}(x)| \le K(nM)^m, \qquad m = 0, 1, 2, \ldots$$

for $x$ in $I$.

(c) Show that the Taylor series for $f$ about any point $x_0$ in $I$ converges to $f$ throughout $I$.

(d) If the function $g$ is a solution of Eq. (1) on an interval $J$ and satisfies the initial conditions $y^{(i)}(x_0) = 0, 0 \le i \le n-1$, show that $g(x) = 0$ for all $x$ in $J$.

(e) If $f_1$ and $f_2$ are both solutions of the equation $P(D) y = F(x)$ on an interval $J$, and if both functions satisfy the same set of initial conditions $y^{(i)}(x_0) = k_i, 0 \le i \le n-1$, show that $f_1(x) = f_2(x)$ for all $x$ in $J$.

2. Prove the lemma in this section for the case when $x < x_0$.

3. Let $u$ and $v$ be nonnegative continuous functions defined on an interval $[a, b]$ and let $c$ be a nonnegative constant. If

$$u(x) \le c + \int_a^x u(t)\, v(t)\, dt, \qquad a \le x \le b,$$

show that

$$u(x) \le c\left[\exp \int_a^x v(t)\, dt\right], \qquad a \le x \le b.$$

Suggestion: let $W(x) = c + \displaystyle\int_a^x u(t)\, v(t)\, dt$. Then $W'(x) - v(x)\, W(x) \le 0$.

**4.** Let the function $f$ be a solution of the equation

$$y'' + [k^2 + a(x)]y = 0 \tag{2}$$

for $x \geq x_0$. The function $a$ is assumed to be continuous for $x \geq x_0$.
(a) Show that $f$ is a solution of the nonhomogeneous equation

$$y'' + k^2 y = -a(x)f(x)$$

for $x \geq x_0$, and hence that

$$f(x) = A \sin(kx + \alpha) + \int_{x_0}^{x} \sin k(t - x)a(t)f(t)dt,$$

where $A$ and $\alpha$ are constants.

(b) If $\int_{x_0}^{\infty} |a(t)| \, dt$ converges, show that every solution of Eq. (2) is bounded

on $[x_0, \infty)$. Suggestion: use the result of Exercise 3.

**5.** Let $\alpha$ and $\beta$ be positive real numbers, with $\alpha \neq \beta$. Let the function $a$ be

continuous on $[x_0, \infty)$. If $\int_{x_0}^{\infty} |a(t)| \, dt$ converges, show that every solution

of the equation

$$[(D + \alpha)(D + \beta) + a(x)] \, y = 0$$

is bounded on $[x_0, \infty)$. Suggestion: follow the procedure used in Exercise 4.

**6.** Consider the equation

$$y'' + a_1(x) \, y' + a_2(x) \, y = 0. \tag{3}$$

(a) Making the change of variable $y = u \, F(x)$, where $F$ is to be determined
show that Eq. (3) becomes

$$Fu'' + (2F' + a_1 F) \, u' + (F'' + a_1 F' + a_2 F) u = 0. \tag{4}$$

(b) If the function $F$ in part (a) is chosen as

$$F(x) = \exp\left[ -\tfrac{1}{2} \int a_1(x) \, dx \right],$$

show that Eq. (4) assumes the form

$$u'' + g(x) u = 0, \tag{5}$$

in which the first derivative is missing.

(c) Put the equation

$$xy'' + y' + xy = 0$$

in the form (5).

7. If the function $f$ is a solution of the equation

$$xy'' + y' + xy = 0$$

on the interval $(0, \infty)$ show that $x^{1/2} f(x)$ is bounded as $x$ becomes infinite and hence that $f(x)$ tends to zero. Use the results of Exercises 4 and 6.

8. Let $u_1$ and $u_2$ be solutions of the differential equation

$$y'' + a_1(x) y' + a_2(x) y = 0$$

on an interval where $a_1$ and $a_2$ are continuous. If $W$ is the Wronskian of $u_1$ and $u_2$, show that

$$W'(x) + a_1(x) W(x) = 0$$

and hence that

$$W(x) = C \exp\left[ - \int a_1(x) \, dx \right].$$

This shows that either $W$ is identically zero ($C = 0$) or never zero ($C \neq 0$).

9. Let $u_1, u_2, \ldots, u_n$ be solutions of the equation

$$y^{(n)} + a_1(x) y^{(n-1)} + \cdots + a_{n-1}(x) y' + a_n(x) y = 0$$

on an interval $I$ where the functions $a_i$ are continuous. Let $W$ be the Wronskian of the functions $u_i$. Use the procedure for differentiating a determinant (Section 2.6) to show that

$$W'(x) + a_1(x) W(x) = 0$$

and hence that

$$W(x) = C \exp\left[ - \int a_1(x) \, dx \right]$$

for $x$ in $I$. This last formula is known as *Abel's formula.*

# VI

Systems
of Differential Equations

## 6.1  INTRODUCTION

In this chapter we shall deal with sets of simultaneous equations that involve several unknown functions and their derivatives. Such a set of equations is called a *system of differential equations*. An example of a system is

$$\frac{d^2x_1}{dt^2} - 2\frac{dx_1}{dt} - \frac{dx_2}{dt} = -3e^{3t},$$

$$(6.1)$$

$$\frac{dx_2}{dt} - 6x_1 = 0.$$

A solution of this system consists of an ordered pair of functions $(x_1, x_2)$ that satisfy both the equations. An example of a solution of the system (6.1) is the pair of functions $(f_1, f_2)$ where

$$f_1(t) = e^{3t}, \qquad f_2(t) = 2e^{3t}$$

for all $t$. To verify this assertion, we observe that

$$f_1''(t) - 2f_1'(t) - f_2'(t) = 9e^{3t} - 6e^{3t} - 6e^{3t} = -3e^{3t},$$
$$f_2'(t) - 6f_1(t) = 6e^{3t} - 6e^{3t} = 0.$$

Systems of differential equations commonly arise in problems of mechanics and electric circuits. For example, let $x_1(t)$, $x_2(t)$, and $x_3(t)$ denote the rectangular coordinates, at time $t$, of the center of mass of a moving object. Then, according to Newton's laws of motion we must have

$$m\mathbf{a} = \mathbf{F}, \tag{6.2}$$

where $m$ is the mass of the object,

$$\mathbf{a} = \ddot{x}_1\mathbf{i} + \ddot{x}_2\mathbf{j} + \ddot{x}_3\mathbf{k}$$

is the acceleration, and

$$\mathbf{F} = F_1\mathbf{i} + F_2\mathbf{j} + F_3\mathbf{k}$$

is the force. Equating corresponding components in the vector equation (6.2), we have

$$m\ddot{x}_1 = F_1, \qquad m\ddot{x}_2 = F_2, \qquad m\ddot{x}_3 = F_3. \tag{6.3}$$

In many instances the components $F_1$, $F_2$, and $F_3$ of the force are known functions of $t$, $x_1$, $x_2$, $x_3$, $\dot{x}_1$, $\dot{x}_2$, and $\dot{x}_3$. If this is the case, we have a system of three differential equations for $x_1$, $x_2$, and $x_3$. Applications of differential equations that involve moving bodies are discussed in Section 6.10. Electric circuits are considered in Section 6.11.

We consider one additional example. Suppose that two tanks each initially contain 100 gal of a solution of a chemical, with 20 lb of the chemical in the first tank and 10 lb in the second. At $t = 0$, water begins to flow into the first tank at the rate of 2 gal/min. The (stirred) mixture runs into the second tank at the rate of 2 gal/min, and the (stirred) mixture in the second tank runs out at the same rate. We wish to find formulas for the amounts of chemical in each tank at time $t$.

Let $x_1(t)$ and $x_2(t)$ denote the amounts in the first and second tanks, respectively. Then we must have

$$\frac{dx_1}{dt} = -\frac{2}{100}x_1,$$
$$\frac{dx_2}{dt} = \frac{2}{100}x_1 - \frac{2}{100}x_2. \tag{6.4}$$

since the rate of change of the amount of chemical in a tank must be equal to the rate at which it enters minus the rate at which it leaves. We also have

$$x_1(0) = 20, \qquad x_2(0) = 10. \tag{6.5}$$

Here the first equation of the system (6.4) involves only the one unknown $x_1$. Solving this equation subject to the first of the conditions (6.5), we find that

$$x_1(t) = 20e^{-t/50}. \tag{6.6}$$

Substituting for $x_1$ in the second equation of the system (6.4), we arrive at the equation

$$\frac{dx_2}{dt} + \frac{1}{50}x_2 = \frac{2}{5}e^{-t/50}$$

for $x_2$. The solution of this equation that satisfies the second of the conditions (6.5) is

$$x_2(t) = (\tfrac{2}{5}t + 10)\, e^{-t/50}. \tag{6.7}$$

Formulas (6.6) and (6.7) describe the desired solution of the system (6.4).

In the examples presented in this section, the number of unknown functions was always the same as the number of differential equations of the systems. This is not a coincidence, as we shall see in the next section. There we begin a more systematic discussion of systems of differential equations.

### Exercises for Section 6.1

1.  Verify that the ordered pair of functions $(f_1, f_2)$, where

$$f_1(t) = t^2, \qquad f_2(t) = 2t$$

for all $t$, is a solution of the system

$$x_1'' - x_1'x_2 + 4x_1 = 2,$$
$$x_1 - x_2' + x_2^2 = 5t^2 - 2.$$

2.  Verify that the ordered triple of functions $(f_1, f_2, f_3)$, where

$$f_1(t) = e^t, \qquad f_2(t) = 2e^t, \qquad f_3(t) = -e^t$$

for all $t$, constitutes a solution of the system

$$x_1' = x_2 + x_3, \qquad x_2' = x_1 - x_3, \qquad x_3' = -x_2 - x_3.$$

Also find another solution by inspection.

**3.** Describe a procedure that could be used for solving a system of the form

$$F(t, x_1, x_1') = 0,$$
$$G(t, x_1, x_2, x_2') = 0,$$
$$H(t, x_1, x_2, x_3, x_3') = 0,$$

where $t$ is the independent variable and $x_1$, $x_2$, and $x_3$ are the unknown functions. (Notice that the first equation involves only one unknown.)

**4.** Two tanks each contain 50 gal of a salt solution initially, with 10 lb of salt in the first tank and 20 lb in the second tank. A salt solution containing 2 lb of salt per gallon runs into the first tank at the rate of 1 gal/min. The mixture runs into the second tank at the rate of 1 gal/min, and the mixture in the second tank runs out at the same rate. Find the amount of salt in each tank after 30 min.

**5.** A radioactive substance A decays at a rate equal to $k_1$ times the amount of A present. Let $k_2$ be the proportion of A that goes into a second radioactive substance B, where B decays at a rate equal to $k_3$ times the amount of B present. Let $x_1(t)$ and $x_2(t)$ be the amounts of A and B present at time $t$, and let $x_1(0) = a$, $x_2(0) = b$.

(a) Find a system of differential equations for $x_1$ and $x_2$.

(b) Find formulas for $x_1$ and $x_2$. (Assume $k_1 \neq k_3$.)

**6.** Two bodies, of masses $m_1$ and $m_2$, are suspended from springs as shown in Fig. 6.1. The spring constants are $k_1$ and $k_2$. If $x_1$ and $x_2$ denote the directed distances downward of the bodies from their equilibrium positions, find a system of differential equations for $x_1$ and $x_2$.

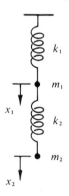

**Figure 6.1**

**7.** Two bodies, with masses $m_1$ and $m_2$, respectively, move along a straight line as shown in Fig. 6.2. Let $x_1$ and $x_2$ denote the respective directed

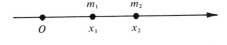

**Figure 6.2**

distances of the bodies from a fixed point on the line of motion. Assume that the bodies attract each other with a force equal to $km_1m_2/r^2$, where $k$ is a positive constant and $r$ is the distance between the bodies. Find a system of differential equations for $x_1$ and $x_2$.

## 6.2   FIRST-ORDER SYSTEMS

A system of differential equations for unknown functions $x_1, x_2, \ldots, x_n$ that is of the form

$$\frac{dx_1}{dt} = f_1(t, x_1, x_2, \ldots, x_n),$$

$$\frac{dx_2}{dt} = f_2(t, x_1, x_2, \ldots, x_n), \tag{6.8}$$

$$\ldots\ldots\ldots\ldots\ldots\ldots\ldots$$

$$\frac{dx_n}{dt} = f_n(t, x_1, x_2, \ldots, x_n),$$

where the functions $f_i$ are given, is called a *first-order system*. Notice that the number of equations is the same as the number of unknowns, and that no derivatives of order higher than 1 appear. A *solution* of the system is an ordered set of functions $(x_1, x_2, \ldots, x_n)$ that satisfies the system on some interval. The set of all solutions is called the *general solution*.

A set of auxiliary conditions of the form

$$x_1(t_0) = k_1, \quad x_2(t_0) = k_2, \ldots, \quad x_n(t_0) = k_n, \tag{6.9}$$

where $k_i$ are given numbers, is called a set of *initial conditions* for the system (6.8). The system (6.8) together with the conditions (6.9) is called an *initial value problem*. It can be shown that, with certain restrictions on the functions $f_i$, the initial value problem possesses a solution and that there is essentially only one solution. (See Section 6.12.) This means that if there were more equations than unknowns, we could not expect to find any solution that satisfied a set of initial conditions, while if there were fewer equations than unknowns we could expect more than one solution. We shall usually be able to find explicit formulas for the solutions of the systems that we encounter in this chapter.

We notice that although the system (6.4) is a first-order system, the system (6.1) is not. In particular, this latter system involves a derivative of order higher than 1. However, most systems that are not of the form (6.8) can be rewritten as first-order systems, as we shall presently show. In developing a theory for systems of differential equations, it is more convenient to work with first-order systems. However, the theory will still be applicable to systems such as (6.1).

Let us show how the system (6.1) can be replaced by a first-order system. We introduce new unknowns $u_1, u_2, u_3$ by setting

$$u_1 = x_1, \qquad u_2 = x_2, \qquad u_3 = \frac{dx_1}{dt}. \tag{6.10}$$

Then

$$\frac{du_1}{dt} = u_3,$$

and from the system (6.1) we have

$$\frac{du_3}{dt} - 2u_3 - \frac{du_2}{dt} = -3e^{3t},$$

$$\frac{du_2}{dt} - 6u_1 = 0.$$

Notice that only *first* derivatives appear in these equations. Upon solving algebraically for $du_2/dt$ and $du_3/dt$, we arrive at the first-order system

$$\frac{du_1}{dt} = u_3,$$

$$\frac{du_2}{dt} = 6u_1, \tag{6.11}$$

$$\frac{du_3}{dt} = 6u_1 + 2u_3 - 3e^{3t}$$

for $u_1, u_2$, and $u_3$. Since initial conditions for the system (6.11) are of the form

$$u_1(t_0) = k_1, \qquad u_2(t_0) = k_2, \qquad u_3(t_0) = k_3, \tag{6.12}$$

appropriate conditions for the original system (6.1) are of the form

$$x_1(t_0) = k_1, \qquad x_2(t_0) = k_2, \qquad x_1'(t_0) = k_3. \qquad (6.13)$$

More generally, suppose we have a system for two unknown functions $x_1$ and $x_2$ that is of the form

$$F[t, x_1, x_1', \ldots, x_1^{(m)}, x_2, x_2', \ldots, x_2^{(n)}] = 0.$$
$$G[t, x_1, x_1', \ldots, x_1^{(m)}, x_2, x_2', \ldots, x_2^{(n)}] = 0. \qquad (6.14)$$

Here $m$ and $n$ are the orders of the highest order derivatives of $x_1$ and $x_2$ that appear in the two equations. Suppose that it is possible to solve algebraically for $x_1^{(m)}$ and $x_2^{(n)}$, so that we can replace the system (6.14) by the equivalent[1] system

$$x_1^{(m)} = f[t, x_1, x_1', \ldots, x_1^{(m-1)}, x_2, x_2', \ldots, x_2^{(n-1)}],$$
$$x_2^{(n)} = g[t, x_1, x_1', \ldots, x_1^{(m-1)}, x_2, x_2', \ldots, x_2^{(n-1)}]. \qquad (6.15)$$

Let us set

$$u_1 = x_1, \qquad u_2 = x_1', \qquad u_3 = x_1'', \ldots, u_m = x_1^{(m-1)}$$
$$u_{m+1} = x_2, \qquad u_{m+2} = x_2', \qquad u_{m+3} = x_2'', \ldots, u_{m+n} = x_2^{(n-1)}. \qquad (6.16)$$

Then we have

$$u_m' = f(t, u_1, u_2, \ldots, u_m, u_{m+1}, \ldots, u_{m+n}),$$
$$u_{m+n}' = g(t, u_1, u_2, \ldots, u_m, u_{m+1}, \ldots, u_{m+n}),$$
$$u_1' = u_2,$$
$$u_2' = u_3,$$
$$\ldots\ldots\ldots\ldots$$
$$u_{m-1}' = u_m,$$
$$u_{m+1}' = u_{m+2},$$
$$\ldots\ldots\ldots\ldots\ldots$$
$$u_{m+n-1}' = u_{m+n}. \qquad (6.17)$$

This is a first-order system for the unknowns $u_1, u_2, \ldots, u_{m+n}$. Our discussion has been about a system (6.14) with two equations and two unknowns, but the general case of $k$ equations and $k$ unknowns can be treated similarly.

[1] Two systems are said to be *equivalent* if they have the same solutions.

An important special case is that of a single differential equation for one unknown function, of the form

$$x^{(n)} = F[t, x, x', \ldots, x^{(n-1)}].$$ (6.18)

Setting

$$u_1 = x, \qquad u_2 = x', \quad u_3 = x'', \ldots, \quad u_n = x^{(n-1)},$$ (6.19)

we have

$$
\begin{aligned}
u_1' &= u_2, \\
u_2' &= u_3, \\
&\cdots\cdots\cdots \\
u_{n-1}' &= u_n, \\
u_n' &= F(t, u_1, u_2, \ldots, u_n).
\end{aligned}
$$ (6.20)

The last equation of this first-order system comes from the differential equation (6.18). If $x$ is a solution of Eq. (6.18) the relations (6.19) yield a solution $(u_1, u_2, \ldots, u_n)$ of the system (6.20). On the other hand, if a solution $(u_1, u_2, \ldots, u_n)$ of the system (6.20) is known, the function $u_1$ is a solution of Eq. (6.18).

As an example, we consider the equation

$$\frac{d^3x}{dt^3} = 3tx - \frac{dx}{dt} + \frac{d^2x}{dt^2}.$$ (6.21)

Setting

$$u_1 = x, \qquad u_2 = x', \qquad u_3 = x'',$$

we have

$$
\begin{aligned}
u_1' &= u_2, \\
u_2' &= u_3, \\
u_3' &= 3tu_1 - u_2 + u_3.
\end{aligned}
$$ (6.22)

Notice that the initial conditions

$$x(t_0) = k_1, \qquad x'(t_0) = k_2, \qquad x''(t_0) = k_3$$

for Eq. (6.21) correspond to the initial conditions

$$u_1(t_0) = k_1, \qquad u_2(t_0) = k_2, \qquad u_3(t_0) = k_3$$

for the first-order system (6.22).

### Exercises for Section 6.2

In Exercises 1–6, rewrite the given system as a first-order system. Also, indicate what quantities must be specified in an appropriate set of auxiliary conditions for the given system at a point $t_0$.

**1.** $x_1' - x_2' - x_1 = \cos t$
$x_2' - 3x_1 = e^t$

**2.** $x_1' - 2x_2' - x_2 = t^2$
$x_1' - 3x_2' = 0$

**3.** $x_1' - x_2' = e^t$
$x_2'' - x_1 - x_2 - x_2' = \sin t$

**4.** $x_1'' = 1$
$x_2'' = 2$

**5.** $x_1''' - x_1' + x_1 x_2' = \sin t$
$x_1'' - x_2 x_1' - x_2'' = \cos t$

**6.** $x_1'' = x_1 \sin t + x_2 x_3$
$x_2'' = x_1^2 x_2' + x_1'$
$x_3' = x_1 x_2 x_3$

In Exercises 7–10, rewrite the differential equation as a first-order system.

**7.** $x'' - tx' + x^2 = \sin t$

**8.** $x'' + x = 0$

**9.** $x''' - x'' + x = e^t$

**10.** $x^{(4)} + x'''x - (x'')^2 = 0$

## 6.3   LINEAR SYSTEMS WITH CONSTANT COEFFICIENTS

A first-order system of the special form

$$\frac{dx_1}{dt} = a_{11}(t)\, x_1 + a_{12}(t)\, x_2 + \cdots + a_{1n}(t)\, x_n + b_1(t),$$

$$\frac{dx_2}{dt} = a_{21}(t)\, x_1 + a_{22}(t)\, x_2 + \cdots + a_{2n}(t)\, x_n + b_2(t), \qquad (6.23)$$

$$\cdots\cdots\cdots\cdots\cdots\cdots\cdots\cdots\cdots\cdots\cdots\cdots$$

$$\frac{dx_n}{dt} = a_{n1}(t)\, x_1 + a_{n2}(t)\, x_2 + \cdots + a_{nn}(t)\, x_n + b_n(t),$$

where the functions $a_{ij}$ and $b_i$ are given is called a first-order *linear* system. The functions $a_{ij}$ are called the *coefficients* of the system. If each of the functions $b_i$ is the zero function, the system is said to be *homogeneous*; otherwise it is said to be *nonhomogeneous*. The system (6.23) can be written more briefly as

$$\frac{dx_i}{dt} = \sum_{j=1}^{n} a_{ij}(t)x_j + b_i(t), \qquad 1 \le i \le n. \tag{6.24}$$

In an initial value problem associated with the system we wish to find a solution that satisfies conditions of the form

$$x_1(t_0) = k_1, \quad x_2(t_0) = k_2, \dots, \quad x_n(t_0) = k_n. \tag{6.25}$$

For example, when $n = 2$ (two equations and two unknowns) the initial value problem has the form

$$\frac{dx_1}{dt} = a_{11}(t)x_1 + a_{12}(t)x_2 + b_1(t),$$

$$\frac{dx_2}{dt} = a_{21}(t)x_1 + a_{22}(t)x_2 + b_2(t),$$

$$x_1(t_0) = k_1, \qquad x_2(t_0) = k_2.$$

In what follows, we restrict our attention to linear systems with constant coefficients. For convenience we use the operator notation

$$Df = f', \qquad Df(t) = f'(t).$$

A first-order linear system with constant coefficients has the form

$$\begin{aligned}
Dx_1 &= a_{11}x_1 + a_{12}x_2 + \cdots + a_{1n}x_n + b_1(t), \\
Dx_2 &= a_{21}x_1 + a_{22}x_2 + \cdots + a_{2n}x_n + b_2(t), \\
&\;\cdots\cdots\cdots\cdots\cdots\cdots\cdots\cdots\cdots\cdots\cdots\cdots \\
Dx_n &= a_{n1}x_1 + a_{n2}x_2 + \cdots + a_{nn}x_n + b_n(t),
\end{aligned} \tag{6.26}$$

where the $a_{ij}$ are constants. We shall also be concerned with more general linear systems with constant coefficients, of the form

$$\begin{aligned}
P_{11}(D)\,x_1 + P_{12}(D)\,x_2 + \cdots + P_{1n}(D)\,x_n &= b_1(t), \\
P_{21}(D)\,x_1 + P_{22}(D)\,x_2 + \cdots + P_{2n}(D)\,x_n &= b_2(t), \\
&\cdots\cdots\cdots\cdots\cdots\cdots\cdots\cdots\cdots \\
P_{n1}(D)\,x_1 + P_{n2}(D)\,x_2 + \cdots + P_{nn}(D)\,x_n &= b_n(t),
\end{aligned} \tag{6.27}$$

where the $P_{ij}(D)$ are polynomial operators.[2] Every system of the form (6.26) is of the form (6.27); but not every system of the form (6.27) is a first-order system. Usually, however, a system such as (6.27) can be rewritten as a first-order system, as was shown in Section 6.2.

One procedure that can be used to solve systems of the type (6.27) is called the *method of elimination.* The theory and technique are reminiscent of the elimination method for solving systems of linear algebraic equations. Two systems of differential equations are said to be *equivalent* if they have the same solutions. In the method of elimination, we replace a given system by an equivalent, but simpler, system that is relatively easier to solve.

The reduction to a simpler system is carried out by performing a sequence of operations, each of which leads to an equivalent system. These operations are of three types. First, we can simply interchange two equations of the system. For example, the two systems

$$Dx_1 = 1, \qquad Dx_2 = 2,$$
$$Dx_2 = 2, \qquad Dx_1 = 1,$$

are equivalent. Notice, however, that the two systems

$$Dx_1 = 1, \qquad Dx_2 = 1,$$
$$Dx_2 = 2, \qquad Dx_1 = 2,$$

are *not* equivalent, because a solution consists of an *ordered* pair of functions. The second type of operation consists of simply multiplying through in one equation of the system by a nonzero constant. In the third type of operation we operate on both members of one equation of the system, say the $i$th with a polynomial operator $Q(D)$ and add the result to another equation of the system, say the $j$th. In the new system, only the $j$th equation has changed. To illustrate, suppose that the $i$th and $j$th equations of the original system are

$$P_{i1}(D) x_1 + \cdots + P_{in}(D) x_n = b_i(t)$$
$$P_{j1}(D) x_1 + \cdots + P_{jn}(D) x_n = b_j(t). \tag{6.28}$$

Operating on both members of the $i$th equation with $Q(D)$ and adding the result to the $j$th equation, we have

$$P_{i1}(D) x_1 + \cdots + P_{in}(D) x_n = b_i(t)$$
$$[P_{j1}(D) + Q(D) P_{i1}(D)]x_1 + \cdots + [P_{jn}(D) + Q(D)P_{in}(D)]x_n \tag{6.29}$$
$$= b_j(t) + Q(D) b_i(t)$$

---

[2] See Section 5.2.

Notice that the equation

$$Q(D) P_{i1}(D) x_1 + \cdots + Q(D) P_{in}(D) = Q(D) b_i(t)$$

does not appear in either system. It is not hard to verify that if an ordered $n$-tuple of functions $(x_1, x_2, \ldots, x_n)$ satisfies the pair of equations (6.28) it also satisfies the pair (6.29) and vice versa.[3]

In order to illustrate the method, we solve the system

$$-(D + 2)x_1 + (D^2 - 4)x_2 = 4t,$$
$$(D + 3)x_1 + (D + 7)x_2 = 0. \tag{6.30}$$

Let us eliminate $x_1$. Adding the second equation to the first, we obtain

$$x_1 + (D^2 + D + 3)x_2 = 4t,$$
$$(D + 3)x_1 + (D + 7)x_2 = 0. \tag{6.31}$$

Next we operate on both members of the first equation with $(D + 3)$, obtaining the equation

$$(D + 3)x_1 + (D + 3)(D^2 + D + 3)x_2 = 4 + 12t.$$

Subtracting this equation from the second equation of the system (6.31), we obtain the system

$$x_1 + (D^2 + D + 3)x_2 = 4t,$$
$$[-(D + 3)(D^2 + D + 3) + (D + 7)]x_2 = -4 - 12t. \tag{6.32}$$

The second equation, which involves only the one unknown $x_2$ can be written as

$$(D + 1)^2(D + 2)x_2 = 4 + 12t.$$

Using the methods of Chapter 5 to solve this equation, we have

$$x_2 = c_1 e^{-t} + c_2 t e^{-t} + c_3 e^{-2t} + 6t - 13. \tag{6.33}$$

From the first equation of the system (6.32) we have

$$x_1 = -(D^2 + D + 3)x_2 + 4t$$

---

[3] The functions $b_i$, $x_1, x_2, \ldots, x_n$ must be sufficiently differentiable, so that all indicated derivatives exist. This is usually the case, in practice.

or

$$x_1 = -3c_1 e^{-t} + c_2(1 - 3t)e^{-t} - 5c_3 e^{-2t} - 14t + 33. \qquad (6.34)$$

Formulas (6.33) and (6.34) describe the general solution of the system (6.30). Notice that three arbitrary constants are involved in the description of the general solution, even though the system involves only two unknowns. This is in accordance with the fact that the system (6.30) can be replaced by a *first-order* system with three unknown functions.

In the general case where we have a system of the form (6.27), we attempt to find an equivalent system of the form[4]

$$Q_{11}(D) x_1 + Q_{12}(D) x_2 + \cdots + Q_{1n}(D) x_n = f_1(t),$$
$$Q_{22}(D) x_2 + \cdots + Q_{2n}(D) x_n = f_2(t), \qquad (6.35)$$
$$\cdots\cdots\cdots\cdots\cdots\cdots\cdots\cdots\cdots\cdots\cdots\cdots$$
$$Q_{n-1,\,n-1}(D) x_{n-1} + Q_{n-1,\,n}(D) x_n = f_{n-1}(t),$$
$$Q_{nn}(D) x_n = f_n(t),$$

where none of the operators $Q_{ii}(D)$, $1 \le i \le n$, is the zero operator. We can solve the $n$th equation for $x_n$, then find $x_{n-1}$ from the $(n-1)$st equation, and so on.

As a final example, we consider the first-order homogeneous system for three unknowns,

$$Dx_1 = 3x_1 + x_2 - 2x_3,$$
$$Dx_2 = -x_1 + 2x_2 + x_3, \qquad (6.36)$$
$$Dx_3 = 4x_1 + x_2 - 3x_3.$$

This system can be written as

$$(D-3)x_1 - x_2 + 2x_3 = 0,$$
$$x_1 + (D-2)x_2 - x_3 = 0, \qquad (6.37)$$
$$-4x_1 - x_2 + (D+3)x_3 = 0.$$

The second equation can be used to eliminate $x_1$ from the first and third equations. First we multiply through in the second equation by 4 and add the result to the third equation. Next we operate on the second equation with $(D-3)$ and subtract the result from the first equation. In this way we arrive at the equivalent system

---

[4] The unknowns may have to be renumbered.

$$(-D^2 + 5D - 7)x_2 + (D - 1)x_3 = 0,$$
$$x_1 + (D - 2)x_2 - x_3 = 0, \tag{6.38}$$
$$(4D - 9)x_2 + (D - 1)x_3 = 0.$$

We next eliminate $x_3$ between the first and third equations. Subtracting the third equation from the first, we have

$$(-D^2 + D + 2)x_2 = 0,$$
$$x_1 + (D - 2)x_2 - x_3 = 0, \tag{6.39}$$
$$(4D - 9)x_2 + (D - 1)x_3 = 0.$$

The first equation of this system, which involves only the one unknown $x_2$, can be written as

$$(D - 2)(D + 1)x_2 = 0.$$

Hence we have

$$x_2 = c_1 e^{-t} + c_2 e^{2t}. \tag{6.40}$$

From the third equation of the system (6.39) we have

$$(D - 1)x_3 = (9 - 4D)x_2 = 13c_1 e^{-t} + c_2 e^{2t}.$$

This is a first-order equation for $x_3$ and we find that

$$x_3 = -\tfrac{13}{2}c_1 e^{-t} + c_2 e^{2t} + c_3 e^t. \tag{6.41}$$

The unknown $x_2$ can now be found from the second equation of the system (6.39). We have

$$x_1 = (2 - D)x_2 + x_3$$

or

$$x_1 = -\tfrac{7}{2}c_1 e^{-t} + c_2 e^{2t} + c_3 e^t. \tag{6.42}$$

The formulas (6.41) and (6.42) become slightly simpler if we set $c_1 = 2c_1'$, where $c_1'$ is an arbitrary constant. With this change the formulas (6.40) (6.41), and (6.42) become

$$x_1 = -7c_1' e^{-t} + c_2 e^{2t} + c_3 e^t,$$
$$x_2 = 2c_1' e^{-t} + c_2 e^{2t}, \tag{6.43}$$
$$x_3 = -13c_1' e^{-t} + c_2 e^{2t} + c_3 e^t.$$

## Exercises for Section 6.3

1. Show that the linear system

$$P_{11}(D)x_1 + P_{12}(D)x_2 = b_1(t),$$
$$P_{21}(D)x_1 + P_{22}(D)x_2 = b_2(t),$$

where

$$P_{ij}(D) = a_{ij} D^2 + b_{ij} D + c_{ij}, \quad i = 1, 2, \quad j = 1, 2,$$

can be replaced by a first-order system if $a_{11}a_{22} - a_{21}a_{12} \neq 0$.

2. Show that the system

$$Dx_1 - x_2 = 0, \qquad -x_1 + Dx_2 = 0$$

is equivalent to the system

$$Dx_1 - x_2 = 0, \qquad (D^2 - 1)x_1 = 0,$$

but that it is not equivalent to the system

$$D^2 x_1 - Dx_2 = 0, \qquad (D^2 - 1)x_1 = 0.$$

In Examples 3–17, find the general solution of the system. If initial conditions are given, also find the solution that satisfies the conditions.

3. $Dx_1 = -4x_1 - 6x_2 + 9e^{-3t}$
   $Dx_2 = x_1 + x_2 - 5e^{-3t} \quad x_1(0) = -9, \quad x_2(0) = 4$

4. $Dx_1 = -2x_1 + x_2$
   $Dx_2 = -3x_1 + 2x_2 + 2 \sin t \quad x_1(0) = 3, \quad x_2(0) = 4$

5. $Dx_1 = -x_1 - 3e^{-2t}$
   $Dx_2 = -2x_1 - x_2 - 6e^{-2t}$

6. $D^2 x_1 + (D + 2)x_2 = 2e^{-2t}$
   $Dx_1 - (D + 2)x_2 = 0 \quad x_1(0) = 4, \quad x_2(0) = 1, \quad x_1'(0) = -2$

7. $(2D + 1)x_1 + (D^2 - 4)x_2 = -7e^{-t}$
   $Dx_1 - (D + 2)x_2 = -3e^{-t}$

**8.**  $(D + 2)x_1 + (D^2 + 2D)x_2 = 5e^{-t}$
$(D + 1)x_1 - (D + 2)x_2 = 0$

**9.**    $(D^2 + 1)x_1 + 2Dx_2 = 0$
$- 3(D^2 + 1)x_1 + 2(D^2 + 2)x_2 = 0 \quad x_1(0) = 1, \quad x_2(0) = 1,$
$x_1'(0) = 0, \quad x_2'(0) = -1$

**10.**  $(D^3 - 2D^2 + 3D)x_1 + (2D^2 - 8)x_2 = 4 - 6t$
$Dx_1 - (D + 2)x_2 = -2t$

**11.**  $Dx_1 = x_1 - 3x_2 + 2x_3$
$Dx_2 = -x_2$
$Dx_3 = -x_2 - 2x_3 \quad x_1(0) = -3, \quad x_2(0) = 0, \quad x_3(0) = 3$

**12.**  $Dx_1 = -x_1 + x_2$          **13.**  $Dx_1 = x_1 - 2x_2 - t^2$
$Dx_2 = 2x_1 - 2x_2 + 2x_3$              $Dx_2 = x_1 + x_3 - 1 - t^2$
$Dx_3 = -x_2 - x_3$                    $Dx_3 = -2x_1 + 2x_2 - x_3 + 2t^2 + 2t$

**14.**  $Dx_1 = x_1 + x_2$          **15.**  $Dx_1 = x_1 - 3x_2 + 2x_3$
$Dx_2 = -2x_1 + x_2 - 2x_3 + e^{2t}$          $Dx_2 = -2x_2 + 2x_3$
$Dx_3 = -x_2 + x_3$                    $Dx_3 = x_1 - 5x_2 + 2x_3$

**16.**  $Dx_1 = -x_1 + x_3$          **17.**  $Dx_1 = -x_1 - x_3 + \cos t$
$Dx_2 = 2x_3$                          $Dx_2 = -x_2 - x_3 + \sin t$
$Dx_3 = x_1 - 2x_2 - 3x_3$              $Dx_3 = -2x_3 + \cos t + 2 \sin t$

## 6.4  MATRIX FORMULATION OF LINEAR SYSTEMS

In this and the next several sections we shall develop a theory for linear systems of differential equations that relies heavily on matrix algebra. This approach yields compact formulas that are convenient for theoretical purposes. We shall also present a practical method of solution for linear systems with constant coefficients that uses matrices. In the examples of the previous section, we dealt with systems involving only two or three unknowns and the coefficients of these systems were integers. In a practical problem, the solver may be faced with a system having a large number of unknowns and coefficients with several significant digits. For such systems, it is desirable to have a method of solution in which the computational work is facilitated.

To begin with, let us show how the first-order linear system

$$\frac{dx_1}{dt} = a_{11}(t)\, x_1 + a_{12}(t)\, x_2 + \cdots + a_{1n}(t)\, x_n + b_1(t),$$

$$\frac{dx_2}{dt} = a_{21}(t)\, x_1 + a_{22}(t)\, x_2 + \cdots + a_{2n}(t)\, x_n + b_2(t), \qquad (6.44)$$

........................................................

$$\frac{dx_n}{dt} = a_{n1}(t)\, x_1 + a_{n2}(t)\, x_2 + \cdots + a_{nn}(t)x_n + b_n(t),$$

can be written in a compact form by the use of matrices. Let $(u_1, u_2, \ldots, u_n)$ be an ordered set of $n$ functions defined on an interval. We write

$$\mathbf{u} = \begin{bmatrix} u_1 \\ u_2 \\ \vdots \\ u_n \end{bmatrix}, \qquad (6.45)$$

and call $\mathbf{u}$ a *vector function* with components $u_1, u_2, \ldots, u_n$. The vector function each of whose components is the zero function is denoted by $\mathbf{0}$. The product $f\mathbf{u}$ of a function $f$ and a vector function $\mathbf{u}$ is defined as

$$f\mathbf{u} = \begin{bmatrix} fu_1 \\ fu_2 \\ \vdots \\ fu_n \end{bmatrix}. \qquad (6.46)$$

The derivative $\mathbf{u}'$ of the vector function $\mathbf{u}$ is defined by the relation

$$\mathbf{u}' = \begin{bmatrix} u'_1 \\ u'_2 \\ \vdots \\ u'_n \end{bmatrix}. \qquad (6.47)$$

More generally we can define *matrix functions*. For example, associated with the system (6.44) is the $n \times n$ matrix function whose elements are the functions $a_{ij}$. Examples of a vector function $\mathbf{u}$ and a matrix function $A$ are

$$\mathbf{u}(t) = \begin{bmatrix} e^t \\ \sin t \end{bmatrix}, \qquad A(t) = \begin{bmatrix} 2t & \cos t \\ t^2 & e^t \end{bmatrix}.$$

The definitions of the sum and product of two matrix functions are similar to those for ordinary matrices, and we shall not bother to write them down. A matrix or vector function is said to be continuous if each of its elements is continuous. The set of all (real) $n$-dimensional vector functions defined on an interval forms a real vector space (Exercise 6).

Let us write

$$\mathbf{x} = \begin{bmatrix} x_1 \\ x_2 \\ \vdots \\ x_n \end{bmatrix}, \qquad \mathbf{b} = \begin{bmatrix} b_1 \\ b_2 \\ \vdots \\ b_n \end{bmatrix},$$

and let $A$ be the $n \times n$ matrix function with elements $a_{ij}$. Then the system (6.44) can be written as

$$\mathbf{x}' = A\mathbf{x} + \mathbf{b} \qquad (6.48)$$

or

$$\frac{d\mathbf{x}}{dt} = A(t)\mathbf{x} + \mathbf{b}(t).$$

To see this, we observe that

$$\begin{bmatrix} a_{11} & a_{12} & \cdots & a_{1n} \\ a_{21} & a_{22} & \cdots & a_{2n} \\ \vdots & & & \vdots \\ a_{n1} & a_{n2} & \cdots & a_{nn} \end{bmatrix} \begin{bmatrix} x_1 \\ x_2 \\ \cdot \\ x_n \end{bmatrix} = \begin{bmatrix} a_{11}x_1 + a_{12}x_2 + \cdots + a_{1n}x_n \\ a_{21}x_1 + a_{22}x_2 + \cdots + a_{2n}x_n \\ \vdots \\ a_{n1}x_1 + a_{n2}x_2 + \cdots + a_{nn}x_n \end{bmatrix}$$

and equate corresponding components in the vector equation (6.48). For example, the system

$$x_1' = 2x_1 - 3x_2 + 3e^t,$$
$$x_2' = -x_1 + x_2 - e^t,$$

can be written as $\mathbf{x}' = A\mathbf{x} + \mathbf{b}$, where

$$A = \begin{bmatrix} 2 & -3 \\ -1 & 1 \end{bmatrix}, \qquad \mathbf{b}(t) = \begin{bmatrix} 3e^t \\ -e^t \end{bmatrix}.$$

A set of initial conditions

$$x_1(t_0) = k_1, \quad x_2(t_0) = k_2, \dots, \quad x_n(t_0) = k_n$$

can be described by the vector equation $\mathbf{x}(t_0) = \mathbf{k}$, where $\mathbf{k}$ is the constant vector with components $k_1, k_2, \ldots, k_n$.

We now consider the special case of a homogeneous equation

$$\mathbf{x}' = A\mathbf{x}, \tag{6.49}$$

where the matix function $A$ is continuous on an inverval $\mathscr{I}$. The set of all real solutions of this equation on $\mathscr{I}$ [5] constitutes a real vector space. To verify this fact, we first observe that if $\mathbf{u}$ is a solution on $\mathscr{I}$ then $c\mathbf{u}$, where $c$ is any real number, is also a solution. For we have

$$(c\mathbf{u})' = c\mathbf{u}' = c(A\mathbf{u}) = A(c\mathbf{u}).$$

Also, if $\mathbf{u}$ and $\mathbf{v}$ are both solutions then $\mathbf{u} + \mathbf{v}$ is also a solution, because since $\mathbf{u}' = A\mathbf{u}$ and $\mathbf{v}' = A\mathbf{v}$ we have

$$(\mathbf{u} + \mathbf{v})' = \mathbf{u}' + \mathbf{v}' = A\mathbf{u} + A\mathbf{v} = A(\mathbf{u} + \mathbf{v}).$$

Thus the set of all solutions is a subspace of the space of all vector functions defined on $\mathscr{I}$. The zero vector function $\mathbf{0}$ is evidently a solution of Eq. (6.49). It is called the *trivial solution*. From properties of a vector space, we see that if $\mathbf{u}_1, \mathbf{u}_2, \ldots, \mathbf{u}_m$ are solutions of Eq. (6.49) and if $c_1, c_2, \ldots, c_m$ are real numbers, then $c_1\mathbf{u}_1 + c_2\mathbf{u}_2 + \cdots + c_m\mathbf{u}_m$ is a solution.

We shall also have occasion to consider complex solutions of a vector differential equation of the form (6.48) or (6.49). Let $\mathbf{w}$ be a *complex vector function* of the form

$$\mathbf{w} = \mathbf{u} + i\mathbf{v},$$

where $\mathbf{u}$ and $\mathbf{v}$ are real vector functions with components $u_1, u_2, \ldots, u_n$ and $v_1, v_2, \ldots, v_n$, respectively. The components of $\mathbf{w}$ are the complex functions $w_j = u_j + iv_j$, $1 \leq j \leq n$. The derivative of $\mathbf{w}$ is defined as $\mathbf{w}' = \mathbf{u}' + i\mathbf{v}'$. If $\mathbf{w}$ is a complex solution of Eq. (6.49), so that $\mathbf{w}' = A\mathbf{w}$, then $\mathbf{u}$ and $\mathbf{v}$ are real solutions of that equation.[6] To see this, observe that

$$(\mathbf{u} + i\mathbf{v})' = A(\mathbf{u} + i\mathbf{v})$$

or

$$\mathbf{u}' + i\mathbf{v}' = A\mathbf{u} + iA\mathbf{v}.$$

---

[5] If $A$ is continuous on $\mathscr{I}$ and if $\mathbf{u}$ is a solution on some interval $\mathscr{J}$ that is contained in $\mathscr{I}$, it can be shown (Chapter 8) that there is a solution $\mathbf{v}$ on $\mathscr{I}$ such that $\mathbf{u}(t) = \mathbf{v}(t)$ for $t$ in $\mathscr{J}$. For this reason we consider only solutions that exist throughout $\mathscr{I}$.

[6] We assume that $A$ is real; that is, that the elements of $A$ are real functions.

Equating real and imaginary parts, we have

$$\mathbf{u}' = A\mathbf{u}, \qquad \mathbf{v}' = A\mathbf{v}.$$

Hence $\mathbf{u}$ and $\mathbf{v}$ are both real solutions. It is left to the reader (Exercise 7) to verify that the set of all complex solutions of Eq. (6.49) is a complex vector space. This space contains all the real solutions of the equation.

## Exercises for Solution 6.4

1.  Write out the equations of the system that corresponds to the vector equation $\mathbf{x}' = A\mathbf{x} + \mathbf{b}$ if $A$ and $\mathbf{b}$ are as given.

   (a)  $A(t) = \begin{bmatrix} 2 & -1 \\ 3 & 0 \end{bmatrix}$,   $\mathbf{b}(t) = \begin{bmatrix} e^t \\ 3e^{-2t} \end{bmatrix}$

   (b)  $A(t) = \begin{bmatrix} t^2 & -e^t \\ 1 & \cos t \end{bmatrix}$,   $\mathbf{b}(t) = \begin{bmatrix} 1-t \\ 0 \end{bmatrix}$

   (c)  $A(t) = \begin{bmatrix} 2 & -1 & 0 \\ 0 & 1 & 1 \\ 3 & 2 & 0 \end{bmatrix}$,   $\mathbf{b}(t) = \begin{bmatrix} e^{2t} \\ 0 \\ e^t \end{bmatrix}$

   (d)  $A(t) = \begin{bmatrix} t+1 & 0 & t \\ -1 & 2-t & 0 \\ t & 0 & 0 \end{bmatrix}$,   $\mathbf{b} = \mathbf{0}$.

2.  Find $A$ and $\mathbf{b}$ if the given system is written as $\mathbf{x}' = A\mathbf{x} + \mathbf{b}$.

   (a)  $x_1' = -2x_1 + x_2 + \cos 2t$
        $x_2' = -x_1 - x_2 - 2\sin 2t$

   (b)  $x_1' = e^t x_1 - e^{-t} x_2$
        $x_2' = 2e^{-t} x_1 + 3e^t x_2$

   (c)  $x_1' = 2x_1 + x_2 - x_3 + 2e^{-t}$
        $x_2' = x_1 - x_2 \qquad\quad - e^{-t}$
        $x_3' = \qquad\quad x_2 + 2x_3$

   (d)  $x_1' = te^t x_1 - x_2 + 1$
        $x_2' = x_3 - e^t$
        $x_3' = x_1 + e^t x_2$

3.  If

$$A(t) = \begin{bmatrix} -2 & 1 \\ -3 & 2 \end{bmatrix}, \qquad \mathbf{b}(t) = \begin{bmatrix} 0 \\ 2\sin t \end{bmatrix},$$

verify that a solution of $\mathbf{x}' = A\mathbf{x} + \mathbf{b}$ is

$$\mathbf{u}(t) = \begin{bmatrix} -\sin t \\ -\cos t - 2\sin t \end{bmatrix}.$$

**4.** If

$$A(t) = \begin{bmatrix} 1 & -3 & 2 \\ 0 & -1 & 0 \\ 0 & -1 & -2 \end{bmatrix},$$

verify that a solution of $\mathbf{x}' = A\mathbf{x}$ is

$$\mathbf{u}(t) = \begin{bmatrix} -2e^{-2t} \\ 0 \\ 3e^{-2t} \end{bmatrix},$$

**5.** If $\mathbf{w} = \mathbf{u} + i\mathbf{v}$ is a complex solution of the equation $\mathbf{x}' = A\mathbf{x} + \mathbf{b}$, where $\mathbf{b} = \mathbf{c} + i\mathbf{d}$, show that $\mathbf{u}$ and $\mathbf{v}$ are real solutions of $\mathbf{x}' = A\mathbf{x} + \mathbf{c}$ and $\mathbf{x}' = A\mathbf{x} + \mathbf{d}$, respectively. (Here $\mathbf{u}$, $\mathbf{v}$, $A$, $\mathbf{c}$, and $\mathbf{d}$ are assumed to be real.)

**6.** Show that the set of all $n$-dimensional real (complex) vector functions defined on an interval forms a vector space over the real (complex) numbers.

**7.** Show that the set of all complex solutions of the equation $\mathbf{x}' = A\mathbf{x}$ on an interval $\mathscr{I}$ is a complex vector space.

## 6.5 FUNDAMENTAL SETS OF SOLUTIONS

It is fairly evident that the set of all real $n$-dimensional vector functions defined on an interval is a real vector space (Exercise 6, Section 6.4). We also found it rather easy to show in the previous section that the set of all real solutions of the vector differential equation

$$\mathbf{x}' = A\mathbf{x} \tag{6.50}$$

on an interval formed a vector space. The additional fact presented in the following theorem is not so apparent.

**Theorem 6.1** Let $A$ be an $n \times n$ matrix function whose elements $a_{ij}$ are continuous functions on an interval $\mathscr{I}$. Then the set of all real solutions of Eq. (6.50) on $\mathscr{I}$ is a real vector space of dimension $n$.

A proof of this theorem that makes use of some results proved in Chapter 8 is deferred until Section 6.12.

In order to solve Eq. (6.50), we must find a basis for the space of solutions. If $\mathbf{u}_1, \mathbf{u}_2, \ldots, \mathbf{u}_n$ are linearly independent solutions, the general solution consists of all vector functions of the form

$$c_1\mathbf{u}_1 + c_2\mathbf{u}_2 + \cdots + c_n\mathbf{u}_n, \tag{6.51}$$

where $c_1, c_2, \ldots, c_n$ are real numbers. A basis for the solution space of the equation (6.50) is called a *fundamental set* of solutions. A criterion for the linear independence of a set of $n$ vector functions with $n$ components is given by the following theorem.

**Theorem 6.2**   Let $\mathbf{u}_1, \mathbf{u}_2, \ldots, \mathbf{u}_n$ be $n$-dimensional vector functions defined on an inverval $\mathscr{I}$. Let $U$ be the matrix function whose column vectors are these vector functions. If the vector functions are linearly dependent then det $U(t) = 0$ for all $t$ in $\mathscr{I}$. Hence if there exists even one point $t_1$ in $\mathscr{I}$ such that det $U(t_1) \neq 0$, the vector functions are linearly independent.

PROOF.    Suppose that the vector functions are linearly dependent. Then there exist numbers $c_1, c_2, \ldots, c_n$, not all zero, such that

$$c_1 \, \mathbf{u}_1(t) + c_2 \, \mathbf{u}_2(t) + \cdots + c_n \, \mathbf{u}_n(t) = 0$$

for all $t$ in $\mathscr{I}$. This equation may be written as

$$U(t) \, \mathbf{c} = \mathbf{0},$$

where $\mathbf{c}$ is the column vector with components $c_1, c_2, \ldots, c_n$. Since $\mathbf{c}$ is not the zero vector, $U(t)$ must be singular for every $t$ in $\mathscr{I}$. If there is a point $t_1$ such that $U(t_1)$ is nonsingular, the vector functions cannot be linearly dependent. Hence they must be linearly independent.

Theorem 6.2 does *not* say that if the determinant, det $U(t)$, of a set of vector functions is zero on an interval then the vector functions are linearly dependent. The vector functions $\mathbf{u}_1$ and $\mathbf{u}_2$, where

$$\mathbf{u}_1(t) = \begin{bmatrix} t \\ 1 \end{bmatrix}, \qquad \mathbf{u}_2(t) = \begin{bmatrix} t^2 \\ t \end{bmatrix}$$

for all $t$, provide an example. The condition

$$c_1 \, \mathbf{u}_1(t) + c_2 \, \mathbf{u}_2(t) = 0$$

is satisfied when $t = 3$ only if $c_1 = c_2 = 0$, so the vector functions are linearly independent. However, det $U(t) = 0$ for all $t$. If the vector functions $\mathbf{u}_1, \mathbf{u}_2, \ldots,$ $\mathbf{u}_n$ are linearly independent and *in addition are solutions of Eq.* (6.50), it can be shown (Section 6.12) that det $U(t)$ is *never* zero.

As an application of the above theory, let us consider the equation $\mathbf{x}' = A\mathbf{x}$, where

$$A = \begin{bmatrix} 3 & 1 & -2 \\ -1 & 2 & 1 \\ 4 & 1 & -3 \end{bmatrix}.$$

(This is the system (6.36), formulated in terms of matrices.) The reader may verify directly (or by the use of the formulas (6.43)) that each of the vector functions $\mathbf{u}_1, \mathbf{u}_2, \mathbf{u}_3$, where

$$\mathbf{u}_1(t) = e^{-t}\begin{bmatrix} -7 \\ 2 \\ -13 \end{bmatrix}, \qquad \mathbf{u}_2(t) = e^{2t}\begin{bmatrix} 1 \\ 1 \\ 1 \end{bmatrix}, \qquad \mathbf{u}_3(t) = e^{t}\begin{bmatrix} 1 \\ 0 \\ 1 \end{bmatrix}$$

for all $t$, is a solution of the equation. If $U(t)$ is the matrix whose column vectors are $\mathbf{u}_1(t), \mathbf{u}_2(t), \mathbf{u}_3(t)$, then

$$\det U(0) = \begin{vmatrix} -7 & 1 & 1 \\ 2 & 1 & 0 \\ -13 & 1 & 1 \end{vmatrix} = 6 \neq 0.$$

By Theorem 6.2, the vectors are linearly independent. By Theorem 6.1, the general solution of the equation may be written as

$$\mathbf{x} = c_1 e^{-t}\begin{bmatrix} -7 \\ 2 \\ -13 \end{bmatrix} + c_2 e^{2t}\begin{bmatrix} 1 \\ 1 \\ 1 \end{bmatrix} + e_3 e^{t}\begin{bmatrix} 1 \\ 0 \\ 1 \end{bmatrix}$$

where $c_1, c_2$, and $c_3$ are arbitrary constants.

The derivative of a vector function was defined by the relation (6.47). We now consider the more general notion of the derivative of a matrix function.[7] If $B$ is an $m \times n$ matrix function whose elements are the functions $b_{ij}$, we define $B'$, the derivative of $B$, as

$$B' = \begin{bmatrix} b'_{11} & b'_{12} & \cdots & b'_{1n} \\ b'_{21} & b'_{22} & \cdots & b'_{2n} \\ \vdots & \vdots & & \vdots \\ b'_{m1} & b'_{m2} & \cdots & b'_{mn} \end{bmatrix}. \qquad (6.52)$$

---

[7] The reader should be careful not to confuse the notion of the derivative of a determinant with that of the derivative of a matrix.

Thus $B'$ is the matrix function whose elements $b'_{ij}$ are the derivatives of the corresponding elements $b_{ij}$ of $B$. It is left to the reader to show that

$$(A + B)' = A' + B', \qquad (AB)' = A'B + AB'. \tag{6.53}$$

If $A$ is an $n \times n$ matrix function defined on an interval $\mathscr{I}$, we can seek to determine an $n \times n$ matrix function $X$ such that

$$X' = AX \tag{6.54}$$

on $\mathscr{I}$. If the column vectors of $X$ are $\mathbf{x}_1, \mathbf{x}_2, \ldots, \mathbf{x}_n$, then the column vectors of $X'$ are $\mathbf{x}'_1, \mathbf{x}'_2, \ldots, \mathbf{x}'_n$. The product $AX$ is an $n \times n$ matrix function whose column vectors are $A\mathbf{x}_1, A\mathbf{x}_2, \ldots, A\mathbf{x}_n$. Hence a matrix function $X$ satisfies Eq. (6.54) if and only if each of its column vectors is a solution of the vector equation

$$\mathbf{x}' = A\mathbf{x}. \tag{6.55}$$

A matrix solution of Eq. (6.54) whose column vectors are linearly independent is called a *fundamental matrix* for Eq. (6.55). Thus a matrix function is a fundamental matrix for Eq. (6.55) if and only if its column vectors constitute a fundamental set of solutions for that equation.

In dealing with an initial value problem

$$\mathbf{x}' = A\mathbf{x}, \qquad \mathbf{x}(t_0) = \mathbf{k}, \tag{6.56}$$

one particular fundamental matrix is especially convenient. Suppose that $\mathbf{u}_1, \mathbf{u}_2, \ldots, \mathbf{u}_n$ are the solutions of the differential equation for which

$$\mathbf{u}_1(t_0) = \begin{bmatrix} 1 \\ 0 \\ 0 \\ \vdots \\ 0 \end{bmatrix}, \qquad \mathbf{u}_2(t_0) = \begin{bmatrix} 0 \\ 1 \\ 0 \\ \vdots \\ 0 \end{bmatrix}, \ldots, \qquad \mathbf{u}_n(t_0) = \begin{bmatrix} 0 \\ 0 \\ 0 \\ \vdots \\ 1 \end{bmatrix}.$$

(It can be shown that such solutions exists. See section 6.12.) If $U$ is the matrix function that has these vector functions as its column vectors, we see that

$$U(t_0) = I, \tag{6.57}$$

where $I$ is the $n \times n$ identity matrix. Since $I$ is nonsingular, $U$ is a fundamental matrix, according to Theorem 6.2. Let $\mathbf{u}$ be defined by the relation

$$\mathbf{u} = U\mathbf{k}.$$

Making use of formula (2.47), we have

$$\mathbf{u} = k_1\mathbf{u}_1 + k_2\mathbf{u}_2 + \cdots + k_n\mathbf{u}_n ;$$

thus $\mathbf{u}$ is a solution of the differential equation. Since $\mathbf{u}(t_0) = \mathbf{k}$, $\mathbf{u}$ is the solution of the initial value problem (6.56). A method for finding the fundamental matrix for a system with constant coefficients that satisfies the condition (6.57) will be developed in Section 6.8.

In order to illustrate the ideas discussed here, let us consider the equation $\mathbf{x}' = A\mathbf{x}$, where

$$A = \begin{bmatrix} -2 & 1 \\ -4 & 3 \end{bmatrix}, \qquad \mathbf{x} = \begin{bmatrix} x_1 \\ x_2 \end{bmatrix}.$$

It may be verified that $\mathbf{u}_1$ and $\mathbf{u}_2$, where

$$\mathbf{u}_1(t) = \begin{bmatrix} e^{-t} \\ e^{-t} \end{bmatrix}, \qquad \mathbf{u}_2(t) = \begin{bmatrix} e^{2t} \\ 4e^{2t} \end{bmatrix},$$

are linearly independent solutions. Then the matrix function $U$, where

$$U(t) = [\mathbf{u}_1(t), \mathbf{u}_2(t)] = \begin{bmatrix} e^{-t} & e^{2t} \\ e^{-t} & 4e^{2t} \end{bmatrix}$$

is a fundamental matrix for the equation. It may also be verified that the matrix function $V$, where

$$V = \tfrac{1}{3}[4\mathbf{u}_1 - \mathbf{u}_2, \quad -\mathbf{u}_1 + \mathbf{u}_2]$$

and

$$V(t) = \frac{1}{3}\begin{bmatrix} 4e^t - e^{2t} & -e^{-t} + e^{2t} \\ 4e^t - 4e^{2t} & -e^{-t} + 4e^{2t} \end{bmatrix}$$

is also a fundamental matrix, with $V(0) = I$. The solution of the equation that satisfies $\mathbf{x}(0) = \mathbf{k}$ is given by

$$\mathbf{x}(t) = V(t)\,\mathbf{k}.$$

## Exercises for Section 6.5

In Exercises 1–3, determine whether or not the given set of vector functions is linearly dependent. The interval of definition is assumed to be the set of all real numbers.

1. (a) $\mathbf{u}_1(t) = \begin{bmatrix} 2t - 1 \\ -t \end{bmatrix},$    $\mathbf{u}_2(t) = \begin{bmatrix} -t + 1 \\ 2t \end{bmatrix}$

   (b) $\mathbf{u}_1(t) = \begin{bmatrix} \cos t \\ \sin t \end{bmatrix},$    $\mathbf{u}_2(t) = \begin{bmatrix} \sin t \\ \cos t \end{bmatrix}$

   (c) $\mathbf{u}_1(t) = \begin{bmatrix} t - 2t^2 \\ -t \end{bmatrix},$    $\mathbf{u}_2(t) = \begin{bmatrix} -2t + 4t^2 \\ 2t \end{bmatrix}$

   (d) $\mathbf{u}_1(t) = \begin{bmatrix} te^t \\ t \end{bmatrix},$    $\mathbf{u}_2(t) = \begin{bmatrix} e^t \\ 1 \end{bmatrix}$

2. (a) $\mathbf{u}_1(t) = \begin{bmatrix} 2 - t \\ t \\ -2 \end{bmatrix},$    $\mathbf{u}_2(t) = \begin{bmatrix} t \\ -1 \\ 2 \end{bmatrix},$    $\mathbf{u}_3(t) = \begin{bmatrix} 2 + t \\ t - 2 \\ 2 \end{bmatrix}$

   (b) $\mathbf{u}_1(t) = \begin{bmatrix} \cos t \\ \sin t \\ 0 \end{bmatrix},$    $\mathbf{u}_2(t) = \begin{bmatrix} \cos t \\ 0 \\ \sin t \end{bmatrix},$    $\mathbf{u}_3(t) = \begin{bmatrix} 0 \\ \cos t \\ \sin t \end{bmatrix}$

   (c) $\mathbf{u}_1(t) = \begin{bmatrix} e^t \\ -e^t \\ e^t \end{bmatrix},$    $\mathbf{u}_2(t) = \begin{bmatrix} -e^t \\ 2e^t \\ -e^t \end{bmatrix},$    $\mathbf{u}_3(t) = \begin{bmatrix} 0 \\ e^t \\ 0 \end{bmatrix}$

   (d) $\mathbf{u}_1(t) = \begin{bmatrix} e^t \\ 0 \\ 0 \end{bmatrix},$    $\mathbf{u}_2(t) = \begin{bmatrix} 0 \\ \cos t \\ \cos t \end{bmatrix},$    $\mathbf{u}_3(t) = \begin{bmatrix} 0 \\ \sin t \\ \sin t \end{bmatrix}$

3. (a) $\mathbf{u}_1(t) = \begin{bmatrix} 2 - t \\ t \end{bmatrix},$    $\mathbf{u}_2(t) = \begin{bmatrix} t + 1 \\ -2 \end{bmatrix},$    $\mathbf{u}_3(t) = \begin{bmatrix} t \\ t + 2 \end{bmatrix}$

   (b) $\mathbf{u}_1(t) = \begin{bmatrix} e^t \\ 0 \end{bmatrix},$    $\mathbf{u}_2(t) = \begin{bmatrix} 0 \\ 0 \end{bmatrix},$    $\mathbf{u}_3(t) = \begin{bmatrix} 0 \\ e^t \end{bmatrix}$

   (c) $\mathbf{u}_1(t) = \begin{bmatrix} t^2 \\ t^4 \end{bmatrix},$    $\mathbf{u}_2(t) = \begin{bmatrix} t|t| \\ t|t^3| \end{bmatrix}$

   (d) $\mathbf{u}_1(t) = \begin{bmatrix} \cos(t + \pi/4) \\ 0 \\ 0 \\ 0 \end{bmatrix},$    $\mathbf{u}_2(t) = \begin{bmatrix} \cos t \\ 0 \\ 0 \\ e^t \end{bmatrix},$    $\mathbf{u}_3(t) = \begin{bmatrix} \sin t \\ 0 \\ 0 \\ e^t \end{bmatrix}$

4. Show that any set of vector functions that contains the zero function is linearly dependent.

5. Let $\mathbf{u}$ and $\mathbf{v}$ be vector functions, where

$$\mathbf{u} = \begin{bmatrix} u_1 \\ u_2 \end{bmatrix}, \qquad \mathbf{v} = \begin{bmatrix} v_1 \\ v_2 \end{bmatrix}.$$

(a) If $u_1$ and $v_1$ are linearly independent functions, is it necessarily true that **u** and **v** are linearly independent?

(b) Suppose that $u_1$ and $v_1$ are linearly dependent and also that $u_2$ and $v_2$ are linearly dependent. Is it necessarily true that **u** and **v** are linearly dependent?

6. Verify the differentiation rules

$$(A + B)' = A' + B' \quad \text{and} \quad (AB)' = A'B + AB'$$

for matrix functions $A$ and $B$.

7. Let $\mathbf{u}_1, \mathbf{u}_2, \ldots, \mathbf{u}_m$ be vector functions each with $n$ components defined on an interval $J$. Let $U(t)$ be the $n \times m$ matrix with column vectors $\mathbf{u}_1, \mathbf{u}_2, \ldots, \mathbf{u}_m$.

(a) If the vector functions are linearly dependent, what can be said about the rank of $U(t)$ for each $t$ in $J$?

(b) If $U(t_0)$ has rank equal to $m$ for some $t_0$ in $J$, what can be said about the linear dependence of the functions $\mathbf{u}_i$, $1 \le i \le m$?

8. For all $t$, let

$$\mathbf{u}_1(t) = \begin{bmatrix} e^{2t} \\ e^{2t} \end{bmatrix}, \qquad \mathbf{u}_2(t) = \begin{bmatrix} 3e^{3t} \\ 2e^{3t} \end{bmatrix}, \qquad A(t) = \begin{bmatrix} 5 & -3 \\ 2 & 0 \end{bmatrix}.$$

(a) Verify that $\mathbf{u}_1$ and $\mathbf{u}_2$ form a fundamental set of solutions for the equation $\mathbf{x}' = A\mathbf{x}$.

(b) If $U$ is the matrix function with $\mathbf{u}_1$ and $\mathbf{u}_2$ as its column vectors, verify that $U' = AU$.

(c) Write down a formula that describes the set of all solutions of the equation $\mathbf{x}' = A\mathbf{x}$.

(d) Find the solution of $\mathbf{x}' = A\mathbf{x}$ for which

$$\mathbf{x}(0) = \begin{bmatrix} 1 \\ 0 \end{bmatrix}.$$

9. Let the matrix function $U$ be defined by the equation

$$U(t) = \begin{bmatrix} \cos 2t & \sin 2t \\ \sin 2t & -\cos 2t \end{bmatrix}$$

for all $t$. Verify that $U$ is a fundamental matrix for the equation $x' = Ax$, where

$$A = \begin{bmatrix} 0 & -2 \\ 2 & 0 \end{bmatrix}.$$

Find the solution of $\mathbf{x}' = A\mathbf{x}$ for which

$$\mathbf{x}(0) = \begin{bmatrix} 2 \\ 3 \end{bmatrix}.$$

**10.** Let

$$A(t) = \begin{bmatrix} -1 & 4 & -4 \\ 0 & -1 & 1 \\ 0 & 0 & 0 \end{bmatrix}, \qquad U(t) = \begin{bmatrix} 0 & 4te^{-t} & e^{-t} \\ 1 & e^{-t} & 0 \\ 1 & 0 & 0 \end{bmatrix}$$

for all $t$.

(a)  Verify that $U$ is a fundamental matrix for the equation $\mathbf{x}' = A\mathbf{x}$.

(b)  Find the solution of $\mathbf{x}' = A\mathbf{x}$ for which

$$\mathbf{x}(0) = \begin{bmatrix} 0 \\ 1 \\ 2 \end{bmatrix}.$$

**11.** Let $U$ and $A$ be $n \times n$ matrix functions such that $U'(t) = A(t)\,U(t)$ for $t$ in an interval $J$.

(a)  Show that

$$\frac{d}{dt} \det U(t) = [a_{11}(t) + a_{22}(t) + \cdots + a_{nn}(t)] \det U(t)$$

and hence that

$$\det U(t) = (\det U(t_0)) \exp\left\{ \int_{t_0}^{t} [a_{11}(s) + \cdots + a_{nn}(s)]\,ds \right\},$$

where $t_0$ is any point in $J$. Suggestion: differentiate $\det U$ by rows.

(b)  Use the result of part (a) to show that if $U$ is a matrix solution of $\mathbf{x}' = A\mathbf{x}$, then either $\det U(t) = 0$ for all $t$ in $J$ or $\det U(t)$ is not zero for any $t$ in $J$.

**12.** Let $U$ be a fundamental matrix for the equation $\mathbf{x}' = A\mathbf{x}$, and let $V = UC$ where $C$ is a square constant matrix.

(a) Show that $V' = AV$.

(b) Show that $V$ is a fundamental matrix for the equation $\mathbf{x}' = A\mathbf{x}$ if and only if $C$ is nonsingular.

(c) If $W = CU$, show that $W' \neq AW$ in general.

## 6.6 SOLUTIONS BY CHARACTERISTIC VALUES

In Chapter 5 we attacked the equation with constant coefficients $P(D) y = 0$ by attempting to find solutions of the form $y = e^{rx}$. Let us consider here the system with constant coefficients,

$$
\begin{aligned}
x_1' &= x_1 - 3x_2, \\
x_2' &= -2x_1 + 2x_2.
\end{aligned}
\tag{6.58}
$$

We shall seek solutions of the form

$$
x_1(t) = k_1 e^{\lambda t}, \qquad x_2(t) = k_2 e^{\lambda t},
\tag{6.59}
$$

where $\lambda$, $k_1$, and $k_2$ are numbers to be determined. Substituting in the system (6.58), we obtain the requirements

$$
\begin{aligned}
\lambda k_1 e^{\lambda t} &= (k_1 - 3k_2)e^{\lambda t} \\
\lambda k_2 e^{\lambda t} &= (-2k_1 + 2k_2)e^{\lambda t}
\end{aligned}
$$

or

$$
\begin{aligned}
(\lambda - 1)k_1 + 3k_2 &= 0, \\
2k_1 + (\lambda - 2)k_2 &= 0.
\end{aligned}
$$

Thus a nontrivial solution of the form (6.59) exists if and only if $\lambda$ is a characteristic value of the matrix

$$
A = \begin{bmatrix} 1 & -3 \\ -2 & 2 \end{bmatrix},
$$

which is the coefficient matrix of the system (6.58). The vector $\mathbf{k} = (k_1, k_2)$ must be a corresponding characteristic vector. The characteristic values of the matrix are $\lambda_1 = -1$ and $\lambda_2 = 4$. Corresponding characteristic vectors are $(3, 2)$ and $(1, -1)$, respectively. Thus the vector functions $\mathbf{x}_1$ and $\mathbf{x}_2$, where

$$\mathbf{x}_1(t) = e^{-t}\begin{bmatrix} 3 \\ 2 \end{bmatrix}, \qquad \mathbf{x}_2(t) = e^{4t}\begin{bmatrix} 1 \\ -1 \end{bmatrix},$$

are nontrivial solutions of the equation $\mathbf{x}' = A\mathbf{x}$. If $X = [\mathbf{x}_1, \mathbf{x}_2]$, we see that $\det X(0) = -5 \neq 0$, so these solutions form a fundamental set. The general solution of the system (6.58) is $\mathbf{x} = c_1 \mathbf{x}_1 + c_2 \mathbf{x}_2$, or

$$x_1 = 3c_1 e^{-t} + c_2 e^{4t}, \qquad x_2 = 2c_1 e^{-t} - c_2 e^{4t}.$$

Let us now consider the general case of an equation

$$\mathbf{x}' = A\mathbf{x}, \tag{6.60}$$

where $A$ is an $n \times n$ constant matrix. A vector function $\mathbf{x}$ of the form

$$\mathbf{x}(t) = e^{\lambda t}\mathbf{k}, \tag{6.61}$$

where $\mathbf{k}$ is a constant vector, is a solution if and only if

$$\lambda e^{\lambda t}\mathbf{k} = e^{\lambda t} A\mathbf{k}$$

or

$$(\lambda I - A)\mathbf{k} = \mathbf{0}.$$

Thus a vector function of the form (6.61) is a nontrivial solution when and only when $\lambda$ is a characteristic value of $A$ and $\mathbf{k}$ is a corresponding characteristic vector. The question now is whether it is possible to find a fundamental set of solutions of the form (6.61).

**Theorem 6.3**   Let the $n \times n$ matrix $A$ possess $n$ real linearly independent characteristic vectors $\mathbf{k}_1, \mathbf{k}_2, \ldots, \mathbf{k}_n$, and let $\lambda_i$ be the real characteristic value that corresponds to $\mathbf{k}_i$. (The numbers $\lambda_1, \lambda_2, \ldots, \lambda_n$ need not all be distinct.) Then the $n$ vector functions $\mathbf{u}_1, \mathbf{u}_2, \ldots, \mathbf{u}_n$, where

$$\mathbf{u}_i(t) = e^{\lambda_i t}\mathbf{k}_i \tag{6.62}$$

for all $t$, constitute a fundamental set of solutions for Eq. (6.60).

PROOF We have already shown that each of the vector functions (6.62) is a solution. If $U = [\mathbf{u}_1, \mathbf{u}_2, \ldots, \mathbf{u}_n]$, then $U(0) = [\mathbf{k}_1, \mathbf{k}_2, \ldots, \mathbf{k}_n]$. By Theorem 3.5, $U(0)$ is nonsingular. By Theorem 6.2, $U$ is a fundamental matrix for Eq. (6.60).

According to Theorem 4.5, the matrix $A$ has $n$ linearly independent characteristic vectors if and only if the maximum number of linearly independent characteristic vectors associated with each characteristic value is equal to the multiplicity of the characteristic value. In particular, $A$ has $n$ linearly independent characteristic vectors if it has $n$ distinct characteristic values.

It may happen that $A$ possesses a set of $n$ linearly independent characteristic vectors, some of which are complex. In this case it is still possible to find a fundamental set of real solutions by using the method of characteristic values. If $\lambda_1 = \alpha + i\beta, \beta \neq 0$, is a characteristic value of $A$, then by Theorem 4.2, so is $\lambda_2 = \alpha - i\beta$. If $\mathbf{k}_1 = \mathbf{r} + i\mathbf{s}$ is a characteristic vector corresponding to $\lambda_1$ then $\mathbf{k}_2 = \mathbf{r} - i\mathbf{s}$ is a characteristic vector corresponding to $\lambda_2$. The vector functions

$$\mathbf{w}_1(t) = (\mathbf{r} + i\mathbf{s})e^{(\alpha + i\beta)t} = \mathbf{u}(t) + i\,\mathbf{v}(t),$$
$$\mathbf{w}_2(t) = (\mathbf{r} - i\mathbf{s})e^{(\alpha - i\beta)t} = \mathbf{u}(t) - i\,\mathbf{v}(t),$$

are complex solutions of Eq. (6.60). The real and imaginary parts,

$$\mathbf{u}(t) = e^{\alpha t}(\mathbf{r}\cos\beta t - \mathbf{s}\sin\beta t),$$
$$\mathbf{v}(t) = e^{\alpha t}(\mathbf{r}\sin\beta t + \mathbf{s}\cos\beta t),$$

are real solutions. Thus to each pair of complex conjugate characteristic vectors corresponds a pair of real vector solutions of the differential equation. It can be shown (Exercise 12) that the $n$ real solutions obtained by this process are linearly independent.

To consider an example, let

$$A = \begin{bmatrix} -3 & -1 \\ 2 & -1 \end{bmatrix}.$$

Then

$$p(\lambda) = \begin{vmatrix} \lambda + 3 & 1 \\ -2 & \lambda + 1 \end{vmatrix} = \lambda^2 + 4\lambda + 5;$$

therefore the characteristic values of $A$ are

$$\lambda_1 = -2 + i, \qquad \lambda_2 = -2 - i.$$

The equation $(\lambda_1 I - A)\mathbf{k} = \mathbf{0}$ corresponds to the system

$$(1 + i)k_1 + k_2 = 0$$
$$-2k_1 + (-1 + i)k_2 = 0.$$

A characteristic vector is

$$\begin{bmatrix} -1 \\ 1 + i \end{bmatrix} = \begin{bmatrix} -1 \\ 1 \end{bmatrix} + i \begin{bmatrix} 0 \\ 1 \end{bmatrix}.$$

Then

$$e^{(-2+i)t} \begin{bmatrix} -1 \\ 1 + i \end{bmatrix} = e^{-2t}\left( \begin{bmatrix} -1 \\ 1 \end{bmatrix} + i \begin{bmatrix} 0 \\ 1 \end{bmatrix} \right)(\cos t + i \sin t)$$

is a complex solution of the equation $\mathbf{x}' = A\mathbf{x}$. The real and imaginary parts are

$$\mathbf{u}(t) = e^{-2t}(\cos t) \begin{bmatrix} -1 \\ 1 \end{bmatrix} - e^{-2t}(\sin t) \begin{bmatrix} 0 \\ 1 \end{bmatrix}$$

and

$$\mathbf{v}(t) = e^{-2t}(\sin t) \begin{bmatrix} -1 \\ 1 \end{bmatrix} + e^{-2t}(\cos t) \begin{bmatrix} 0 \\ 1 \end{bmatrix},$$

respectively. The functions $\mathbf{u}$ and $\mathbf{v}$ are linearly independent real solutions.

When the $n \times n$ matrix $A$ does not possess $n$ linearly independent characteristic vectors, we cannot find a fundamental set of solutions of the form (6.61). It is possible to develop a complete theory that parallels that for a single equation with constant coefficients. However, in the next two sections, we shall adopt a different approach. The method of solution that we shall describe yields the general solution in all cases, and it is particularly convenient in finding the specific solution of an initial value problem.

### Exercises for Section 6.6

In Exercises 1–11, using the method of this section, find the general solution of the equation $\mathbf{x}' = A\mathbf{x}$, where $A$ is the given matrix. If an initial condition is given, also find the solution that satisfies the condition.

1. $\begin{bmatrix} -2 & 4 \\ 1 & 1 \end{bmatrix}$,   $\mathbf{x}(0) = \begin{bmatrix} -2 \\ 3 \end{bmatrix}$

2. $\begin{bmatrix} -3 & 2 \\ 1 & -2 \end{bmatrix}.$

3. $\begin{bmatrix} 2 & 4 \\ -2 & -2 \end{bmatrix}$, $\quad \mathbf{x}(0) = \begin{bmatrix} 1 \\ 3 \end{bmatrix}$.

4. $\begin{bmatrix} -1 & 2 \\ -1 & -3 \end{bmatrix}$.

5. $\begin{bmatrix} -2 & 0 \\ 0 & -2 \end{bmatrix}$.

6. $\begin{bmatrix} 3 & 0 & -1 \\ -2 & 2 & 1 \\ 8 & 0 & -3 \end{bmatrix}$, $\quad \mathbf{x}(0) = \begin{bmatrix} -1 \\ 2 \\ -8 \end{bmatrix}$

7. $\begin{bmatrix} -2 & 2 & 1 \\ 0 & -1 & 0 \\ 2 & -2 & -1 \end{bmatrix}$

8. $\begin{bmatrix} 3 & -4 & 4 \\ 4 & -5 & 4 \\ 4 & -4 & 3 \end{bmatrix}$, $\quad \mathbf{x}(0) = \begin{bmatrix} 2 \\ 1 \\ -1 \end{bmatrix}$

9. $\begin{bmatrix} -3 & 0 & -3 \\ 1 & -2 & 3 \\ 1 & 0 & 1 \end{bmatrix}$

10. $\begin{bmatrix} 0 & 4 & 0 \\ -1 & 0 & 0 \\ 1 & 4 & -1 \end{bmatrix}$

11. $\begin{bmatrix} 5 & -5 & -5 \\ -1 & 4 & 2 \\ 3 & -5 & -3 \end{bmatrix}$

12. Let $\mathbf{w}_1 = \mathbf{u} + i\mathbf{v}$ and $\mathbf{w}_2 = \mathbf{u} - i\mathbf{v}$ be linearly independent elements of the space of complex vector functions.

(a) Show that $\mathbf{u}$ and $\mathbf{v}$ are linearly independent elements of the space of real vector functions.

(b) Show that any linearly independent set of complex vector functions that contains $\mathbf{w}_1$ and $\mathbf{w}_2$ remains linearly independent when $\mathbf{w}_1$ and $\mathbf{w}_2$ are replaced by $\mathbf{u}$ and $\mathbf{v}$.

13. Let $P(r) = r^n + a_1 r^{n-1} + \cdots + a_{n-1} r + a_n$. If the differential equation $P(D) y = 0$ is replaced by a first-order system, show that the characteristic polynomial of the matrix of that system is $P$.

## 6.7   THE EXPONENTIAL MATRIX FUNCTION

In Section 4.6 we defined $e^A$, for a square matrix $A$, by the formula[8]

$$e^A = I + \sum_{k=1}^{\infty} \frac{1}{k!} A^k. \tag{6.63}$$

In a few cases we can find simple expressions for the elements of $e^A$ directly from this definition. If $A$ is the zero matrix we evidently have

$$e^0 = I. \tag{6.64}$$

---

[8] The formula (6.63) is also used to define $e^A$ when $A$ has complex elements. In this case an understanding of the formula requires a knowledge of the theory of sequences and series of complex numbers.

If

$$A = \begin{bmatrix} 2 & 0 \\ 0 & -3 \end{bmatrix},$$

then

$$e^A = \begin{bmatrix} 1 & 0 \\ 0 & 1 \end{bmatrix} + \begin{bmatrix} 2 & 0 \\ 0 & -3 \end{bmatrix} + \frac{1}{2!} \begin{bmatrix} 2^2 & 0 \\ 0 & (-3)^2 \end{bmatrix} + \frac{1}{3!} \begin{bmatrix} 2^3 & 0 \\ 0 & (-3)^3 \end{bmatrix} + \cdots$$

$$= \begin{bmatrix} e^2 & 0 \\ 0 & e^{-3} \end{bmatrix}.$$

More generally, if $D = \text{diag}(d_1, d_2, \ldots, d_n)$ then (Exercise 2)

$$e^D = \text{diag}(e^{d_1}, e^{d_2}, \ldots, e^{d_n}). \tag{6.65}$$

Another special case of interest is that where $A$ is a scalar matrix, $A = dI$. It is left as an exercise to show that

$$e^{dI} = e^d I. \tag{6.66}$$

If $A$ and $B$ are square matrices of the same size, it is natural to inquire whether $e^{A+B}$ is equal to $e^A e^B$. We know that

$$e^{A+B} = I + (A + B) + \tfrac{1}{2}(A + B)^2 + \cdots \tag{6.67}$$

$$e^B = I + B + \tfrac{1}{2}B^2 + \cdots \tag{6.68}$$

$$e^A = I + A + \tfrac{1}{2}A^2 + \cdots. \tag{6.69}$$

If we multiply the series for $e^A$ and $e^B$ in the manner employed for the multiplication of ordinary power series (a procedure that we have certainly not justified) we find that

$$e^A e^B = I + (A + B) + \tfrac{1}{2}(A^2 + 2AB + B^2) + \cdots. \tag{6.70}$$

Comparing the terms of second degree here with those in formula (6.67), we see that the matrices

$$A^2 + 2AB + B^2$$

and

$$(A + B)^2 = (A + B)(A + B) = A^2 + AB + BA + B^2$$

are not equal unless $BA = AB$. It can be shown, however, that

$$e^{A+B} = e^A e^B \qquad (6.71)$$

if $A$ and $B$ commute.

In particular, since $A$ and $-A$ commute, we have

$$e^A e^{-A} = e^{A-A} = e^0 = I.$$

Hence $e^A$ is nonsingular and

$$(e^A)^{-1} = e^{-A}. \qquad (6.72)$$

If $A$ is a square matrix then $tA$ is a square matrix for every real number $t$. Then the formula

$$e^{tA} = I + tA + \frac{1}{2!}t^2 A^2 + \cdots + \frac{1}{k!}t^k A^k + \cdots \qquad (6.73)$$

defines a matrix function. By using facts about the termwise differentiation of ordinary power series, it is possible to show that termwise differentiation of the series in Eq. (6.73) is valid. Then

$$\frac{de^{tA}}{dt} = A + \frac{2}{2!}tA^2 + \cdots + \frac{k}{k!}t^{k-1}A^k + \cdots$$

$$= A\left[I + tA + \cdots + \frac{1}{(k-1)!}t^{k-1}A^{k-1} + \cdots\right]$$

or

$$\frac{de^{tA}}{dt} = Ae^{tA}. \qquad (6.74)$$

The importance of the exponential matrix function (6.73) in the theory of linear systems of differential equations with constant coefficients is given by the following theorem.

**Theorem 6.4**  Let the matrix function $U$ be defined by the formula

$$U(t) = e^{tA} \qquad (6.75)$$

for all $t$. Then $U$ is a fundamental matrix for the differential equation

$$\mathbf{x}' = A\mathbf{x}.\tag{6.76}$$

Also, the solution of Eq. (6.76) that satisfies the initial condition

$$\mathbf{x}(0) = \mathbf{k}\tag{6.77}$$

is given by the formula

$$\mathbf{x}(t) = U(t)\,\mathbf{k} = e^{tA}\mathbf{k}.\tag{6.78}$$

PROOF   It follows from the relation

$$U'(t) = Ae^{tA} = A\,U(t)$$

that $U$ is a matrix solution. Since

$$\det U(0) = \det I = 1,$$

$U$ is a fundamental matrix for Eq. (6.76). Each of the column vectors $\mathbf{u}_1, \mathbf{u}_2, \ldots, \mathbf{u}_n$ of $U$ is a solution of Eq. (6.76). If $\mathbf{k} = (k_1, k_2, \ldots, k_n)$ then

$$U\mathbf{k} = k_1\mathbf{u}_1 + k_2\mathbf{u}_2 + \cdots + k_n\mathbf{u}_n.$$

Hence $\mathbf{x} = U\mathbf{k}$ is a solution of the differential equation and

$$\mathbf{x}(0) = U(0)\mathbf{k} = I\mathbf{k} = \mathbf{k}.$$

Thus $U\mathbf{k}$ is the solution that satisfies the initial condition (6.77).
    The proof of the following corollary is left as an exercise.

**Corollary**   The solution of the initial value problem

$$\mathbf{x}' = A\mathbf{x}, \qquad \mathbf{x}(t_0) = \mathbf{k}$$

is given by the formula

$$\mathbf{x}(t) = e^{(t-t_0)A}\mathbf{k}.$$

It is difficult to determine the behavior of the elements of $e^{tA}$ directly from the series definition (6.73), except in certain cases. In the next section we shall show how $e^{tA}$ can be expressed in a simple way in terms of ordinary polynomials and exponential functions.

## Exercises for Section 6.7

1. Show that $e^{dI} = e^d I$, where $d$ is a number and $I$ is the identity matrix.

2. If $D = \operatorname{diag}(d_1, d_2, \ldots, d_n)$, show that

$$e^D = \operatorname{diag}(e^{d_1}, e^{d_2}, \ldots, e^{d_n}).$$

3. Use the result of Exercise 2 to find $e^{tA}$ if $A$ is

   (a) $\begin{bmatrix} 4 & 0 \\ 0 & -1 \end{bmatrix}$   (b) $\begin{bmatrix} 0 & 0 & 0 \\ 0 & 3 & 0 \\ 0 & 0 & 2 \end{bmatrix}$

4. Show that $(e^A)^2 = e^{2A}$, and that $(e^A)^m = e^{mA}$ for every positive integer $m$.

5. Use properties of ordinary power series to show that $de^{tA}/dt = Ae^{tA}$.

6. If $\mathbf{u}$ is a solution of the equation $\mathbf{x}' = A\mathbf{x}$, where $A$ is a constant matrix, show that $\mathbf{v}$, where $\mathbf{v}(t) = \mathbf{u}(t - t_0)$, is also a solution.

7. Prove the corollary to Theorem 6.4 (Use the result of Exercise 6).

8. Use Theorem 6.4 to solve the initial value problem $\mathbf{x}' = A\mathbf{x}$, $\mathbf{x}(0) = \mathbf{k}$, where $A$ and $\mathbf{k}$ are as given.

   (a) $A = \begin{bmatrix} 1 & 0 \\ 0 & 2 \end{bmatrix}$,   $\mathbf{k} = \begin{bmatrix} 3 \\ 5 \end{bmatrix}$

   (b) $A = \begin{bmatrix} 2 & 0 & 0 \\ 0 & -1 & 0 \\ 0 & 0 & -1 \end{bmatrix}$,   $\mathbf{k} = \begin{bmatrix} 4 \\ 5 \\ 6 \end{bmatrix}$

9. Find the solution of the problem $\mathbf{x}' = A\mathbf{x}$, $\mathbf{x}(t_0) = \mathbf{k}$, where $A$ and $\mathbf{k}$ are as in Exercise 8.

10. Find the elements of $e^{tA}$ if

$$A = \begin{bmatrix} 0 & 1 & -1 \\ 0 & 0 & 2 \\ 0 & 0 & 0 \end{bmatrix}.$$

Suggestion: $A^3 = 0$.

## 6.8   A MATRIX METHOD

If $A$ is an $n \times n$ matrix, each column vector of $U$, where $U(t) = e^{tA}$, is a solution of the differential equation

$$\mathbf{x}' = A\mathbf{x}. \tag{6.79}$$

But if $A$ has $n$ linearly independent characteristic vectors $\mathbf{k}_1, \mathbf{k}_2, \ldots, \mathbf{k}_n$ corresponding to the characteristic values $\lambda_1, \lambda_2, \ldots, \lambda_n$, then every solution of Eq. (6.79) is of the form

$$\mathbf{x}(t) = c_1 e^{\lambda_1 t}\mathbf{k}_1 + c_2 e^{\lambda_2 t}\mathbf{k}_2 + \cdots + c_n e^{\lambda_n t}\mathbf{k}_n,$$

where $c_1, c_2, \ldots, c_n$ are constants. Consequently there must exist constant matrices $B_1, B_2, \ldots, B_n$ such that

$$e^{tA} = e^{\lambda_1 t}B_1 + e^{\lambda_2 t}B_2 + \cdots + e^{\lambda_n t}B_n.$$

We shall now derive a formula for $e^{tA}$ in the general case where $A$ need not have $n$ linearly independent characteristic vectors. For a general treatment of matrix functions defined by power series, the reader is referred to Smiley (1965).

Let $\lambda_1, \lambda_2, \ldots, \lambda_k$ be the *distinct* characteristic values of the $n \times n$ matrix $A$, with corresponding multiplicities $m_1, m_2, \ldots, m_k$. Then

$$p(\lambda) = \det(\lambda I - A) = (\lambda - \lambda_1)^{m_1}(\lambda - \lambda_2)^{m_2} \cdots (\lambda - \lambda_k)^{m_k}. \tag{6.80}$$

We pause for a moment to introduce the product notation

$$\prod_{j=1}^{m} c_j = c_1 c_2 \cdots c_m.$$

We also use the notation

$$\prod_{\substack{j=1 \\ j \neq i}}^{m} c_j = c_1 c_2 \cdots c_{i-1} c_{i+1} \cdots c_m.$$

We now define polynomials $p_i$, $1 \leq i \leq k$, by means of the formula

$$p_i(\lambda) = \prod_{\substack{j=1 \\ j \neq i}}^{k} (\lambda - \lambda_j)^{m_j}. \tag{6.81}$$

(If $k = 1$, we define $p_1(\lambda) = 1$.) Thus

$$p_1(\lambda) = (\lambda - \lambda_2)^{m_2}(\lambda - \lambda_3)^{m_3} \cdots (\lambda - \lambda_k)^{m_k},$$
$$p_2(\lambda) = (\lambda - \lambda_1)^{m_1}(\lambda - \lambda_3)^{m_3} \cdots (\lambda - \lambda_k)^{m_k},$$

and so on.

If we expand $1/p(\lambda)$ by partial fractions, the terms corresponding to $\lambda_i$ are

$$\frac{c_{i1}}{\lambda - \lambda_i} + \frac{c_{i2}}{(\lambda - \lambda_i)^2} + \cdots + \frac{c_{im_i}}{(\lambda - \lambda_i)^{m_i}} = \frac{a_i(\lambda)}{(\lambda - \lambda_i)^{m_i}},$$

where the $c_{ij}$ are constants and $a_i$ is a polynomial of degree less than $m_i$. Then

$$\frac{1}{p(\lambda)} = \frac{a_1(\lambda)}{(\lambda - \lambda_1)^{m_1}} + \frac{a_2(\lambda)}{(\lambda - \lambda_2)^{m_2}} + \cdots + \frac{a_k(\lambda)}{(\lambda - \lambda_k)^{m_k}}. \tag{6.82}$$

Multiplying through in this equation by $p(\lambda)$, we have

$$1 = a_1(\lambda)\, p_1(\lambda) + a_2(\lambda)\, p_2(\lambda) + \cdots + a_k(\lambda)\, p_k(\lambda). \tag{6.83}$$

Hence we have

$$I = a_1(A)\, p_1(A) + a_2(A)\, p_2(A) + \cdots + a_k(A)\, p_k(A). \tag{6.84}$$

For each fixed integer $i$, $1 \le i \le k$, we may write[9]

$$e^{tA} = e^{\lambda_i t} e^{t(A - \lambda_i I)}$$

and hence

$$e^{tA} = e^{\lambda_i t} \sum_{j=0}^{\infty} \frac{1}{j!} t^j (A - \lambda_i I)^j.$$

Premultiplying both members of this equation by $a_i(A)\, p_i(A)$, and observing that

$$p_i(A)\, (A - \lambda_i I)^{m_i} = p(A) = 0,$$

and hence that

$$p_i(A)\, (A - \lambda_i I)^j = 0$$

---

[9] The formulas here are valid when $\lambda_i$ is complex. See the footnote at the beginning of the previous section.

for all $j \geq m_i$, we have

$$a_i(A)\, p_i(A)\, e^{tA} = e^{\lambda_i t} a_i(A)\, p_i(A) \sum_{j=0}^{m_i-1} \frac{t^j}{j!}(A - \lambda_i I)^j.$$

Summing from $i = 1$ to $i = k$, and using the relation (6.84), we arrive at the formula

$$e^{tA} = \sum_{i=1}^{k} e^{\lambda_i t} a_i(A)\, p_i(A) \sum_{j=0}^{m_i-1} \frac{t^j}{j!}(A - \lambda_i I)^j. \tag{6.85}$$

In the special case where $A$ has $n$ distinct characteristic values, so that $k = n$, we have $m_1 = m_2 = \cdots = m_k = 1$ and the $a_i$ are constants. Then formula (6.85) becomes

$$e^{tA} = \sum_{i=1}^{n} e^{\lambda_i t} a_i\, p_i(A). \tag{6.86}$$

Equation (6.83) becomes

$$1 = a_1 p_1(\lambda) + a_2 p_2(\lambda) + \cdots + a_n p_n(\lambda).$$

Upon setting $\lambda = \lambda_i$, we see that

$$1 = a_i\, p_i(\lambda_i)$$

or

$$a_i = \frac{1}{p_i(\lambda_i)}, \quad 1 \leq i \leq n. \tag{6.87}$$

**Example 1**   We consider the initial value problem

$$x_1' = -x_1 - 2x_2, \qquad x_1(0) = 2$$
$$x_2' = 3x_1 + 4x_2, \qquad x_2(0) = -1$$

Here

$$A = \begin{bmatrix} -1 & -2 \\ 3 & 4 \end{bmatrix}$$

and

$$p(\lambda) = \begin{vmatrix} \lambda + 1 & 2 \\ -3 & \lambda - 4 \end{vmatrix} = (\lambda - 1)(\lambda - 2).$$

Since $n = k = 2$ we may use formulas (6.86) and (6.87). Let $\lambda_1 = 1$ and $\lambda_2 = 2$.
Since

$$p_1(\lambda) = \lambda - 2, \qquad p_2(\lambda) = \lambda - 1$$

we have

$$a_1 = \frac{1}{p_1(\lambda_1)} = -1, \qquad a_2 = \frac{1}{p_2(\lambda_2)} = 1$$

and

$$p_1(A) = A - 2I, \qquad p_2(A) = A - I.$$

Then

$$e^{tA} = -(A - 2I) e^t + (A - I) e^{2t}.$$

Since

$$-(A - 2I) = \begin{bmatrix} 3 & 2 \\ -3 & -2 \end{bmatrix}, \qquad A - I = \begin{bmatrix} -2 & -2 \\ 3 & 3 \end{bmatrix},$$

we have

$$e^{tA} = \begin{bmatrix} 3 & 2 \\ -3 & -2 \end{bmatrix} e^t + \begin{bmatrix} -2 & -2 \\ 3 & 3 \end{bmatrix} e^{2t}.$$

The solution of the initial value problem is given by

$$x(t) = e^{tA} \mathbf{k},$$

where

$$\mathbf{k} = \begin{bmatrix} 2 \\ -1 \end{bmatrix}.$$

It is a matter of routine calculation to show that

$$x(t) = \begin{bmatrix} 4 \\ -4 \end{bmatrix} e^t + \begin{bmatrix} -2 \\ 3 \end{bmatrix} e^{2t}.$$

**Example 2**  This time we consider the system $\mathbf{x}' = A\mathbf{x}$, where

$$A = \begin{bmatrix} 3 & 1 & -1 \\ -1 & 2 & 1 \\ 2 & 1 & 0 \end{bmatrix}.$$

Here

$$p(\lambda) = \begin{vmatrix} \lambda - 3 & -1 & 1 \\ 1 & \lambda - 2 & -1 \\ -2 & -1 & \lambda \end{vmatrix} = (\lambda - 1)(\lambda - 2)^2 .$$

Setting $\lambda_1 = 1$ and $\lambda_2 = 2$ we have $m_1 = 1$ and $m_2 = 2$. Then

$$p_1(\lambda) = (\lambda - 2)^2 = \lambda^2 - 4\lambda + 4, \qquad p_2(\lambda) = \lambda - 1.$$

Using partial fractions, we have

$$\frac{1}{p(\lambda)} = \frac{1}{\lambda - 1} + \frac{-\lambda + 3}{(\lambda - 2)^2}$$

so

$$a_1(\lambda) = 1, \qquad a_2(\lambda) = -\lambda + 3.$$

From the general formula (6.85), with $k = 2$, we have

$$e^{tA} = e^t (A - 2I)^2 + e^{2t}(-A + 3I)(A - I)[I + t(A - 2I)]$$

or

$$e^{tA} = (A^2 - 4A + 4I) e^t + (-A^2 + 4A - 3I) e^{2t} + (A^2 - 3A + 2I) te^{2t}.$$

Here we have used the Cayley–Hamilton theorem to simplify the coefficient of $te^{2t}$. Since

$$A^2 = \begin{bmatrix} 6 & 4 & -2 \\ -3 & 4 & 3 \\ 5 & 4 & -1 \end{bmatrix}$$

some calculation yields the formula

$$e^{tA} = \begin{bmatrix} -2 & 0 & 2 \\ 1 & 0 & -1 \\ -3 & 0 & 3 \end{bmatrix} e^t + \begin{bmatrix} 3 & 0 & -2 \\ -1 & 1 & 1 \\ 3 & 0 & -2 \end{bmatrix} e^{2t} + \begin{bmatrix} -1 & 1 & 1 \\ 0 & 0 & 0 \\ -1 & 1 & 1 \end{bmatrix} te^{2t}.$$

Suppose that $\lambda_1 = \alpha + i\beta$, $\lambda_2 = \alpha - i\beta$, $\beta \neq 0$, are a pair of complex conjugate characteristic values of the matrix $A$. Then in the expansion (6.82) it turns out that

$$a_2(\lambda) = \overline{a_1(\lambda)}$$

for all real $\lambda$. Since $A$ is real,

$$a_2(A) = \overline{a_1(A)}.$$

Consequently the sum of the terms in formula (6.85) that correspond to $\lambda_1$ and $\lambda_2$ is equal to twice the real part of the sum of the terms that correspond to $\lambda_1$.

**Example 3** We examine the equation $\mathbf{x}' = A\mathbf{x}$, where

$$A = \begin{bmatrix} -1 & -1 & 0 \\ 0 & -4 & -1 \\ 0 & 5 & 0 \end{bmatrix}.$$

We find that

$$\begin{aligned} p(\lambda) &= \det(\lambda I - A) \\ &= (\lambda + 1)(\lambda^2 + 4\lambda + 5) \\ &= (\lambda + 1)[\lambda - (-2 + i)][\lambda - (-2 - i)]. \end{aligned}$$

Setting $\lambda_1 = -1$, $\lambda_2 = -2 + i$, $\lambda_3 = -2 - i$, we have

$$\begin{aligned} p_1(\lambda) &= \lambda^2 + 4\lambda + 5, \\ p_2(\lambda) &= (\lambda + 1)(\lambda + 2 + i), \\ p_3(\lambda) &= (\lambda + 1)(\lambda + 2 - i) \end{aligned}$$

and

$$a_1 = \frac{1}{2}, \qquad a_2 = -\frac{1-i}{4}, \qquad a_3 = \overline{a_2} = -\frac{1+i}{4}.$$

In formula (6.86), the expression corresponding to $\lambda_3$ will be the complex conjugate of that corresponding to $\lambda_2$. Then

$$\begin{aligned} e^{tA} &= \tfrac{1}{2}(A^2 + 4A + 5I)e^{-t} \\ &\quad + 2\,\mathrm{Re}\{-\tfrac{1}{4}(1 - i)(A + I)[A + (2 + i)I]\,e^{(-2+i)t}\} \end{aligned}$$

or

$$\begin{aligned} e^{tA} &= \tfrac{1}{2}(A^2 + 4A + 5I)\,e^{-t} - \tfrac{1}{2}(A^2 + 4A + 3I)\,e^{-2t}\cos t \\ &\quad - \tfrac{1}{2}(A^2 + 2A + I)e^{-2t}\sin t. \end{aligned}$$

## Exercises for Section 6.8

In Exercises 1–12, find (a) $e^{tA}$, and (b) the solution of the initial value problem $\mathbf{x}' = A\mathbf{x}$, $\mathbf{x}(0) = \mathbf{k}$.

1. $A = \begin{bmatrix} 0 & 1 \\ 1 & 0 \end{bmatrix}$,   $\mathbf{k} = \begin{bmatrix} 3 \\ 1 \end{bmatrix}$

2. $A = \begin{bmatrix} 1 & -2 \\ -2 & 1 \end{bmatrix}$,   $\mathbf{k} = \begin{bmatrix} 2 \\ 0 \end{bmatrix}$

3. $A = \begin{bmatrix} 0 & -1 \\ 1 & 2 \end{bmatrix}$,   $\mathbf{k} = \begin{bmatrix} 2 \\ -3 \end{bmatrix}$

4. $A = \begin{bmatrix} -1 & -1 \\ 1 & -3 \end{bmatrix}$,   $\mathbf{k} = \begin{bmatrix} 2 \\ 1 \end{bmatrix}$

5. $A = \begin{bmatrix} 1 & -5 \\ 2 & -1 \end{bmatrix}$,   $\mathbf{k} = \begin{bmatrix} 3 \\ 0 \end{bmatrix}$

6. $A = \begin{bmatrix} 3 & -2 \\ 1 & 1 \end{bmatrix}$,   $\mathbf{k} = \begin{bmatrix} 4 \\ 1 \end{bmatrix}$

7. $A = \begin{bmatrix} 2 & 1 & 0 \\ -2 & -1 & 2 \\ 1 & 1 & 1 \end{bmatrix}$,   $\mathbf{k} = \begin{bmatrix} 3 \\ 0 \\ -3 \end{bmatrix}$

8. $A = \begin{bmatrix} 1 & 2 & -1 \\ -1 & 1 & 1 \\ 2 & 2 & -2 \end{bmatrix}$,   $\mathbf{k} = \begin{bmatrix} 0 \\ 1 \\ 0 \end{bmatrix}$

9. $A = \begin{bmatrix} 3 & -2 & 1 \\ 2 & -1 & 1 \\ -4 & 4 & 1 \end{bmatrix}$,   $\mathbf{k} = \begin{bmatrix} 1 \\ 1 \\ 0 \end{bmatrix}$

10. $A = \begin{bmatrix} -3 & 0 & 2 \\ 2 & -1 & -1 \\ -4 & 0 & 3 \end{bmatrix}$,   $\mathbf{k} = \begin{bmatrix} 1 \\ 0 \\ 1 \end{bmatrix}$

11. $A = \begin{bmatrix} -1 & -1 & 1 \\ 0 & 2 & 0 \\ -3 & -1 & 3 \end{bmatrix}$,   $\mathbf{k} = \begin{bmatrix} 0 \\ 2 \\ 2 \end{bmatrix}$

12. $A = \begin{bmatrix} -1 & 0 & 0 \\ -1 & 0 & 3 \\ 3 & -3 & 0 \end{bmatrix}$,   $\mathbf{k} = \begin{bmatrix} 0 \\ 1 \\ 1 \end{bmatrix}$

13. If $\lambda_2 = \bar{\lambda}_1$, show that $a_2(\lambda) = \overline{a_1(\lambda)}$ for all real $\lambda$ in formula (6.82).

In Exercises 14 and 15, solve the initial value problem by converting the differential equation to a first-order system and then using the method of this section.

14. $y'' - 2y' + y = 0$,   $y(0) = -2$,   $y'(0) = 3$

15. $y'' + 3y' + 2y = 0$,   $y(0) = 1$,   $y'(0) = 2$

## 6.9 NONHOMOGENEOUS LINEAR SYSTEMS

The theory for nonhomogeneous linear systems parallels that for non-homogeneous single linear equations. We write our nonhomogeneous linear system in vector form as

$$\mathbf{x}' = A\mathbf{x} + \mathbf{b}, \tag{6.88}$$

where $\mathbf{x}$ and $\mathbf{b}$ are vector functions with $n$ components and $A$ is an $n \times n$ matrix function. Associated with the nonhomogeneous equation (6.88) is the homogeneous equation

$$\mathbf{x}' = A\mathbf{x}. \tag{6.89}$$

The connection between the two equations is provided by the following theorem.

**Theorem 6.5**   Let $\mathbf{u}_1, \mathbf{u}_2, \ldots, \mathbf{u}_n$ constitute a fundamental set of solutions for the homogeneous equation (6.89) on an interval $\mathscr{I}$ and let $\mathbf{u}_p$ be any particular solution of the nonhomogeneous equation (6.88) on $\mathscr{I}$. Then the general solution of the equation (6.88) on $\mathscr{I}$ consists of the set of all vector functions of the form

$$c_1\mathbf{u}_1 + c_2\mathbf{u}_2 + \cdots c_n\mathbf{u}_n + \mathbf{u}_p, \tag{6.90}$$

where $c_1, c_2, \ldots, c_n$ are constants.

PROOF   We first verify that every function of the form (6.90) is a solution of equation (6.88). Since

$$\mathbf{u}_p' = A\mathbf{u}_p + \mathbf{b}, \qquad \mathbf{u}_i' = A\mathbf{u}_i, \qquad 1 \le i \le n,$$

we have

$$
\begin{aligned}
(c_1\mathbf{u}_1 + \cdots + c_n\mathbf{u}_n + \mathbf{u}_p)' &= c_1\mathbf{u}_1' + \cdots + c_n\mathbf{u}_n' + \mathbf{u}_p' \\
&= c_1 A\mathbf{u}_1 + \cdots + c_n A\mathbf{u}_n + A\mathbf{u}_p + \mathbf{b} \\
&= A(c_1\mathbf{u}_1 + \cdots + c_n\mathbf{u}_n + \mathbf{u}_p) + \mathbf{b}.
\end{aligned}
$$

Hence $c_1\mathbf{u}_1 + \cdots + c_n\mathbf{u}_n + \mathbf{u}_p$ is a solution.

We must next show that every solution of Eq. (6.88) is of the form (6.90). Let $\mathbf{u}$ be any solution. Then $\mathbf{u}' = A\mathbf{u}$ and

$$(\mathbf{u} - \mathbf{u}_p)' = \mathbf{u}' - \mathbf{u}_p' = (A\mathbf{u} + \mathbf{b}) - (A\mathbf{u}_p + \mathbf{b}) = A(\mathbf{u} - \mathbf{u}_p).$$

Hence $\mathbf{u} - \mathbf{u}_p$ is a solution of the homogeneous equation (6.89). By Theorem 6.1, there exist numbers $c_1, c_2, \ldots, c_n$ such that

$$\mathbf{u} - \mathbf{u}_p = c_1 \mathbf{u}_1 + c_2 \mathbf{u}_2 + \cdots + c_n \mathbf{u}_n$$

or

$$\mathbf{u} = c_1 \mathbf{u}_1 + c_2 \mathbf{u}_2 + \cdots + c_n \mathbf{u}_n + \mathbf{u}_p.$$

Thus $\mathbf{u}$ is of the form (6.90).

Our next theorem also resembles a previous theorem for single linear equations. Its proof is left as an exercise.

**Theorem 6.6**   Let $\mathbf{u}$ and $\mathbf{v}$ be solutions of the equations

$$\mathbf{x}' = A\mathbf{x} + \mathbf{b}, \qquad \mathbf{x}' = A\mathbf{x} + \mathbf{c},$$

respectively. Then $\mathbf{u} + \mathbf{v}$ is a solution of the equation

$$\mathbf{x}' = A\mathbf{x} + \mathbf{b} + \mathbf{c}.$$

Let us consider briefly a special case. In the equation

$$\mathbf{x}' = A\mathbf{x} + \mathbf{b} \tag{6.91}$$

suppose that $A$ is a constant matrix and that $\mathbf{b}$ is of the form

$$\mathbf{b}(t) = \mathbf{h}e^{rt}, \tag{6.92}$$

where $\mathbf{h}$ is a constant vector and $r$ is a constant. The general solution of the associated homogeneous equation can be found, if not by the method of Section 6.6, then by that of Section 6.8. It is natural here to look for a particular solution $\mathbf{x}_p$ of Eq. (6.91) that is of the form

$$\mathbf{x}_p(t) = \mathbf{k}e^{rt}, \tag{6.93}$$

where $\mathbf{k}$ is a constant vector. Substituting into the differential equation, we obtain the requirement that

$$r\mathbf{k}e^{rt} = A\mathbf{k}e^{rt} + \mathbf{h}e^{rt}$$

or

$$(rI - A)\mathbf{k} = \mathbf{h}.$$

If $\det(rI - A) \neq 0$, $\mathbf{k}$ is uniquely determined. Thus Eq. (6.91) possesses a particular solution of the form (6.93) if $r$ is not a characteristic value of the matrix $A$.

As an example, we consider the system,

$$x_1' = -3x_1 + 2x_2 + 3e^{2t}$$
$$x_2' = -4x_1 + 3x_2 . \tag{6.94}$$

Here the coefficient matrix

$$A = \begin{bmatrix} -3 & 2 \\ -4 & 3 \end{bmatrix}$$

has characteristic values $\lambda_1 = 1$ and $\lambda_2 = -1$. Since $r = 2$ is not a characteristic value, the system possesses a solution of the form

$$x_1(t) = k_1 e^{2t}, \qquad x_2(t) = k_2 e^{2t} .$$

Substituting these expressions into the equations of the system, we obtain the conditions

$$2k_1 = -3k_1 + 2k_2 + 3, \qquad 2k_2 = -4k_1 + 3k_2$$

for $k_1$ and $k_2$. We find easily that $k_1 = -1$ and $k_2 = -4$. Hence a particular solution is

$$x_1(t) = -e^{2t}, \qquad x_2(t) = -4e^{2t} .$$

The general solution of the system (6.94) is described by the formula

$$\mathbf{x}(t) = c_1 \begin{bmatrix} 1 \\ 2 \end{bmatrix} e^t + c_2 \begin{bmatrix} 1 \\ 1 \end{bmatrix} e^{-t} + \begin{bmatrix} -1 \\ -4 \end{bmatrix} e^{2t},$$

where $c_1$ and $c_2$ are arbitrary constants.

It is possible to develop a method of undetermined coefficients for certain linear systems with constant coefficients. We shall content ourselves here with a statement of the facts. (See, however, Exercise 9.) Let

$$\mathbf{b}(t) = t^s e^{rt} \mathbf{h}, \tag{6.95}$$

where $s$ is a nonnegative integer, $\mathbf{h}$ is a constant vector, and $r$ is a number, real or complex. If $r$ is a characteristic value, of multiplicity $m$, of the matrix $A$, then the equation

$$\mathbf{x}' = A\mathbf{x} + \mathbf{b}$$

possesses a particular solution of the form

$$\mathbf{x}_p(t) = e^{rt} \sum_{j=0}^{m+s} t^j \mathbf{k}_j,$$    (6.96)

where the $\mathbf{k}_j$ are constant vectors. If $r$ is not a characteristic value of $A$, formula (6.96) is valid with $m = 0$.

The *method of variation of parameters* for systems, which we shall describe shortly, enables us to construct a particular solution of the nonhomogeneous equation (6.88) whenever a fundamental matrix for the associated homogeneous equation (6.89) is known. In the special case when the matrix function $A$ in equation (6.88) is constant, we can always find such a fundamental matrix. As in the case of a single linear equation, the method of variation of parameters is extremely important for theoretical purposes.

In what follows, we shall need the notion of the integral of a vector function. If $\mathbf{u} = (u_1, u_2, \ldots, u_n)$ is a vector function, we define

$$\int_a^b \mathbf{u}(t)\, dt = \begin{bmatrix} \int_a^b u_1(t)\, dt \\ \int_a^b u_2(t)\, dt \\ \vdots \\ \int_a^b u_n(t)\, dt \end{bmatrix}.$$    (6.97)

Thus the integral of $\mathbf{u}$ is the vector whose components are the integrals of the corresponding components of $\mathbf{u}$. For example, if

$$\mathbf{u}(t) = \begin{bmatrix} 3t^2 \\ -2 \\ 2t + 1 \end{bmatrix},$$

then

$$\int_0^1 \mathbf{u}(t)\, dt = \begin{bmatrix} 1 \\ -2 \\ 2 \end{bmatrix}, \qquad \int_0^t \mathbf{u}(s)\, ds = \begin{bmatrix} t^3 \\ -2t \\ t^2 + t \end{bmatrix}.$$

We now proceed with a description of the method of variation of parameters.

Let $\mathbf{u}_1, \mathbf{u}_2, \ldots, \mathbf{u}_n$ constitute a fundamental set of solutions for the equation

$$\mathbf{x}' = A\mathbf{x}$$    (6.98)

on an interval $\mathcal{I}$. We attempt to find functions $c_1, c_2, \ldots, c_n$ such that

$$c_1 \mathbf{u}_1 + c_2 \mathbf{u}_2 + \cdots + c_n \mathbf{u}_n$$

is a solution of the nonhomogeneous equation

$$\mathbf{x}' = A\mathbf{x} + \mathbf{b} \tag{6.99}$$

on $\mathcal{I}$. Let $U$ be the matrix function whose column vectors are $\mathbf{u}_1, \mathbf{u}_2, \ldots, \mathbf{u}_n$ and let $\mathbf{c}$ be the vector function with components $c_1, c_2, \ldots, c_n$. Since

$$U\mathbf{c} = c_1 \mathbf{u}_1 + c_2 \mathbf{u}_2 + \cdots + c_n \mathbf{u}_n,$$

we are seeking to determine a vector function $\mathbf{c}$ such that

$$(U\mathbf{c})' = A(U\mathbf{c}) + \mathbf{b}$$

or

$$U'\mathbf{c} + U\mathbf{c}' = AU\mathbf{c} + \mathbf{b}.$$

Since $U' = AU$ we have $U'\mathbf{c} = AU\mathbf{c}$. The requirement for $\mathbf{c}$ becomes

$$U\mathbf{c}' = \mathbf{b}. \tag{6.100}$$

Since $U$ is a fundamental matrix for Eq. (6.98) on $\mathcal{I}$, $U(t)$ is nonsingular for every $t$ in $\mathcal{I}$. Then $U^{-1}(t)$ exists for each $t$ and

$$\mathbf{c}'(t) = U^{-1}(t)\,\mathbf{b}(t).$$

We may choose

$$\mathbf{c}(t) = \int_{t_0}^t U^{-1}(s)\,\mathbf{b}(s)\,ds,$$

where $t_0$ is any point in $\mathcal{I}$. Then a particular solution $\mathbf{u}_p = U\mathbf{c}$ of Eq. (6.99) is defined by the formula

$$\mathbf{u}_p(t) = U(t) \int_{t_0}^t U^{-1}(s)\,\mathbf{b}(s)\,ds. \tag{6.101}$$

Note that $\mathbf{u}_p(t_0) = \mathbf{0}$.

In the special case where $A$ is a constant matrix, we may choose

$$U(t) = e^{(t-t_0)A}.$$

Then $U^{-1}(t) = e^{-(t-t_0)A}$ and formula (6.101) becomes

$$\mathbf{u}_p(t) = \int_{t_0}^t e^{(t-s)A}\, \mathbf{b}(s)\, ds. \tag{6.102}$$

Recalling that $e^{0A} = I$, we can verify that the solution of the initial value problem

$$\mathbf{x}' = A\mathbf{x} + \mathbf{b}, \qquad \mathbf{x}(t_0) = \mathbf{k}, \tag{6.103}$$

where $A$ and $\mathbf{k}$ are constant, is

$$\mathbf{x}(t) = e^{(t-t_0)A}\mathbf{k} + \int_{t_0}^t e^{(t-s)A}\, \mathbf{b}(s)\, ds. \tag{6.104}$$

As an example, we consider the equation $\mathbf{x}' = A\mathbf{x} + \mathbf{b}$ where

$$A = \begin{bmatrix} 4 & -3 \\ 2 & -1 \end{bmatrix}, \qquad \mathbf{b}(t) = \begin{bmatrix} 0 \\ (e^{-t} + 1)^{-1} \end{bmatrix}.$$

We find, by using the methods of Section 6.8, that

$$e^{tA} = A_1 e^t + A_2 e^{2t},$$

where

$$A_1 = \begin{bmatrix} -2 & 3 \\ -2 & 3 \end{bmatrix}, \qquad A_2 = \begin{bmatrix} 3 & -3 \\ 2 & -2 \end{bmatrix}.$$

Then

$$e^{(t-s)A} = A_1 e^{t-s} + A_2 e^{2(t-s)}$$

and a particular solution is

$$\mathbf{u}_p(t) = \int_0^t e^{(t-s)A}\, \mathbf{b}(s)\, ds = A_1 \int_0^t e^{t-s}\, \mathbf{b}(s)\, ds + A_2 \int_0^t e^{2(t-s)}\, \mathbf{b}(s)\, ds.$$

Since

$$\int_0^t e^{t-s}(e^{-s}+1)^{-1}\,ds = -e^t \ln \frac{e^{-t}+1}{2}$$

and

$$\int_0^t e^{2(t-s)}(e^{-s}+1)^{-1}\,ds = -e^t + e^{2t} + e^{2t}\ln \frac{e^{-t}+1}{2},$$

we have

$$\mathbf{u}_p(t) = A_1 \begin{bmatrix} 0 \\ -e^t \ln \dfrac{e^{-t}+1}{2} \end{bmatrix} + A_2 \begin{bmatrix} 0 \\ -e^t + e^{2t} + e^{2t}\ln \dfrac{e^{-t}+1}{2} \end{bmatrix}.$$

## Exercises for Section 6.9

In Exercises 1–8, find (a) a particular solution of the form (6.92) and (b) the general solution of the system. The characteristic values of the coefficient matrix are as indicated.

**1.** $\begin{aligned} x_1' &= 3x_1 - 2x_2 - 2e^{-t} \\ x_2' &= x_1 \qquad\quad - 2e^{-t} \end{aligned}$  $\quad \lambda_1 = 1, \quad \lambda_2 = 2$

**2.** $\begin{aligned} x_1' &= 3x_1 - x_2 + 2e^{3t} \\ x_2' &= 3x_1 - x_2 + 5e^{3t} \end{aligned}$  $\quad \lambda_1 = 0, \quad \lambda_2 = 2$

**3.** $\begin{aligned} x_1' &= -4x_2 - 2e^{-2t} \\ x_2' &= \quad\; x_1 - 5e^{-2t} \end{aligned}$  $\quad \lambda_1 = 2i, \quad \lambda_2 = -2i$

**4.** $\begin{aligned} x_1' &= -x_2 + 20 \cos 2t \\ x_2' &= 2x_1 - 3x_2 \end{aligned}$  $\quad \lambda_1 = -1, \quad \lambda_2 = -2$

Suggestion: replace $\cos 2t$ by $e^{2it}$ and take the real part of a complex solution.

**5.** $\begin{aligned} x_1' &= -2x_1 + x_2 + 2\sin t \\ x_2' &= -3x_1 + 2x_2 + 2\sin t \end{aligned}$  $\quad \lambda_1 = 1, \quad \lambda_2 = -1$

Suggestion: replace $\sin t$ by $e^{it}$ and take the imaginary part of a complex solution.

**6.**  $x_1' = -2x_1 + x_2 + 2e^{-2t}$

$\quad x_2' = \quad x_1 - 2x_2 + 3e^{-4t}$    $\qquad \lambda_1 = -1, \quad \lambda_2 = -3$

**7.**  $x_1' = 2x_2$

$\quad x_2' = -x_1 + x_2 + x_3 - 2e^{2t}$

$\quad x_3' = \quad x_1 + 2x_2 - x_3 + 3e^{2t}$    $\qquad \lambda_1 = 0, \quad \lambda_2 = 1, \quad \lambda_3 = -1$

**8.**  $x_1' = 2x_1 + 2x_3 + e^{-t}$

$\quad x_2' = \quad x_2$    $\qquad \lambda_1 = 0, \quad \lambda_2 = \lambda_3 = 1$

$\quad x_3' = -x_1 - x_3 - e^{-t}$

**9.**  Let the quantities $P_{ij}(D)$, $1 \le i \le n$, $1 \le j \le n$, be polynomial operators. We define a matrix differential operator $M(D)$,

$$M(D) = \begin{bmatrix} P_{11}(D) \dots P_{1n}(D) \\ \vdots \qquad\qquad \vdots \\ P_{n1}(D) \dots P_{nn}(D) \end{bmatrix}$$

such that, if $U$ is an $n \times q$ matrix function then $M(D) U = V$, where $V$ is the $n \times q$ matrix function with elements

$$v_{ij} = \sum_{k=1}^{n} P_{ik}(D) u_{kj}.$$

In particular, we write

$$I Q(D) = \text{diag}(Q(D), Q(D), \dots, Q(D)).$$

(a) Show that the equation $\mathbf{x}' = A\mathbf{x}$, where $A$ is a constant matrix, can be written as $(ID - A)\mathbf{x} = \mathbf{0}$. Then show that the $i$th component $x_i$ of every solution satisfies the equation $p(D) x_i = 0$, where $p$ is the characteristic polynomial of $A$. Suggestion: find $M(D)$ such that $M(D)(ID - A) = I p(D)$.

(b) Show that the $i$th component $x_i$ of a solution of the equation $\mathbf{x}' = A\mathbf{x} + \mathbf{b}$, where $\mathbf{b}(t) = t^s e^{rt}\mathbf{h}$, satisfies an equation of the form

$$p(D) x_i = \sum_{j=0}^{s} k_{ij} t^j e^{rt},$$

where the $k_{ij}$ are constants. Then deduce formula (6.96).

In Exercises 10–14, use formula (6.96) to find (a) a particular solution and (b) the general solution.

10. $x_1' = 3x_1 - 2x_2 + 3e^t$

$x_2' = x_1$

Suggestion: let $x_1 = (A_1 + A_2 t)e^t$, $\qquad x_2 = (B_1 + B_2 t)e^t$.

11. $x_1' = -3x_1 - x_2 + e^{-2t}$

$x_2' = 6x_1 + 4x_2 + 4e^{-2t}$

12. $x_1' = -2x_1 + 3x_2$

$x_2' = -x_1 + 2x_2 + 4te^{-t}$

13. $x_1' = -x_2 + 2 \sin t$

$x_2' = x_1 + 4 \sin t$

14. $x_1' = 3e^t$

$x_2' = 2x_1 + x_2 - 2x_3$

$x_3' = x_1 - x_3 + e^t$

In Exercises 15–22 use the method of variation of parameters to find a particular solution $\mathbf{x}_p$ of the equation $\mathbf{x}' = A\mathbf{x} + \mathbf{b}$, where $A$ and $\mathbf{b}$ are as given.

15. $A = \begin{bmatrix} 2 & 1 \\ -3 & -2 \end{bmatrix}$, $\qquad \mathbf{b}(t) = \begin{bmatrix} 2e^t \\ 4e^t \end{bmatrix}$

16. $A = \begin{bmatrix} 0 & 1 \\ -1 & 2 \end{bmatrix}$, $\qquad \mathbf{b}(t) = \begin{bmatrix} e^t/(t+1) \\ 0 \end{bmatrix}$

17. $A = \begin{bmatrix} 2 & 2 \\ -3 & -3 \end{bmatrix}$, $\qquad \mathbf{b}(t) = \begin{bmatrix} 1 \\ 2t \end{bmatrix}$

18. $A = \begin{bmatrix} 0 & -1 \\ 2 & 3 \end{bmatrix}$, $\qquad \mathbf{b}(t) = \begin{bmatrix} 0 \\ 2te^t \end{bmatrix}$

19. $A = \begin{bmatrix} 3 & 2 \\ -4 & -3 \end{bmatrix}$, $\qquad \mathbf{b}(t) = \begin{bmatrix} 2 \cos t \\ 2 \sin t \end{bmatrix}$

20. $A = \begin{bmatrix} 0 & 1 \\ -1 & 0 \end{bmatrix}$, $\qquad \mathbf{b}(t) = \begin{bmatrix} 4 \sin t \\ 0 \end{bmatrix}$

21. $A = \begin{bmatrix} 1 & -1 & 1 \\ 0 & 0 & 1 \\ 0 & -1 & 2 \end{bmatrix}$, $\qquad \mathbf{b}(t) = \begin{bmatrix} 2e^t \\ 6te^t \\ 0 \end{bmatrix}$

22. $A = \begin{bmatrix} 1 & 1 & 1 \\ 0 & -1 & 0 \\ -2 & -1 & -2 \end{bmatrix}$, $\qquad \mathbf{b}(t) = \begin{bmatrix} 1 \\ 0 \\ 2e^{-t} \end{bmatrix}$

23. If $\mathbf{u}$ is a vector function such that $|\mathbf{u}(t)| \le g(t)$ for $a \le t \le b$, show that

$$\left| \int_a^b \mathbf{u}(t) \, dt \right| \le \int_a^b g(t) \, dt.$$

24. If every characteristic value of $A$ has a negative real part, show that there exist positive numbers $K$ and $r$ such that

$$|e^{tA}| \le Ke^{-rt}, \quad t \ge 0$$

$$|e^{(t-s)A}| \le Ke^{-r(t-s)}, \quad t \ge s.$$

25. Let $\mathbf{u}$ be the solution of the initial value problem $\mathbf{x}' = A\mathbf{x} + \mathbf{b}$, $\mathbf{x}(0) = \mathbf{k}$. Assume that every characteristic value of $A$ has a negative real part.

    (a)   If there is a positive number $M$ such that $|\mathbf{b}(t)| \le M$ for $t \ge 0$, show that $|\mathbf{u}|$ is bounded for $t \ge 0$.

    (b)   If $|\mathbf{b}(t)| \le g(t)$ for $t \ge 0$ and $\int_0^{\infty} g(t)\, dt$ converges, show that $|\mathbf{u}|$ is bounded for $t \ge 0$.

    Suggestion: use formula (6.104) and the results of Exercises 23 and 24.

## 6.10   MECHANICAL SYSTEMS

In this section we consider some problems that involve the motion of one or more objects. The mathematical description of such a problem involves a system of differential equations in which the unknown functions are the co-ordinates of the centers of mass of the moving objects.

For our first example we consider the configuration of Fig. 6.3. Two bodies with masses $m_1$ and $m_2$, respectively, are suspended by springs as shown in the figure. The downward displacements of the bodies from their equilibrium

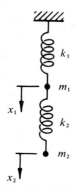

**Figure 6.3**

positions are denoted by $x_1$ and $x_2$ respectively. At equilibrium, the forces due to gravity are balanced by forces due to a stretching of the springs. We denote by $F_1$ and $F_2$ the net forces acting on the moving bodies with masses $m_1$ and $m_2$, respectively. Suppose first that $x_1 > 0$ and that $x_2 > x_1$. Then the spring with constant $k_2$ is stretched by a distance $x_2 - x_1$. The net force acting on the object with mass $m_1$ (positive direction downward) is

$$F_1 = -k_1 x_1 + k_2(x_2 - x_1) \qquad (6.105)$$

while the net force acting on the second body is

$$F_2 = -k_2(x_2 - x_1). \qquad (6.106)$$

If $x_1 < 0$, the first spring is compressed and it exerts a downward force on the first object. If $x_2 - x_1 < 0$, the second spring is compressed. It then exerts an upward force on the first object and a downward force on the second. But if either or both of these situations occurs, it can be seen that the formulas (6.105) and (6.106) are still valid. Consequently our equations of motion become

$$m_1 \ddot{x}_1 = -k_1 x_1 + k_2(x_2 - x_1),$$
$$m_2 \ddot{x}_2 = -k_2(x_2 - x_1). \qquad (6.107)$$

This is a linear system with constant coefficients. The nature of the solutions is discussed in Exercise 1.

We next consider the motion of a single body in a plane, subject to a constant gravitational force. It is convenient to work in a rectangular coordinate system in the plane. We choose the origin to be at the surface of the earth, with one axis vertical, as shown in Fig. 6.4. According to Newton's law of motion,

$$m\mathbf{a} = \mathbf{F}, \qquad (6.108)$$

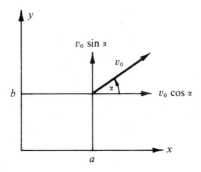

**Figure 6.4**

where $m$ is the mass of the body,

$$\mathbf{a} = \ddot{x}\mathbf{i} + \ddot{y}\mathbf{j} \tag{6.109}$$

is the acceleration, and

$$\mathbf{F} = 0\mathbf{i} - mg\mathbf{j} \tag{6.110}$$

is the force acting on the body. Then Eq. (6.108) becomes

$$m(\ddot{x}\mathbf{i} + \ddot{y}\mathbf{j}) = 0\mathbf{i} - mg\mathbf{j}.$$

Equating corresponding components, we arrive at the system

$$m\ddot{x} = 0, \qquad m\ddot{y} = -mg. \tag{6.111}$$

We denote by $(a, b)$ the position of the body at $t = 0$. We assume that the initial velocity vector $\mathbf{v}_0$ has magnitude $v_0$ and is inclined at an angle $\alpha$ with the horizontal. Thus

$$\mathbf{v}_0 = v_0(\mathbf{i} \cos \alpha + \mathbf{j} \sin \alpha). \tag{6.112}$$

The initial conditions for the system (6.111) are

$$x(0) = a, \qquad y(0) = b, \qquad \dot{x}(0) = v_0 \cos \alpha, \qquad \dot{y}(0) = v_0 \sin \alpha. \tag{6.113}$$

Some specific cases are considered in the exercises.

In some applications that involve the motion of a body in a plane, it is more convenient to work with the polar coordinates $(r, \theta)$ of the center of mass of the body. This is the case, for example, when attempting to describe the motion of a satellite about the earth. We introduce the unit vectors $\mathbf{e}_r$ and $\mathbf{e}_\theta$, where

$$\begin{aligned} \mathbf{e}_r &= \mathbf{i} \cos \theta + \mathbf{j} \sin \theta, \\ \mathbf{e}_\theta &= -\mathbf{i} \sin \theta + \mathbf{j} \cos \theta. \end{aligned} \tag{6.114}$$

The vector $\mathbf{e}_r$ points away from the origin in the direction of increasing $r$. The vector $\mathbf{e}_\theta$ is perpendicular to $\mathbf{e}_r$ and points in the direction of increasing $\theta$. The situation is illustrated in Fig. 6.5. From formulas (6.114), we deduce the relations

$$\frac{d}{d\theta} \mathbf{e}_r = \mathbf{e}_\theta, \qquad \frac{d}{d\theta} \mathbf{e}_\theta = -\mathbf{e}_r. \tag{6.115}$$

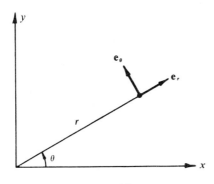

**Figure 6.5**

We resolve the force **F** acting on the body into radial and circumferential components, writing

$$\mathbf{F} = F_r \mathbf{e}_r + F_\theta \mathbf{e}_\theta. \tag{6.116}$$

In many important cases, $F_\theta = 0$. For instance, when the origin is at the center of the earth the force exerted on a satellite of mass $m$ by the earth is

$$\mathbf{F} = -\frac{mgR^2}{r^2}\,\mathbf{e}_r, \tag{6.117}$$

where $R$ is the radius of the (spherical) earth.

Since

$$\mathbf{i}x + \mathbf{j}y = \mathbf{i}r\cos\theta + \mathbf{j}r\sin\theta = r\mathbf{e}_r,$$

the velocity of the moving body is

$$\mathbf{v} = \dot{r}\mathbf{e}_r + r\frac{d\mathbf{e}}{d\theta}\frac{d\theta}{dt} = \dot{r}\mathbf{e}_r + r\dot{\theta}\mathbf{e}_\theta \tag{6.118}$$

and the acceleration is

$$\mathbf{a} = \ddot{r}\mathbf{e}_r + \dot{r}\dot{\theta}\mathbf{e}_\theta + (r\ddot{\theta} + \dot{r}\dot{\theta})\mathbf{e}_\theta - r\dot{\theta}^2\mathbf{e}_r$$

or

$$\mathbf{a} = (\ddot{r} - r\dot{\theta}^2)\mathbf{e}_r + (r\ddot{\theta} + 2\dot{r}\dot{\theta})\mathbf{e}_\theta. \tag{6.119}$$

Then, by equating corresponding components in the equation

$$ma = F,$$

we arrive at the system

$$m(\ddot{r} - r\dot{\theta}^2) = F_r,$$
$$m(r\ddot{\theta} + 2\dot{r}\dot{\theta}) = F_\theta. \qquad (6.120)$$

In the special case where $F$ is given by formula (6.117), we have

$$\ddot{r} - r\dot{\theta}^2 = -gR^2 r^{-2}, \qquad (6.121)$$
$$r\ddot{\theta} + 2\dot{r}\dot{\theta} = 0.$$

## Exercises for Section 6.10

1.  (a)  Show that the system (6.107) is equivalent to the system

$$P(D)x_1 = 0, \qquad x_2 = \frac{1}{k_2}(m_1 D^2 + k_1 + k_2)x_1,$$

where

$$P(D) = m_1 m_2 D^4 + [m_2(k_1 + k_2) + m_1 k_2]D^2 + k_1 k_2.$$

(b)  Show that the polynomial $P$ of part (a) has four distinct pure imaginary zeros, and hence that $x_1$ and $x_2$ are bounded functions.

2.  Find the equations of motion for the mechanical system of Fig. 6.6.

3.  Find the equations of motion for the mechanical system of Fig. 6.7.

4.  Find the equations of motion for the mechanical system of Fig. 6.8.

5.  A ball is thrown horizontally from the top of a tower of height $h$, with a velocity $v_0$. Neglect air resistance.
    (a)  How far from the base of the tower will the ball land?
    (b)  Find the $xy$ equation of the path of the ball and show that it is a parabola.

6.  A ball is thrown horizontally from the top of a tower of height $h$, with a velocity $v_0$. Assume that the force due to air resistance is $F = -c\mathbf{v}$, where $c$ is a positive constant and $\mathbf{v}$ is the velocity vector.
    (a)  Find the coordinates, $x(t)$ and $y(t)$, of the ball.
    (b)  Find the $xy$ equation of the path.

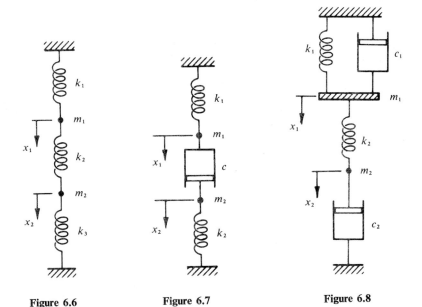

Figure 6.6                Figure 6.7                Figure 6.8

7.  A projectile is fired from a gun located at the origin of the coordinate
    system in Fig. 6.1. The gun is inclined at an angle $\alpha$ with the horizontal
    and its muzzel velocity is $v_0$. Neglect air resistance.

    (a)  How far from the gun will the projectile land?

    (b)  Find the $xy$ equation of the path of the projectile and show that it is
    a parabola.

8.  In Exercise 7, suppose that the force due to air resistance is $\mathbf{F} = -c\mathbf{v}$,
    where $c$ is a positive constant and $\mathbf{v}$ is the velocity vector.

    (a)  Find the coordinates, $x(t)$ and $y(t)$, of the projectile.

    (b)  Find the $xy$ equation of the path.

9.  An earth satellite of mass $m$ is attracted toward the center of the earth
    by a force of magnitude $mkr^{-2}$, where $k$ is a positive constant and $r$ is
    the distance to the earth's center. Using a rectangular $xy$ coordinate
    system with origin at the center of the earth, find the equations of motion
    of the satellite.

10.  (a)  Deduce from the second equation of the system (6.121) that
    $r^2\dot{\theta} = c$ where $c$ is a constant. Then show that $r$ satisfies the equation

$$\ddot{r} = c^2 r^{-3} - gR^2 r^{-2}.$$

(b)   Setting $u = r^{-1}$, show that

$$\frac{d^2u}{d\theta^2} + u = \frac{gR^2}{c^2} .$$

(c)   From the result of part (b) deduce that the path is a conic section.

11.   Two bodies of masses $m_1$ and $m_2$ move in a plane. The force of attraction between them has magnitude $km_1m_2/r^2$, where $k$ is a constant and $r$ is the distance between the centers of mass. Find a system of differential equations for the rectangular coordinates $(x_1, y_1)$ and $(x_2, y_2)$ of the centers of mass.

## 6.11   ELECTRIC CIRCUITS

Figure 6.9 illustrates an electrical network that involves two loops. Our aim is to formulate a system of differential equations and initial conditions for the two unknown loop currents $I_1$ and $I_2$. We assume that the switch is closed at time $t = 0$, and that the charge on the capacitance is zero before this time.

Each loop has been oriented by assigning positive directions to the currents. The currents flowing from node 2 to node 1 must be $I_1 - I_2$. This follows from the law of Kirchhoff that says that the current leaving a node must be equal to the current entering it. Kirchhoff's other law says that the sum of the voltage drops around each loop must be equal to the applied voltage. Applying this law to each loop in turn, we arrive at the system of equations

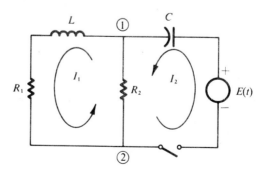

**Figure 6.9**

$$L\frac{dI_1}{dt} + R_1 I_1 + R_2(I_1 - I_2) = 0,\tag{6.122}$$

$$R_2(I_2 - I_1) + \frac{1}{C}Q = E(t).\tag{6.123}$$

Differentiating through in Eq. (6.123) with respect to $t$ and noting that $dQ/dt = I_2$, we obtain the system of differential equations

$$L\frac{dI_1}{dt} + (R_1 + R_2)I_1 - R_2 I_2 = 0,\tag{6.124}$$

$$R_2\frac{dI_2}{dt} - R_1\frac{dI_1}{dt} + \frac{1}{C}I_2 = E'(t)\tag{6.125}$$

for $I_1$ and $I_2$. It is not hard to see that this system is equivalent to a first-order system for $I_1$ and $I_2$. We must therefore specify $I_1(0)$ and $I_2(0)$ in our initial conditions. Because of the presence of the inductance in the loop for $I_1$ we must have

$$I_1(0) = 0.\tag{6.126}$$

From Eq. (6.123) (or by inspection of Fig. 6.9) we see that

$$I_2(0) = \frac{E(0)}{R_2}.\tag{6.127}$$

In the network of Fig. 6.10 there are three loops. However, it is possible to formulate a system of two differential equations for the voltage drops $E_1 - E_0$ and $E_2 - E_0$. It is convenient to set $E_0 = 0$ and work with $E_1$ and $E_2$.

We use the principle that the current leaving a node is equal to the current entering it. The net current leaving node 1 is

$$C\frac{dE_1}{dt} + \frac{1}{R_1}E_1 + \frac{1}{L}\int_0^t [E_1(s) - E_2(s)]\, ds = 0.\tag{6.128}$$

The current entering the second node is $I(t)$. We must have

$$\frac{1}{R_2}E_2 + \frac{1}{L}\int_0^t [E_2(s) - E_1(s)]\, ds = I(t).\tag{6.129}$$

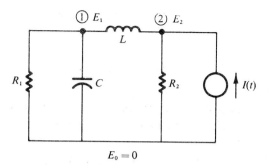

**Figure 6.10**

Differentiating through in Eqs. (6.128) and (6.129), we arrive at the system

$$C\frac{d^2E_1}{dt^2} + \frac{1}{R_1}\frac{dE_1}{dt} + \frac{1}{L}(E_1 - E_2) = 0, \tag{6.130}$$

$$\frac{1}{R_2}\frac{dE_2}{dt} + \frac{1}{L}(E_2 - E_1) = I'(t). \tag{6.131}$$

Because of the presence of the capacitance we must have

$$E_1(0) = 0. \tag{6.132}$$

From Eq. (6.129) (or from the figure),

$$E_2(0) = R_2 I(0). \tag{6.133}$$

From Eq. (6.128),

$$E_1'(0) = 0. \tag{6.134}$$

### Exercises for Section 6.11

In Exercises 1–6, formulate a system of differential equations and initial conditions for the loop currents, assuming that all initial charges are zero and that the switch is closed at $t = 0$.

**1.** The circuit of Fig. 6.11.

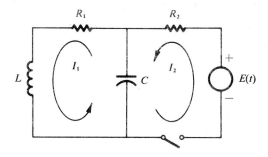

**Figure 6.11**

2.  The circuit of Fig. 6.12.

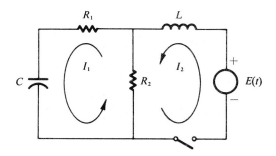

**Figure 6.12**

3.  The circuit of Fig. 6.13. Find the loop currents if $E = 6$ volts, $R_1 = 2$ ohms, $R_2 = 1$ ohm, and $L = 4$ henrys.

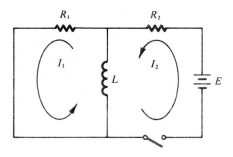

**Figure 6.13**

4. The circuit of Fig. 6.14. Find the *steady-state* loop currents if $R = 1.0$ ohms, $C = 0.5$ farad, $L = 0.5$ henry, and $E(t) = 2\cos(t/2)$ volts.

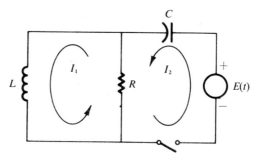

**Figure 6.14**

5. The circuit of Fig. 6.15.

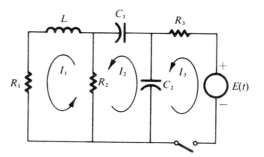

**Figure 6.15**

6. The circuit of Fig. 6.16.

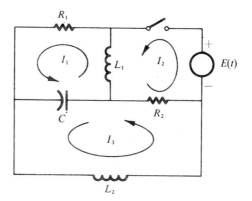

**Figure 6.16**

In Exercises 7 and 8, formulate a system of differential equations and initial conditions for the node voltages. Assume that all currents are zero at $t = 0$.

7.  The circuit of Fig. 6.17.

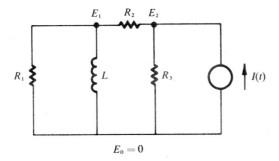

$$E_0 = 0$$

**Figure 6.17**

8.  The circuit of Fig. 6.18.

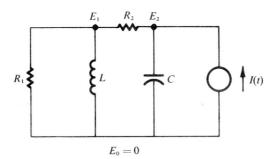

$$E_0 = 0$$

**Figure 6.18**

## 6.12 THEORY OF LINEAR SYSTEMS

In this section we shall discuss some questions about the existence and uniqueness of solutions of first-order linear systems. We consider the initial value problem

$$\mathbf{x}' = A\mathbf{x} + \mathbf{b}, \qquad \mathbf{x}(t_0) = \mathbf{k}, \tag{6.135}$$

where $A$ is a matrix function that is not necessarily constant. In terms of components, problem (6.135) becomes

$$\frac{dx_i}{dt} = \sum_{j=1}^{n} a_{ij} x_j + b_i, \qquad x_i(t_0) = k_i, \qquad 1 \le i \le n. \tag{6.136}$$

The fundamental theorem is as follows.

**Theorem 6.7**   Let the functions $a_{ij}$ and $b_i$ be continuous on an interval $\mathscr{I}$ that contains $t_0$. Then the initial value problem (6.135) possesses a solution that exists throughout $\mathscr{I}$. Any two solutions that exist on an interval that contains $t_0$ are identical on that interval.

In the case of a system with constant coefficients, we are able to write down a formula that describes the solution of the initial value problem (6.135). In fact, as shown in Section 6.9, the solution is

$$\mathbf{x}(t) = e^{(t-t_0)A}\mathbf{k} + \int_{t_0}^{t} e^{(t-s)A}\mathbf{b}(s)\,ds. \tag{6.137}$$

Theorem 6.7 asserts the existence of a solution for much more general problems, in which the system of differential equations need not have constant coefficients. A description of the proof that a solution exists is given in the final chapter of this book. A proof of the uniqueness of a solution follows.

In proving uniqueness we shall make use of two facts about vector functions. First, if $\mathbf{u}$ is a continuous[10] vector function then the function $|\mathbf{u}|$, where

$$|\mathbf{u}|(t) = |\mathbf{u}(t)|,$$

is continuous.[11] Second is the inequality

$$\left| \int_a^b \mathbf{u}(t)\,dt \right| \le \left| \int_a^b |\mathbf{u}(t)|\,dt \right|. \tag{6.138}$$

Proofs of both properties are left as exercises.

Now suppose that $\mathbf{u}$ and $\mathbf{v}$ are both solutions of the problem (6.135) on an interval $\mathscr{I}$ that contains $t_0$. Setting $\mathbf{w} = \mathbf{u} - \mathbf{v}$, we see that

$$\mathbf{w}' = A\mathbf{w}, \tag{6.139}$$

$$\mathbf{w}(t_0) = \mathbf{0}. \tag{6.140}$$

Let $\mathscr{J}$ be a closed interval that contains $t_0$ and that is contained in $\mathscr{I}$. (Each point in $\mathscr{I}$ belongs to such an interval.) Then, since the elements of $A$ are continuous, there exists a number $M$ such that $|A(t)| \le M$ for $t$ in $\mathscr{J}$. Integrating both members of Eq. (6.139) from $t_0$ to $t$, and taking into account the conditions (6.140), we have

$$\mathbf{w}(t) = \int_{t_0}^{t} A(s)\,\mathbf{w}(s)\,ds.$$

---

[10] A vector function is said to be continuous if its components are all continuous.

[11] Notice that $|\mathbf{u}|$ is a scalar function whose value at $t$ is $|\mathbf{u}(t)|$; $|\mathbf{u}|$ is *not* the vector function with components $|u_i|$.

By the use of the inequality (6.138) we see that

$$|\mathbf{w}(t)| \le nM \left| \int_{t_0}^t |\mathbf{w}(s)| \, ds \right|.$$

By the lemma of Section 5.11, $\mathbf{w}(t) = \mathbf{0}$ and $\mathbf{u}(t) = \mathbf{v}(t)$ for all $t$ in $\mathscr{I}$.

We previously saw that the set of all solutions of the equation

$$\mathbf{x}' = A\mathbf{x}, \tag{6.141}$$

where $A$ is an $n \times n$ matrix function, formed a vector space. We wish to show that the dimension of this space is $n$.

First of all, there exists a set of $n$ linearly independent elements of the space. For let $\mathbf{k}_1, \mathbf{k}_2, \ldots, \mathbf{k}_n$ be linearly independent constant vectors and let $\mathbf{u}_1, \mathbf{u}_2, \ldots, \mathbf{u}_n$ be the solutions of Eq. (6.141) for which

$$\mathbf{u}_i(t_0) = \mathbf{k}_i, \qquad 1 \le i \le n.$$

Theorem 6.7 guarantees the existence of such solution. If $U$ is the matrix function whose column vectors are the $\mathbf{u}_i$, then $U(t_0)$ is nonsingular. Hence the column vectors of $U$ are linearly independent vector functions.

We next show that $U(t)$ is nonsingular for all $t$. Suppose that $\det U(t_1) = 0$ for some $t_1$. Then there exist constants $c_1, c_2, \ldots, c_n$, not all zero, such that

$$c_1 \mathbf{u}_1(t_1) + c_2 \mathbf{u}_2(t_1) + \cdots + c_n \mathbf{u}_n(t_1) = \mathbf{0}.$$

Let $\mathbf{v} = c_1 \mathbf{u}_1 + c_2 \mathbf{u}_2 + \cdots + c_n \mathbf{u}_n$. Then $\mathbf{v}$ is a solution of equation (6.141) and $\mathbf{v}(t_1) = \mathbf{0}$. But the zero vector function is also a solution so $\mathbf{v}(t) = \mathbf{0}$ for all $t$. That is,

$$c_1 \mathbf{u}_1 + c_2 \mathbf{u}_2 + \cdots + c_n \mathbf{u}_n = \mathbf{0}.$$

But this is impossible, since the $\mathbf{u}_i$ are linearly independent. Hence $U(t)$ must be nonsingular for all $t$.

Our final result is essentially Theorem 6.1.

**Theorem 6.8** Let $\mathbf{u}_1, \mathbf{u}_2, \ldots, \mathbf{u}_n$ be linearly independent solutions of Eq. (6.141) on an interval $\mathscr{I}$. If $\mathbf{v}$ is any solution on $\mathscr{I}$, there exist constants $c_1, c_2, \ldots, c_n$ such that

$$\mathbf{v} = c_1 \mathbf{u}_1 + c_2 \mathbf{u}_2 + \cdots + c_n \mathbf{u}_n.$$

PROOF    Let $U$ be the matrix function whose column vectors are $\mathbf{u}_1, \mathbf{u}_2, \ldots,$ $\mathbf{u}_n$. Let $t_0$ be any point in $\mathscr{I}$. Since $\det U(t_0) \neq 0$, there exist constants $c_1, c_2, \ldots, c_n$ such that

$$c_1 \mathbf{u}_1(t_0) + c_2 \mathbf{u}_2(t_0) + \cdots + c_n \mathbf{u}_n(t_0) = \mathbf{v}(t_0).$$

Then $c_1\mathbf{u}_1 + c_2\mathbf{u}_2 + \cdots + c_n\mathbf{u}_n$ and $\mathbf{v}$ are both solutions of the Eq. (6.141) and they have the same values at $t_0$. By Theorem 6.7 the solutions are identical; that is,

$$\mathbf{v} = c_1\mathbf{u}_1 + c_2\mathbf{u}_2 + \cdots + c_n\mathbf{u}_n.$$

## Exercises for Section 6.12

1.  If $\mathbf{u}$ is a continuous vector function, show that $|\mathbf{u}|$ is continuous. Suggestion: show that

$$\big||\mathbf{u}(t+h)| - |\mathbf{u}(t)|\big| \le |\mathbf{u}(t+h) - \mathbf{u}(t)|.$$

2.  Prove the inequality (6.138).

3.  Use Theorem 6.7 to prove Theorem 5.6.

4.  Use the results of this section to prove Theorem 5.8.

5.  Let $A$ be an $n \times n$ constant matrix and let $B$ be an $n \times n$ matrix function that is defined and continuous on $[0, \infty)$. Let $\mathbf{u}$ be a solution of the differential equation $\mathbf{x}' = A\mathbf{x} + B\mathbf{x}$ on $[0, \infty)$.
    (a)  Show that

$$\mathbf{u}(t) = e^{tA}\,\mathbf{c} + \int_0^t e^{(t-s)A}B(s)\,\mathbf{u}(s)\,ds$$

for $t \geq 0$, where $\mathbf{c} = \mathbf{u}(0)$.
    (b)  Let every characteristic value of $A$ have a negative real part. Show that there exist positive numbers $K$ and $\alpha$ such that

$$|\mathbf{u}(t)| \le K\,|\mathbf{c}|\,e^{-\alpha t} + nK \int_0^t e^{-\alpha(t-s)}\,|B(s)|\,|\mathbf{u}(s)|\,ds$$

for $t \geq 0$. If $\displaystyle\int_0^\infty |B(s)|\,ds$ converges, show that $|\mathbf{u}|$ is bounded on $[0, \infty)$. (Use the result of Exercise 2, Section 5.11.)

# VII

## Series Solutions

## 7.1 POWER SERIES

The simplest functions in calculus are the polynomials, and those functions "built" from polynomials by a finite number of operations. These operations include the four arithmetic operations, raising to a positive integral power, and root-taking. Functions of this class are known as *algebraic functions*. Examples of such functions are $f$ and $g$, where

$$f(x) = \left(\frac{x^2 - 3x + 1}{2x + 5}\right)^3, \quad g(x) = \sqrt{x^3 + 1}.$$

Still other functions are too complicated to be defined by the procedures described above. Some can be defined in terms of integrals of simpler functions. For instance, the formulas

$$\ln x = \int_0^x \frac{1}{t}\,dt, \quad x > 0,$$

$$\operatorname{arc\,sin} x = \int_0^x (1 - t^2)^{-1/2}\,dt, \quad |x| \le 1$$

can be used to define the natural logarithm and inverse sine functions. Other functions can be defined by infinite sequences or series of simpler functions. Thus we may define

$$e^x = \sum_{n=0}^{\infty} \frac{x^n}{n!}, \quad \sin x = \sum_{n=1}^{\infty} (-1)^n \frac{x^{2n-1}}{(2n-1)!}$$

*353*

for all $x$. In doing computational work manually with any of these functions, or even with the square-root function, a table of values of the functions would be convenient. A knowledge of the basic properties of the function is also useful. For instance, we know that the exponential function is nonnegative and increasing. The sine function is periodic with period $2\pi$ and its values vary between $-1$ and $1$. Actually we may feel more comfortable in working with some of these functions than with many of the more complicated algebraic functions.

Even simple differential equations can have complicated solutions. Thus the solutions of the equation $y' = y$ are not algebraic functions but are multiples of the exponential function. The solutions of the equation $y'' = xy$ are not algebraic functions, nor can they be expressed in a simple way in terms of the standard exponential and trigonometric functions of our library of functions. Rather, they become new functions that can be added to our library if we find them sufficiently important.

In this chapter we consider the possibility of representing the solutions of certain linear differential equations in terms of power series. These "series solutions" form a starting point for the investigation of the general properties of the solutions and for the computation of their values. We begin with a review of power series.

A *power series* is a series of functions which at every real number $x$ has the form

$$\sum_{n=0}^{\infty} a_n(x - x_0)^n. \tag{7.1}$$

Here $x_0$ is a fixed number, called the *center of expansion* (we speak of a power series *about the point* $x_0$) and the numbers $a_i$ are the *coefficients* of the power series. The series (7.1) always converges at $x = x_0$ to the sum $a_0$. It may converge only at this point. It may converge for all values of $x$. If neither of these extreme situations occurs, there is a positive number $R$ such that the series converges absolutely for $|x - x_0| < R$ and diverges when $|x - x_0| > R$. The number $R$ is called the *radius of convergence* of the series. (If the series converges only at $x_0$ we set $R = 0$; if it converges everywhere we write $R = \infty$.) The interval $(x_0 - R, x_0 + R)$ is called the *interval of convergence* of the series.

The radius of convergence of a power series can sometimes be found by the following method, which is based on the ratio test for convergence of series. We leave the proof as an exercise.

**Theorem 7.1**  The radius of convergence of the series (7.1) is given by

$$R = \lim_{n \to \infty} \left| \frac{a_n}{a_{n+1}} \right|,$$

provided that the limit exists. (If $|a_n/a_{n+1}| \to \infty$, then $R = \infty$.)

**Example 1**  For the series

$$\sum_{n=0}^{\infty} \frac{2^n(x-1)^n}{n!}$$

we have $a_n = 2^n/n!$ and $a_{n+1} = 2^{n+1}/(n+1)!$ so $a_n/a_{n+1} = (n+1)/2$. Since this ratio becomes infinite as $n$ increases, we have $R = \infty$. The series converges everywhere.

**Example 2**  The series

$$\sum_{n=0}^{\infty} n!\, x^n$$

converges only at the center of expansion $x = 0$ because

$$R = \lim_{n \to \infty} \left| \frac{a_n}{a_{n+1}} \right| = \lim \frac{1}{n+1} = 0.$$

A power series defines a function $f$ whose domain is the interval of convergence (plus perhaps one or both endpoints of the interval). The value $f(x)$ of the function $f$ at each point $x$ in this interval is the sum of the series at the point $x$.

In the next two theorems we state, without proof, some facts about functions defined by power series. Proofs are given in most books on advanced calculus. See, for example, Taylor (1955).

**Theorem 7.2**  Let

$$f(x) = \sum_{n=0}^{\infty} a_n(x - x_0)^n, \qquad g(x) = \sum_{n=0}^{\infty} b_n(x - x_0)^n$$

for $|x - x_0| < r$. Then for $|x - x_0| < r$,

$$cf(x) = \sum_{n=0}^{\infty} ca_n(x - x_0)^n$$

for every real number $c$,

$$f(x) + g(x) = \sum_{n=0}^{\infty} (a_n + b_n)(x - x_0)^n,$$

and

$$f(x)\,g(x) = \sum_{n=0}^{\infty} c_n(x - x_0)^n,$$

where

$$c_n = a_0\,b_n + a_1 b_{n-1} + \cdots + a_n\,b_0 = \sum_{k=0}^{n} a_k\,b_{n-k} = \sum_{k=0}^{n} a_{n-k}\,b_k.$$

**Example 3**   Let

$$f(x) = \sum_{n=0}^{\infty} (n + 1)x^n, \qquad g(x) = \sum_{n=0}^{\infty} x^n$$

for $|x| < 1$. Then

$$f(x)\,g(x) = \sum_{n=0}^{\infty} c_n x^n,$$

where

$$c_n = \sum_{k=0}^{n} (k + 1)(1) = 1 + 2 + \cdots + (n + 1) = \frac{(n + 1)(n + 2)}{2}.$$

The next theorem concerns facts about differentiation and integration of power series.

**Theorem 7.3**   Let

$$f(x) = \sum_{n=0}^{\infty} a_n(x - x_0)^n$$

for $|x - x_0| < R$. Then $f$ is differentiable on the interval $(x_0 - R, x_0 + R)$, and

$$f'(x) = \sum_{n=1}^{\infty} na_n(x - x_0)^{n-1}$$

for $|x - x_0| < R$.

If $a$ and $b$ are any points in $(x_0 - R, x_0 + R)$, then

$$\int_a^b f(x)\,dx = \sum_{n=0}^{\infty} \frac{a_n}{n + 1} [(b - x_0)^{n+1} - (a - x_0)^{n+1}].$$

In particular,

$$\int_{x_0}^{x} f(t)\, dt = \sum_{n=0}^{\infty} \frac{a_n}{n+1}(x-x_0)^{n+1}, \quad |x-x_0| < R.$$

The first part of the theorem says, roughly speaking, that the derivative of the sum of a power series is equal to the series of derivatives. Since the series of derivatives is again a power series, we may differentiate termwise again to find the second derivative. For instance, if

$$f(x) = \sum_{n=0}^{\infty} \frac{n+1}{n+2} x^n, \quad |x| < 1,$$

then

$$f'(x) = \sum_{n=1}^{\infty} \frac{n(n+1)}{n+2} x^{n-1}, \quad |x| < 1,$$

$$f''(x) = \sum_{n=2}^{\infty} \frac{n(n+1)(n-1)}{n+2} x^{n-2}, \quad |x| < 1,$$

and so on. The series for $f'$ (and the one for $f''$) can be rewritten in the form (7.1) by making a shift in the index of summation. If we set

$$k = n - 1$$

then as $n$ runs over the sequence $(1, 2, 3, \ldots)$, $k$ varies over $(0, 1, 2, \ldots)$. Thus

$$\sum_{n=1}^{\infty} \frac{n(n+1)}{n+2} x^{n-1} = \sum_{k=0}^{\infty} \frac{(k+1)(k+2)}{(k+3)} x^k.$$

This is also the same as

$$\sum_{n=0}^{\infty} \frac{(n+1)(n+2)}{n+3} x^n,$$

since the sum of the series depends only on $x$, and not on either the index of summation $k$ or the index $n$. The procedure is much like making a "change of variable" in an integral. If in the series for $f''$ we set $n = k + 2$, we see that

$$\sum_{n=2}^{\infty} \frac{n(n+1)(n-1)}{n+2} x^{n-2} = \sum_{k=0}^{\infty} \frac{(k+2)(k+3)(k+1)}{k+4} x^k.$$

In working with power series with a center of expansion $x_0$ different from zero, it is often convenient to make a change of variable $z = x - x_0$. Then

$$\sum_{n=0}^{\infty} a_n(x - x_0)^n = \sum_{n=0}^{\infty} a_n z^n.$$

## Exercises for Section 7.1

In Exercises 1–8, find the interval of convergence of the power series.

1. $\displaystyle\sum_{n=0}^{\infty} \frac{n + 2}{n + 1}(x - 1)^n$

2. $\displaystyle\sum_{n=0}^{\infty} \frac{n + 1}{2^n}(x + 3)^n$

3. $\displaystyle\sum_{n=1}^{\infty} n^3(x - 2)^n$

4. $\displaystyle\sum_{n=0}^{\infty} \frac{(n!)^2}{(2n)!} x^n$

5. $\displaystyle\sum_{n=0}^{\infty} \frac{1 \cdot 3 \cdot 5 \cdots (2n + 1)}{(2n)!} x^n$

6. $\displaystyle\sum_{n=1}^{\infty} n^n(x + 5)^n$

7. $\displaystyle\sum_{n=1}^{\infty} (-1)^n \frac{n}{\ln(n + 1)} x^n$

8. $\displaystyle\sum_{n=1}^{\infty} \frac{n!}{n^n} x^n$

In Exercises 9–12, let $f$ and $g$ be defined by the indicated power series. Find power series expansions for $f + g$ and $fg$.

9. $f(x) = \displaystyle\sum_{n=0}^{\infty} (n + 1)(x - 2)^n$,     $g(x) = \displaystyle\sum_{n=0}^{\infty} (x - 2)^n$

10. $f(x) = \displaystyle\sum_{n=0}^{\infty} \frac{x^n}{n!}$,     $g(x) = \displaystyle\sum_{n=1}^{\infty} n x^n$

11. $f(x) = \displaystyle\sum_{n=0}^{\infty} (n + 1)x^n$,     $g(x) = \displaystyle\sum_{n=0}^{\infty} \frac{x^n}{n + 1}$

12. $f(x) = \displaystyle\sum_{n=0}^{\infty} \frac{n + 1}{n + 2} x^n$,     $g(x) = \displaystyle\sum_{n=0}^{\infty} \frac{n + 2}{n + 3} x^n$

In Exercises 13–16, write the series in the form

$$\sum_{n=n_0}^{\infty} a_n x^n,$$

where $n_0$ and $a_n$ are to be determined.

13. $\displaystyle\sum_{n=2}^{\infty} n(n - 1)x^{n-2}$

14. $\displaystyle\sum_{n=3}^{\infty} \frac{n}{n + 1} x^{n-1}$

**15.** $\sum_{n=0}^{\infty} (n^2 + 2)x^{n+1}$          **16.** $\sum_{n=1}^{\infty} (2n + 1)(2n + 3)x^{n+2}$

In Exercises 17–18, find power series that represent $f'(x)$ and $f''(x)$, and determine an interval on which these formulas are valid.

**17.** $f(x) = \sum_{n=0}^{\infty} \dfrac{(-1)^n x^n}{2^n (n + 1)}$          **18.** $f(x) = \sum_{n=0}^{\infty} \dfrac{(x + 1)^{2n}}{n!(n + 1)!}$

**19.** Prove Theorem 7.1.

## 7.2  TAYLOR SERIES

Let $f$ be a function defined by a power series on its interval of convergence. Thus

$$f(x) = \sum_{n=0}^{\infty} a_n(x - x_0)^n, \quad |x - x_0| < R. \tag{7.2}$$

Differentiating and using Theorem 7.3, we have

$$f'(x) = \sum_{n=1}^{\infty} na_n(x - x_0)^{n-1},$$

$$f''(x) = \sum_{n=2}^{\infty} n(n - 1)a_n(x - x_0)^{n-2},$$

$$\dots\dots\dots\dots\dots\dots\dots\dots\dots\dots\dots\dots\dots\dots\dots\dots$$

$$f^{(k)}(x) = \sum_{n=k}^{\infty} n(n - 1) \cdots (n - k + 1)a_n(x - x_0)^{n-k},$$

$$\dots\dots\dots\dots\dots\dots\dots\dots\dots\dots\dots\dots\dots\dots\dots$$

for $|x - x_0| < R$. Setting $x = x_0$ in these formulas, we see that

$$f'(x_0) = a_1, \quad f''(x_0) = 2 \cdot 1a_2, \dots, \quad f^{(k)}(x_0) = k(k - 1) \cdots 2 \cdot 1a_k.$$

Thus if Eq. (7.2) holds we must have

$$a_k = \frac{f^{(k)}(x_0)}{k!} \tag{7.3}$$

for $k = 0, 1, 2, \dots$. On the other hand, suppose that $f$ is any function that possesses derivatives of all orders at a point $x_0$. Then we can form the series

$$\sum_{n=0}^{\infty} \frac{f^{(n)}(x_0)}{n!} (x - x_0)^n, \tag{7.4}$$

called the *Taylor series* of $f$ at $x_0$. (In the special case when $x_0 = 0$, the series (7.4) is also referred to as the *Maclaurin series* for $f$.) If the series (7.4) converges to $f(x)$ for all $x$ in some interval $(x_0 - r, x_0 + r)$, $r > 0$, we say that $f$ is *analytic* at $x_0$.

As an example, let $f(x) = e^x$ for all $x$. Then $f^{(n)}(x) = e^x$ and $f^{(n)}(0) = 1$ for every positive integer $n$. The Maclaurin series for $f$ is

$$\sum_{n=0}^{\infty} \frac{x^n}{n!}.$$

It is easily shown, by application of Theorem 7.1, that this series converges for all $x$. It is more difficult to show that the sum of the series is $e^x$. This can be established by the use of Taylor's formula with remainder, which is discussed in most books on calculus.

For convenience of reference we list a number of Maclaurin series whose validity has been established.

$$e^x = \sum_{n=0}^{\infty} \frac{x^n}{n!}, \quad \text{for all } x, \tag{7.5}$$

$$\cos x = \sum_{n=0}^{\infty} \frac{(-1)^n}{(2n)!} x^{2n} \quad \text{for all } x, \tag{7.6}$$

$$\sin x = \sum_{n=1}^{\infty} \frac{(-1)^{n+1}}{(2n-1)!} x^{2n-1}, \quad \text{for all } x, \tag{7.7}$$

$$\frac{1}{1-x} = \sum_{n=0}^{\infty} x^n, \quad |x| < 1, \tag{7.8}$$

$$(1+x)^m = 1 + \sum_{n=1}^{\infty} \frac{m(m-1)\cdots(m-n+1)}{n!} x^n, \quad |x| < 1. \tag{7.9}$$

The last series is called the *binomial series*. If $m$ is a nonnegative integer there are only a finite number of nonzero terms in the series and it is valid for all $x$. The series (7.8) is called the *geometric series*.

By means of these formulas and Theorems 7.2 and 7.3, it is possible to establish the validity of many other Taylor series expansions rather easily. For example, we have

$$\ln(1+x) = \int_0^x \frac{1}{1+t}\, dt, \quad x > -1.$$

Using the geometric series (7.8) (with $x$ replaced by $-t$) we have

$$\ln(1 + x) = \int_0^x \left[ \sum_{n=0}^{\infty} (-1)^n t^n \right] dt, \quad |x| < 1.$$

An application of Theorem 7.3 yields the formula

$$\ln(1 + x) = \sum_{n=0}^{\infty} \frac{(-1)^n}{n + 1} x^{n+1}, \quad |x| < 1.$$

As a second example, let us find the Maclaurin series for the function $f$, where

$$f(x) = \frac{x}{2x + 3}.$$

Writing

$$f(x) = x \frac{1}{3} \frac{1}{1 + \frac{2}{3}x}$$

and observing from formula (7.8) that

$$\frac{1}{1 + \frac{2}{3}x} = \sum_{n=0}^{\infty} (-1)^n \left( \frac{2}{3}x \right)^n, \quad |x| < \frac{3}{2},$$

we have

$$f(x) = \frac{1}{2} \sum_{n=0}^{\infty} (-1)^n (\tfrac{2}{3}x)^{n+1}, \quad |x| < \tfrac{3}{2}.$$

## Exercises for Section 7.2

In Exercises 1–6, find the Taylor series for the function $f$ about the point $x_0$ by calculating the derivatives of the function at $x_0$.

1.  $f(x) = 2x^3 - 3x^2 + x - 3, \quad x_0 = 2$
2.  $f(x) = -1/x, \quad x_0 = -1$
3.  $f(x) = \ln x, \quad x_0 = 1$
4.  $f(x) = \cos x, \quad x_0 = \pi/4$

**5.**  $f(x) = (1 + x)^{1/2}$,   $x_0 = 0$

**6.**  $f(x) = e^{3x}$,   $x_0 = 0$

In Exercises 7–10, use the geometric and binomial series to find the Taylor series for the given function $f$ about the point $x_0$. Indicate an interval on which the series converges to the function.

**7.**  $\dfrac{4}{4 + x}$,   $x_0 = 0$        **8.**  $\dfrac{1}{x}$,   $x_0 = 2$

**9.**  $\dfrac{-2}{(x-1)(x+2)}$,   $x_0 = 0$        **10.**  $(1 - x^2)^{-1/2}$,   $x_0 = 0$

Exercises 11–14, find the Maclaurin series of the given function by differentiation or integration of another series.

**11.**  $f(x) = \dfrac{1}{(1-x)^2}$        **12.**  $f(x) = \tan^{-1} x$

**13.**  $f(x) = \tanh^{-1} x$        **14.**  $f(x) = \sin^{-1} x$

In Exercises 15–18, express the sum of the series in terms of elementary functions

**15.**  $\displaystyle\sum_{n=0}^{\infty} \frac{n+1}{n!} x^{n+1}$        **16.**  $\displaystyle\sum_{n=1}^{\infty} n^2 x^n$

**17.**  $\displaystyle\sum_{n=0}^{\infty} \frac{n+2}{n+1} x^{n+1}$        **18.**  $\displaystyle\sum_{n=1}^{\infty} n^2 x^{2n}$

## 7.3   ORDINARY POINTS

The linear homogeneous differential equation

$$y^{(n)} + a_1 y^{(n-1)} + \cdots + a_{n-1} y' + a_n y = 0 \qquad (7.10)$$

is said to have an *ordinary point* at $x_0$ if each of the functions $a_i$ is analytic at $x_0$. A point that is not an ordinary point is called a *singular point* for the equation. For the equation

$$y'' + \frac{2x}{(2x - 1)(x + 2)} y' + \frac{\cos x}{x^2} y = 0$$

the singular points are $\tfrac{1}{2}$, 0, and $-2$. All other points are ordinary points.

The basic result about series solutions at an ordinary point is as follows. (A proof is given in Coddington (1961).)

**Theorem 7.4** Let the functions $a_1, a_2, \ldots, a_n$ in Eq. (7.10) be analytic at $x_0$, and let each of these functions be represented by its Taylor series at $x_0$ on the interval $\mathcal{I} = (x_0 - R, x_0 + R)$. Then every solution of Eq. (7.10) on $\mathcal{I}$ is analytic at $x_0$ and is represented on $\mathcal{I}$ by its Taylor series at $x_0$.

Our interest is mainly in second-order equations, and we consider two examples.

**Example 1** The equation

$$(2x + 1)y'' + y' + 2y = 0, \tag{7.11}$$

which may be written as

$$y'' + \frac{1}{2x + 1} y' + \frac{2}{2x + 1} y = 0, \tag{7.12}$$

has an ordinary point at $x = 0$. Here

$$a_1(x) = \frac{1}{1 + 2x} = \sum_{n=0}^{\infty} (-2)^n x^n, \quad |x| < \tfrac{1}{2},$$

$$a_2(x) = \frac{2}{1 + 2x} = \sum_{n=0}^{\infty} 2(-2)^n x^n, \quad |x| < \tfrac{1}{2}.$$

According to Theorem 7.4, the Maclaurin series for every solution converges at least for $|x| < \tfrac{1}{2}$. The solution for which

$$y(0) = A_0, \qquad y'(0) = A_1$$

has a Maclaurin series of the form

$$y = \sum_{n=0}^{\infty} A_n x^n, \tag{7.13}$$

where $A_2$, $A_3$, and so on, must be determined. Now

$$y' = \sum_{n=1}^{\infty} n A_n x^{n-1}, \qquad y'' = \sum_{n=2}^{\infty} n(n - 1) A_n x^{n-2},$$

so if we substitute in Eq. (7.11) we obtain the requirement

$$2 \sum_{n=2}^{\infty} n(n - 1) A_n x^{n-1} + \sum_{n=2}^{\infty} n(n - 1) A_n x^{n-2} + \sum_{n=1}^{\infty} n A_n x^{n-1} + 2 \sum_{n=0}^{\infty} A_n x^n = 0.$$

(It is convenient to substitute in Eq. (7.11) instead of Eq. (7.12) because the coefficients of the former are polynomials.) The first series here starts with the first power of $x$ while the last three start with $x^0$. Collecting the constant terms, we may write

$$(2A_2 + A_1 + 2A_0) + 2\sum_{n=2}^{\infty} n(n-1)A_n x^{n-1} + \sum_{n=3}^{\infty} n(n-1)A_n x^{n-2}$$

$$+ \sum_{n=2}^{\infty} nA_n x^{n-1} + 2\sum_{n=1}^{\infty} A_n x^n = 0. \tag{7.14}$$

Now all four series start with the first power of $x$. In order to combine the series, we shift the indices of summation in the first, third, and fourth, writing

$$2\sum_{n=2}^{\infty} n(n-1)A_n x^{n-1} = 2\sum_{n=3}^{\infty} (n-1)(n-2)A_{n-1} x^{n-2}$$

$$\sum_{n=2}^{\infty} nA_n x^{n-1} = \sum_{n=3}^{\infty} (n-1)A_{n-1} x^{n-2}$$

$$2\sum_{n=1}^{\infty} A_n x^n = 2\sum_{n=3}^{\infty} A_{n-2} x^{n-2}.$$

Equation (7.14) may now be written as

$$(2A_2 + A_1 + 2A_0) + \sum_{n=3}^{\infty} [n(n-1)A_n + 2(n-1)(n-2)A_{n-1}$$
$$+ (n-1)A_{n-1} + 2A_{n-2}]x^{n-2} = 0.$$

Since the coefficients in the Maclaurin series of the zero function are all zero, we must have[1]

$$2A_2 + A_1 + 2A_0 = 0 \tag{7.15}$$

and

$$n(n-1)A_n + (n-1)(2n-3)A_{n-1} + 2A_{n-2} = 0 \tag{7.16}$$

for $n \geq 3$. This last relation is called a *recurrence relation* for the coefficients $A_i$. From Eq. (7.15) we can find $A_2$ in terms of $A_0$ and $A_1$. By using the

---

[1] The power series representation of a function is unique. If a function $f$ can be represented in the form (7.2) then the coefficients in the series must be given by formula (7.3).

recurrence relation (7.16) with $n = 3$ we may express $A_3$ in terms of $A_0$, $A_1$, and $A_2$ and hence in terms of $A_0$ and $A_1$. In fact, each coefficient $A_i$ can be expressed in terms of $A_0$ and $A_1$.

From Eq. (7.15) we have

$$A_2 = -A_0 - \tfrac{1}{2}A_1. \tag{7.17}$$

Equation (7.16) may be written as

$$A_n = -\frac{2n - 3}{n} A_{n-1} - \frac{2}{n(n-1)} A_{n-2}, \quad n \geq 3 \tag{7.18}$$

For $n = 3$, we have

$$A_3 = -A_2 - \tfrac{1}{3}A_1$$

Substituting from Eq. (7.17) for $A_2$ we find that

$$A_3 = -(A_0 - \tfrac{1}{2}A_1) - \tfrac{1}{3}A_1$$

or

$$A_3 = A_0 + \tfrac{1}{6}A_1.$$

Next, setting $n = 4$ in the relation (7.18), we see that

$$A_4 = -\tfrac{1}{6}A_2 - \tfrac{5}{4}A_3$$

$$= -\tfrac{1}{6}(-A_0 - \tfrac{1}{2}A_1) - \tfrac{5}{4}(A_0 + \tfrac{1}{6}A_1)$$

$$= -\tfrac{13}{12}A_0 - \tfrac{1}{8}A_1.$$

The first few terms in the Maclaurin series expansion of the solution are

$$y(x) = A_0 + A_1 x + (-A_0 - \tfrac{1}{2}A_1)x^2 + (A_0 + \tfrac{1}{6}A_1)x^3$$
$$+ (-\tfrac{13}{12}A_0 - \tfrac{1}{8}A_1)x^4 + \cdots. \tag{7.19}$$

Collecting terms that involve $A_0$ and $A_1$, we have

$$y(x) = A_0(1 - x^2 + x^3 - \tfrac{13}{12}x^4 + \cdots) + A_1(x - \tfrac{1}{2}x^2 + \tfrac{1}{6}x^3 - \tfrac{1}{8}x^4 + \cdots).$$

The series included in parentheses are obtained by setting $A_0 = 1$ and $A_1 = 0$, or $A_0 = 0$ and $A_1 = 1$ in formula (7.19). Thus each converges at least for $|x| < \frac{1}{2}$. If we set

$$y_1(x) = 1 - x^2 + x^3 - \tfrac{13}{12}x^4 + \cdots,$$
$$y_2(x) = x - \tfrac{1}{2}x^2 + \tfrac{1}{6}x^3 - \tfrac{1}{8}x^4 + \cdots,$$

then the general solution on $(-\frac{1}{2}, \frac{1}{2})$ consists of all functions of the form

$$y(x) = A_0 y_1(x) + A_1 y_2(x),$$

where $A_0$ and $A_1$ are arbitrary constants.

**Example 2**

$$y'' - 2(x - 1)y' - y = 0. \tag{7.20}$$

Suppose that we wish to find the Taylor series expansions of the solutions at $x = 1$. For convenience we make the change of variable

$$t = x - 1. \tag{7.21}$$

Then $x = 1$ corresponds to $t = 0$ and Eq. (7.20) becomes

$$\frac{d^2 y}{dt^2} - 2t \frac{dy}{dt} - y = 0. \tag{7.22}$$

Seeking solutions of the form

$$y = \sum_{n=0}^{\infty} A_n t^n,$$

we obtain the requirement

$$\sum_{n=2}^{\infty} n(n - 1)A_n t^{n-2} - 2 \sum_{n=1}^{\infty} n A_n t^n - \sum_{n=0}^{\infty} A_n t^n = 0.$$

By proceeding as in the previous example, we see that this equation can be written as

$$(2A_2 - A_0) + \sum_{n=3}^{\infty} \{n(n - 1)A_n - [2(n - 2) + 1]A_{n-2}\}t^{n-2} = 0.$$

Then

$$A_2 = \tfrac{1}{2}A_0$$

and

$$A_n = \frac{2n - 3}{(n-1)n} A_{n-2}, \quad n \geq 3$$

From these relations we see that

$$A_2 = \tfrac{1}{2}A_0,$$

$$A_4 = \frac{5}{3 \cdot 4} A_2 = \frac{1 \cdot 5}{2 \cdot 3 \cdot 4} A_0,$$

$$A_6 = \frac{9}{5 \cdot 6} A_4 = \frac{1 \cdot 5 \cdot 9}{6!} A_0,$$

$$\cdots\cdots\cdots\cdots\cdots\cdots\cdots\cdots\cdots\cdots\cdots$$

$$A_{2m} = \frac{1 \cdot 5 \cdot 9 \cdots (4m - 3)}{(2m)!} A_0, \quad m \geq 1$$

and

$$A_3 = \frac{3}{2 \cdot 3} A_1,$$

$$A_5 = \frac{7}{4 \cdot 5} A_3 = \frac{3 \cdot 7}{2 \cdot 3 \cdot 4 \cdot 5} A_1$$

$$\cdots\cdots\cdots\cdots\cdots\cdots\cdots\cdots\cdots\cdots\cdots$$

$$A_{2m-1} = \frac{3 \cdot 7 \cdots (4m - 5)}{(2m - 1)!} A_1, \quad m \geq 2.$$

The general solution of Eq. (7.21) (for all $x$) is

$$y(x) = A_0 \left[ 1 + \sum_{m=1}^{\infty} \frac{1 \cdot 5 \cdot 9 \cdots (4m - 3)}{(2m)!} (x - 1)^{2m} \right]$$

$$+ A_1 \left[ (x - 1) + \sum_{m=2}^{\infty} \frac{3 \cdot 7 \cdots (4m - 5)}{(2m - 1)!} (x - 1)^{2m-1} \right].$$

In the general case of an equation

$$p(x) y'' + q(x) y' + r(x) y = 0,$$

where $p$, $q$, and $r$ are analytic at $x_0$ and $p(x_0) \neq 0$, it can be shown (see Coddington (1961) and Rabenstein (1966)) that substitution of the series

$$y = \sum_{n=0}^{\infty} A_n(x - x_0)^n$$

into the equation always leads to a recurrence relation that completely determines the coefficients $A_i$, $i \geq 2$, in terms of $A_0$ and $A_1$.

## Exercises for Section 7.3

1.  Locate all singular points of the given differential equation.

(a)  $y'' + \dfrac{x}{(x-1)(x+2)} y' + \dfrac{1}{x(x-1)^2} y = 0$

(b)  $x(x + 3)y'' + x^2 y' - y = 0$

(c)  $y'' + e^x y' + (\cos x)y = 0$

(d)  $(\sin x)y'' - y = 0$

In Exercises 2–10 verify that $x = 0$ is an ordinary point for the differential equation and express the general solution in terms of power series about this point. Discuss the interval of convergence of the series.

2.  $y'' + xy' + y = 0$                    3.  $2y'' - xy' - 2y = 0$

4.  $(1 - x^2)y'' - 5xy' - 3y = 0$          5.  $(2 + x^2)y'' + 5xy' + 4y = 0$

6.  $y'' - xy = 0$                          7.  $y'' - x^2 y' - 2xy = 0$

8.  $y'' - (x + 1)y' - y = 0$               9.  $(1 + x)y'' - y = 0$

10.  $y'' + e^x y' + y = 0$

In Exercises 11–14, express the general solution of the differential equation in terms of power series about the indicated point $x_0$. Suggestion: make the change of variable $t = x - x_0$.

11.  $y'' + (x - 1)y' + y = 0$,  $x_0 = 1$

12.  $(x^2 + 2x)y'' + (x + 1)y' - 4y = 0$,  $x_0 = -1$

13.  $(3 - 4x + x^2)y'' - 6y = 0$,  $x_0 = 2$

14.  $y'' - (x^2 + 6x + 9)y' - 3(x + 3)y = 0$,  $x_0 = -3$

## 7.4  SINGULAR POINTS

The Cauchy–Euler equation

$$2x^2 y'' + 3xy' - y = 0$$

has a singular point at $x = 0$. Its general solution is

$$y = c_1 |x|^{1/2} + c_2 x^{-1},$$

and from this formula we see that no nontrivial solution is analytic at $x = 0$. On the other hand, the occurrence of a singular point may not preclude the existence of analytic solutions. The general solution of the equation

$$x^2 y'' - 2xy' + 2y = 0$$

is

$$y = c_1 x + c_2 x^2,$$

so *every* solution is analytic at $x = 0$.

We notice that every equation of the form

$$b_0 x^2 y'' + b_1 xy' + b_2 y = 0, \tag{7.23}$$

· $b_0 \neq 0$, possesses at least one solution of the form

$$y = x^s,$$

where $s$ is a number that may be complex. Our concern in this section is with a generalization of the class of equations (7.23). An equation that can be written in the form

$$(x - x_0)^2 y'' + (x - x_0) P(x) y' + Q(x) y = 0 \tag{7.24}$$

is said to have a *regular singular point* at $x_0$ if $P$ and $Q$ are analytic at $x_0$.[2] Such an equation may be written as

$$y'' + p(x) y' + q(x) y = 0 \tag{7.25}$$

where

$$p(x) = \frac{P(x)}{x - x_0}, \qquad q(x) = \frac{Q(x)}{(x - x_0)^2}.$$

[2] More generally, the $n$th-order equation

$$(x - x_0)^n y^{(n)} + (x - x_0)^{n-1} b_1(x) y^{(n-1)} + \cdots + (x - x_0) b_{n-1}(x) y' + b_n(x) y = 0$$

is said to have a regular singular point at $x_0$ if $b_1, b_2, \ldots, b_n$ are analytic at $x_0$.

Thus Eq. (7.25) has a regular singular point at $x_0$ if and only if the functions

$$(x - x_0) p(x), \qquad (x - x_0)^2 q(x)$$

are analytic at $x_0$. It turns out that Eq. (7.24) possesses at least one, and sometimes two solutions of the form

$$y = (x - x_0)^s \sum_{n=0}^{\infty} A_n(x - x_0)^n, \quad A_0 \neq 0, \tag{7.26}$$

where $s$ is a number[3] that need not be an integer. The procedure for finding solutions of the type (7.26) is known as the *method of Frobenius*. We illustrate the method with some examples.

**Example 1**   The equation

$$2x^2 y'' + 3xy' - (1 + x)y = 0 \tag{7.27}$$

has a regular singular point at $x = 0$. It can be written as

$$y'' + p(x)y' + q(x)y = 0,$$

with

$$p(x) = \frac{3}{2x}, \qquad q(x) = -\frac{1 + x}{2x^2},$$

and $x\, p(x)$ and $x^2 q\, (x)$ are analytic at $x = 0$. If

$$y = x^s \sum_{n=0}^{\infty} A_n x^n = \sum_{n=0}^{\infty} A_n x^{n+s},$$

then

$$y' = \sum_{n=0}^{\infty} (n + s)A_n x^{n+s-1}$$

and

$$y'' = \sum_{n=0}^{\infty} (n + s)(n + s - 1)A_n x^{n+s-2}.$$

---

[3] It is possible that $s$ may be complex, but this does not happen in the classical equations in which we are most interested.

Substituting into the differential equation (7.27), we obtain the requirement

$$2 \sum_{n=0}^{\infty} (n + s)(n + s - 1)A_n x^{n+s} + 3 \sum_{n=0}^{\infty} (n + s)A_n x^{n+s}$$

$$- \sum_{n=0}^{\infty} A_n x^{n+s} - \sum_{n=0}^{\infty} A_n x^{n+s+1} = 0.$$

The last sum in this equation begins with a term involving $x^{s+1}$; the remaining series start with a term involving $x^s$. In order to combine terms with like powers of $x$, we separate out those terms with $x^s$ and make a shift of index $(n \to n - 1)$ in the last series. The result is

$$[2s(s - 1) + 3s - 1]A_0 x^s + \sum_{n=1}^{\infty} \{[2(n + s)(n + s - 1)$$

$$+ 3(n + s) - 1]A_n - A_{n-1}\}x^{n+s} = 0.$$

Since $A_0 \neq 0$, we see that $s$ must be a root of the quadratic equation

$$2s(s - 1) + 3s + 1 = 0$$

or

$$(2s - 1)(s + 1) = 0. \tag{7.28}$$

Thus $s$ must have one of the values $s_1 = \frac{1}{2}$ or $s_2 = -1$. In either case the coefficients $A_i$ must satisfy the recurrence relation

$$[2(n + s)(n + s - 1) + 3(n + s) - 1]A_n = A_{n-1}, \quad n \geq 1. \tag{7.29}$$

When $s = \frac{1}{2}$ this becomes

$$n(2n + 3)A_n = A_{n-1}$$

or

$$A_n = \frac{1}{n(2n + 3)} A_{n-1}, \quad n \geq 1.$$

Then

$$A_1 = \frac{1}{1 \cdot 5} A_0,$$

$$A_2 = \frac{1}{2 \cdot 7} A_1 = \frac{1}{(1 \cdot 2)(5 \cdot 7)} A_0,$$

$$\cdots\cdots\cdots\cdots\cdots\cdots\cdots\cdots\cdots\cdots$$

$$A_n = \frac{1}{n! \, 5 \cdot 7 \cdots (2n + 3)} A_0.$$

Taking $A_0 = 1$, we arrive at the specific "series solution"

$$y_1(s) = x^{1/2}\left[1 + \sum_{n=1}^{\infty} \frac{x^n}{n!\, 5 \cdot 7 \cdots (2n + 3)}\right]. \tag{7.30}$$

When $s = -1$, the recurrence relation (7.29) becomes

$$n(2n - 3)A_n = A_{n-1}$$

or

$$A_n = \frac{1}{n(2n - 3)} A_{n-1}, \quad n \ge 1.$$

From this relation we find that

$$A_1 = \frac{1}{1 \cdot (-1)} A_0,$$

$$A_2 = \frac{1}{2 \cdot (1)} A_1 = \frac{1}{1 \cdot 2(-1)(1)} A_0,$$

$$A_3 = \frac{1}{3 \cdot 3} A_2 = \frac{1}{1 \cdot 2 \cdot 3(-1) \cdot 1 \cdot 3} A_0,$$

$$\dots\dots\dots\dots\dots\dots\dots\dots\dots\dots\dots\dots$$

$$A_n = \frac{1}{n!(-1)1 \cdot 3 \cdots (2n - 3)} A_0.$$

Thus a second series solution of Eq. (7.27) is

$$y_2(x) = x^{-1}\left[1 - x - \sum_{n=2}^{\infty} \frac{x^n}{n!\, 1 \cdot 3 \cdots (2n - 3)}\right]. \tag{7.31}$$

We have seen that a function of the form

$$f(x) = x^s \sum_{n=0}^{\infty} A_n x^n, \tag{7.32}$$

where the power series converges in some interval $(-R, R)$, is a solution of the differential equation (7.27) if and only if the exponent $s$ and the coefficients $A_i$ satisfy the relations (7.28) and (7.29). Now the power series in formulas (7.30) and (7.31) converge everywhere, as can be verified by the use of Theorem 7.1.

Consequently the functions $y_1$ and $y_2$ are both solutions on the interval $(0, \infty)$. These solutions are linearly independent. For if

$$c_1 \, y_1(x) + c_2 \, y_2(x) = 0$$

for $x > 0$, then, letting $x \to 0$, we see that $c_2$ must be zero. This is because $y_2(x) \to \infty$ as $x \to 0$. We now have

$$c_1 \, y_1(x) = 0$$

for $x > 0$. But $y_1$ is not the zero function so $c_1 = 0$ also. Hence $y_1$ and $y_2$ are linearly independent.

Let us consider briefly the general second-order equation with a regular singular point at zero. Such an equation has the form

$$Ly = x^2 y'' + x P(x) y' + Q(x) y = 0, \tag{7.33}$$

where

$$P(x) = \sum_{n=0}^{\infty} P_n x^n, \qquad Q(x) = \sum_{n=0}^{\infty} Q_n x^n$$

for $|x| < r$. If $f$ is a function of the form

$$f(x) = x^s \sum_{n=0}^{\infty} A_n x^n,$$

it can be verified that

$$L f(x) = [s(s-1) + s P_0 + Q_0] A_0 x^s + \sum_{n=1}^{\infty} B_n x^{n+s},$$

where $B_n$ depends on $s$ and on $A_0, A_1, \ldots, A_n$ in general. The roots of the quadratic equation

$$s(s-1) + s P_0 + Q_0 = 0 \tag{7.34}$$

are called the *exponents* of the differential equation. The requirement

$$B_n = 0$$

yields a recurrence relation involving $A_0, A_1, \ldots, A_n$. If $s$ is an exponent and if the coefficients $A_i$ of a power series

$$\sum_{n=0}^{\infty} A_n x^n$$

satisfy the corresponding recurrence relation, then it can be shown that the power series converges at least for $|x| < r$. Hence the function defined by the relation

$$y(x) = x^s \sum_{n=0}^{\infty} A_n x \tag{7.35}$$

is a solution, at least for $0 < x < r$.

## Exercises for Section 7.4

1. Locate all regular singular points of the given differential equations.

   (a)  $y'' + \dfrac{1 - x}{x(x + 1)(x + 2)} y' + \dfrac{x + 3}{x^2(x + 2)^3} = 0$

   (b)  $(x - 1)^2(x - 2)y'' + xy' + y = 0$

   (c)  $(2x + 1)(x - 2)^2 y'' + (x + 2)y' = 0$

   (d)  $y'' + \dfrac{\sin x}{x^2} y' + \dfrac{e^x}{x + 1} y = 0$

In Exercises 2–10, verify that $x = 0$ is a regular singular point of the differential equation. If possible, express the general solution in terms of series of the form (7.35).

2.  $2x^2 y'' - 3xy' + (3 - x)y = 0$    3.  $2x^2 y'' + xy' - (x + 1)y = 0$

4.  $2x^2 y'' + (x - x^2)y' - y = 0$    5.  $3xy'' + 2y' + y = 0$

6.  $2xy'' + 3y' - xy = 0$

7.  $3x^2 y'' + (5x + 3x^3)y' + (3x^2 - 1)y = 0$

8.  $2x^2 y'' + 5xy' + (1 - x^3)y = 0$

9.  $(2x^2 - x^3)y'' + (7x - 6x^2)y' + (3 - 6x)y = 0$

10.  $(2x - 2x^2)y'' + (1 + x)y' + 2y = 0$

In Exercises 11 and 12, verify that $x_0$ is a regular singular point for the differential equation. If possible, express the general solution in terms of series of the form (7.26). Suggestion: make the change of variable $t = x - x_0$.

11.  $9(x - 1)^2 y'' + [9(x - 1) - 3(x - 1)^2]y' + (4x - 5)y = 0, \quad x_0 = 1$

12.  $2(x + 1)y'' - (1 + 2x)y' + 7y = 0, \quad x_0 = -1$

13.  Show that the change of variable $x = 1/t$ in the differential equation

$$\frac{d^2 y}{dx^2} + P(x)\frac{dy}{dx} + Q(x)\,y = 0, \tag{1}$$

leads to the equation

$$\frac{d^2y}{dt^2} + \left[\frac{2}{t} - \frac{1}{t^2}P\left(\frac{1}{t}\right)\right]\frac{dy}{dt} + \frac{1}{t^4}Q\left(\frac{1}{t}\right)y = 0. \tag{2}$$

Equation (1) is said to have an ordinary point or a regular singular point at infinity if Eq. (2) has an ordinary point or regular singular point at zero.

In Exercises 14 and 15, use the result of Exercise 13 to express the general solution in terms of series of powers of $1/x$ that converge for large $|x|$.

**14.** $x^4y'' + (2x^3 + x)y' - y = 0$

**15.** $2x^3y'' + (5x^2 - 2x)y' + (x + 3)y = 0$

## 7.5 SOLUTIONS AT A REGULAR SINGULAR POINT

Not every second-order equation with a regular singular point at zero has two linearly independent solutions of the form (7.35). For example, the equation

$$4x^2y'' + y = 0$$

has as solutions

$$y_1(x) = x^{1/2}, \qquad y_2(x) = x^{1/2}\ln x$$

for $x > 0$. Now $y_2(x)$ is not of the form (7.35). If it were we would have $x^{1/2}\ln x = x^s f(x)$, where $f$ is analytic at $x = 0$. But then $f(x) = x^{-s+1/2}\ln x$, and such a function is not analytic at $x = 0$.

Let us denote the exponents of the differential equation

$$x^2y'' + xP(x)y' + Q(x)y = 0 \tag{7.36}$$

by $s_1$ and $s_2$. If $s_1 = s_2$, Eq. (7.36) has only one linearly independent solution of the form (7.35),

$$y_1(x) = x^{s_1}\sum_{n=0}^{\infty} A_n x^n. \tag{7.37}$$

It can be shown (see Coddington (1961) and Rabenstein (1966)) that there exists a second solution of the form

$$y_2(x) = y_1(x)\ln x + x^{s_1}\sum_{n=1}^{\infty} B_n x^n. \tag{7.38}$$

The power series with coefficients $B_n$, $n \geq 1$, converges at least for $|x| < r$ if the Maclaurin series for $P$ and $Q$ are both valid for $|x| < r$.

In case the exponents $s_1$ and $s_2$ are distinct, we find two linearly independent solutions

$$y_1(x) = x^{s_1} \sum_{n=0}^{\infty} A_n x^n, \tag{7.39}$$

$$y_2(x) = x^{s_2} \sum_{n=0}^{\infty} B_n x^n, \tag{7.40}$$

*except possibly when $s_1 - s_2$ is an integer.* (Again, see Coddington (1961), Kaplan (1958) and Rabenstein (1966).) If $s_1 - s_2 = N$, where $N$ is a positive integer, there is always a solution of the form (7.39) associated with the larger exponent $s_1$. There may or may not be a solution of the form (7.40) associated with the smaller exponent. If there is not, it can be shown that there is a solution corresponding to $s_2$ of the form

$$y_2(x) = y_1(x) \ln x + x^{s_2} \sum_{n=0}^{\infty} B_n x^n, \tag{7.41}$$

where $B_0 \neq 0$ and where the power series multiplying $x^{s_2}$ converges at least for $|x| < r$. We shall consider examples that illustrate the two situations.

**Example 1**

$$x^2 y'' + (x - x^2)y' - y = 0. \tag{7.42}$$

If

$$y = x^s \sum_{n=0}^{\infty} A_n x^n, \tag{7.43}$$

we find, after some calculation, that

$$(s^2 - 1)A_0 x^s + \sum_{n=1}^{\infty} [(n + s - 1)(n + s + 1)A_n - (n + s - 1)A_{n-1}]x^{n+s} = 0.$$

The exponents $s_1 = 1$ and $s_2 = -1$ differ by an integer. The recurrence relation is

$$(n + s - 1)(n + s + 1)A_n = (n + s - 1)A_{n-1}, \quad n \geq 1 \tag{7.44}$$

For $s = 1$, this becomes

$$n(n + 2)A_n = nA_{n-1}$$

or

$$A_n = \frac{1}{n+2} A_{n-1}, \qquad n \geq 1.$$

From this relation we find that

$$A_1 = \frac{1}{3} A_0, \quad A_2 = \frac{1}{4} A_1 = \frac{1}{3 \cdot 4} A_0,$$

$$A_n = \frac{2}{(n+2)!} A_0.$$

Hence a solution corresponding to the larger exponent $s_1$ is

$$y_1(x) = x \left[ 1 + 2 \sum_{n=1}^{\infty} \frac{x^n}{(n+2)!} \right].$$

When $s = -1$, the recurrence relation (7.44) becomes

$$(n-2)nA_n = (n-2)A_{n-1}, \qquad n \geq 1. \qquad (7.45)$$

Notice that we cannot simply divide both members of this equation by $n - 2$, because this amounts to dividing by zero when $n$ is 2. For $n = 1$, we have

$$-A_1 = -A_0,$$

therefore $A_1 = A_0$. When $n = 2$, relation (7.45) becomes

$$0 \cdot A_2 = 0 \cdot A_1$$

or

$$0 \cdot A_2 = 0.$$

This condition is satisfied for every choice of $A_2$. A convenient choice is $A_2 = 0$ because

$$A_n = \frac{A_{n-1}}{n}, \qquad n \geq 3,$$

and if $A_2 = 0$, then $A_n = 0$ for $n \geq 3$. Thus we have a second solution of the form (7.43),

$$y_2(x) = x^{-1}(1 + x).$$

**Example 2**

$$x^2 y'' + xy' - (1 + x)y = 0.$$

If

$$y(x) = x^s \sum_{n=0}^{\infty} A_n x^n, \tag{7.46}$$

we must have

$$(s^2 - 1)A_0 x^s + \sum_{n=1}^{\infty} [(n + s - 1)(n + s + 1)A_n - A_{n-1}]x^{n+s} = 0.$$

Thus $s$ must be a root of the equation

$$s^2 - 1 = 0,$$

and the coefficients $A_i$ must satisfy the recurrence relation

$$(n + s - 1)(n + s + 1)A_n = A_{n-1}, \quad n \geq 1. \tag{7.47}$$

The exponents are $s_1 = 1$ and $s_2 = -1$. For $s = 1$, the recurrence relation becomes

$$n(n + 2)A_n = A_{n-1}, \quad n \geq 1.$$

A solution corresponding to $s = 1$ is found to be

$$y_1(x) = x \left[ 1 + 2 \sum_{n=1}^{\infty} \frac{x^n}{n!(n + 2)!} \right].$$

When $s = -1$ the recurrence relation (7.47) is

$$(n - 2)nA_n = A_{n-1}, \quad n \geq 1.$$

For $n = 1$ we have $-A_1 = A_0$ or $A_1 = -A_0$; but when $n = 2$, we have

$$0 \cdot A_2 = A_1$$

or

$$0 \cdot A_2 = -A_0.$$

Since $A_0 \neq 0$, there is no possible value for $A_2$ for which this condition is satisfied. Hence there is no solution of the form (7.46) with $s = -1$. However, there is a solution of the form

$$y_2(x) = y_1(x) \ln x + x^{-1} \sum_{n=0}^{\infty} B_n x^n,$$

where $B_0 \neq 0$.

## Exercises for Section 7.5

In Exercises 1–6, verify that the exponents, relative to $x = 0$, are equal. Find one solution of the type (7.37) and indicate the *form* assumed by a second independent solution.

1. $x^2 y'' - xy' + (1 - x)y = 0$
2. $x^2 y'' + (3x + x^2)y' + y = 0$
3. $xy'' + (1 - x^2)y' + 4xy = 0$
4. $x^2 y'' - (3x + x^2)y' + (4 - x)y = 0$
5. $(x^2 - x^3)y'' + 5xy' + (4 + 12x)y = 0$
6. $x^2 y'' - 5xy' + (9 + 2x)y = 0$

In Exercises 7–16, find a solution of the form (7.39) corresponding to the larger exponent $s_1$. If a solution of the form (7.40) corresponding to $s_2$ exists, find it. If not, indicate the *form* of a solution that corresponds to the smaller exponent $s_2$.

7. $xy'' + xy' + 2y = 0$
8. $x^2 y'' - 2xy' + (2 - x)y = 0$
9. $(x^2 - x^3)y'' + xy' + (6x - 1)y = 0$
10. $x^2 y'' + (2x - 2)y = 0$
11. $(x^2 - x^4)y'' + 4xy' + (2 + 20x^2)y = 0$
12. $x^2 y'' + x^2 y' - 2y = 0$
13. $x^2 y'' + (3x - x^2)y' - xy = 0$
14. $x^2 y'' + (x + x^3)y' + (x^2 - 1)y = 0$

**15.** $(x^2 - x^3)y'' + (3x - 5x^2)y' - 3y = 0$

**16.** $(x^2 + x^3)y'' + (x + x^2)y' - (4 + x)y = 0$

## 7.6   LEGENDRE POLYNOMIALS

In this and the next section, we consider two differential equations that are important in a variety of applications and that cannot be solved by elementary methods. Properties of the solutions of these equations have been extensively investigated and tables of values for some of the solutions have been compiled.

The first of these equations, *Legendre's equation*, is

$$(1 - x^2)y'' - 2xy' + \alpha y = 0, \tag{7.48}$$

where $\alpha$ is a constant. This equation has regular singular points at $x = 1$ and $x = -1$, and no other singular points. We shall seek series solutions that are valid near $x = 1$. Making the change of variable $t = x - 1$, we arrive at the equation

$$(2t + t^2)\frac{d^2y}{dt^2} + 2(1 + t)\frac{dy}{dt} - \alpha y = 0. \tag{7.49}$$

A function of the form

$$y = t^s \sum_{n=0}^{\infty} A_n t^n \tag{7.50}$$

is a solution only if the condition

$$2s^2 A_0 t^{s-1} + \sum_{n=1}^{\infty} \{2(n + s)^2 A_n + [(n + s)(n + s - 1) - \alpha]A_{n-1}\}t^{n+s-1} = 0$$

is satisfied. The exponents are $s_1 = s_2 = 0$, so there is only one solution of the form (7.50). The recurrence relation is

$$A_n = \frac{\alpha - (n - 1)n}{2n^2} A_{n-1}, \quad n \geq 1.$$

From this relation we see that

$$A_1 = \frac{\alpha - 0}{2 \cdot 1^2} A_0,$$

$$A_2 = \frac{\alpha - 1 \cdot 2}{2 \cdot 2^2} A_1 = \frac{(\alpha - 0)(\alpha - 1 \cdot 2)}{2^2 \cdot 1^2 \cdot 2^2} A_0,$$

$$\cdots\cdots\cdots\cdots\cdots\cdots\cdots\cdots\cdots\cdots\cdots\cdots$$

$$A_n = \frac{\alpha(\alpha - 1 \cdot 2)(\alpha - 2 \cdot 3) \cdots [\alpha - (n-1)n]}{2^n(n!)^2} A_0.$$

Thus a solution of the original equation (7.48) is

$$y_1(x) = 1 + \sum_{n=1}^{\infty} \frac{\alpha(\alpha - 1 \cdot 2)(\alpha - 2 \cdot 3) \cdots [\alpha - (n-1)n]}{2^n(n!)^2}(x-1)^n. \quad (7.51)$$

Notice that when $\alpha$ is of the form

$$\alpha = m(m+1),$$

where $m$ is a nonnegative integer, the series (7.51) terminates. The corresponding solution is denoted by $P_m$, and we have

$$P_m(x) =$$

$$1 + \sum_{n=1}^{m} \frac{m(m+1)[m(m+1) - 1 \cdot 2] \cdots [m(m+1) - (n-1)n]}{2^n(n!)^2}(x-1)^n.$$

$$(7.52)$$

This solution $P_m$ is a polynomial of degree $m$; it is known as the *Legendre polynomial* of degree $m$. The corresponding differential equation

$$(1 - x^2)y'' - 2xy' + m(m+1)y = 0$$

is known as *Legendre's differential equation of order m*.

From the definition (7.52), we find that

$$P_0(x) = 1, \qquad P_1(x) = x. \qquad (7.53)$$

It can be shown (see Kreider *et al.* (1966, 1968) and Rabenstein (1966)) that the Legendre polynomials satisfy the recurrence relation

$$n\,P_n(x) = (2n - 1)x\,P_{n-1}(x) - (n - 1)\,P_{n-2}(x), \quad n \geq 2. \qquad (7.54)$$

From this relation we find that

$$2\,P_2(x) = 3x\,P_1(x) - P_0(x)$$

or

$$P_2(x) = \tfrac{3}{2}x^2 - \tfrac{1}{2}.$$

Also, by repeated application of the recurrence relation, we obtain the formulas

$$P_3(x) = \frac{5}{2}\,x^3 - \frac{3}{2}\,x,$$

$$P_4(x) = \frac{35}{8}\,x^4 - \frac{15}{4}\,x^2 + \frac{3}{8},$$

$$P_5(x) = \frac{63}{8}\,x^5 - \frac{35}{4}\,x^3 + \frac{15}{8}\,x.$$

It follows from formulas (7.53) and (7.54) that $P_n(x)$ involves only even powers of $x$ when $n$ is even and only odd powers when $n$ is odd.

Another important property of Legendre polynomials is the *orthogonality*[4] *relation*

$$\int_{-1}^{1} P_m(x)\,P_n(x)\,dx = 0, \qquad m = n. \qquad (7.55)$$

To derive this relation, we start with the fact that $P_n$ and $P_m$ satisfy the equations

$$[(1 - x^2)\,P_n'(x)]' + n(n + 1)\,P_n(x) = 0,$$
$$[(1 - x^2)\,P_m'(x)]' + m(m + 1)\,P_m(x) = 0,$$

----

[4] Two functions $f$ and $g$ are said to be orthogonal on the interval $(a, b)$ if

$$\int_b^a f(x)\,g(x)\,dx = 0.$$

respectively. Multiplying through in the first equation by $P_m(x)$ and in the second by $P_n(x)$, and then subtracting, we find that

$$[n(n + 1) - m(m + 1)] P_m(x) P_n(x)$$
$$= P_n(x)[(1 - x^2)P'_m(x)]' - P_m(x)[(1 - x^2)P'_n(x)]'$$
$$= \{(1 - x^2)[P'_m(x) P_n(x) - P'_n(x) P_m(x)]\}'.$$

Integrating from $-1$ to 1, we see that

$$(n - m)(n + m + 1) \int_{-1}^{1} P_m(x) P_n(x) \, dx = 0.$$

Since $m \neq n$, the orthogonality relation (7.55) holds.

Two other useful properties of Legendre polynomials are

$$\int_{-1}^{1} [P_n(x)]^2 \, dx = \frac{2}{2n + 1}, \quad n \geq 0 \tag{7.56}$$

and

$$P_n(x) = \frac{1}{2^n n!} \frac{d^n}{dx^n} (x^2 - 1)^n, \quad n \geq 0. \tag{7.57}$$

This last formula is known as *Rodrigues' formula.*

### Exercises for Section 7.6

1.  Given that $P_0(x) = 1$ and $P_1(x) = x$, find formulas for $P_3$ and $P_4$ from the recurrence relation (7.54).

2.  Derive the formula (7.56) by using Rodrigues' formula (7.57) and repeated integration by parts.

3.  Show that under the change of independent variable $x = \cos \phi$ Legendre's equation assumes the form

$$\frac{d^2 y}{d\phi^2} + \frac{dy}{d\phi} \cot \phi + \alpha(\alpha + 1)y = 0.$$

4.  For nonnegative integers $m$ and $n$, with $m \leq n$, let

$$p_n^m(x) = \frac{d^m}{dx^m} P_n(x).$$

Show that the function $p_n^m$ is a solution of the differential equation

$$(1 - x^2)y'' - 2(m + 1)xy' + (n - m)(n + m + 1)y = 0.$$

5. Let $f$ and $g$ be solutions of the equations

$$[p(x)y']' + \lambda r(x)y = 0,$$
$$[p(x)y']' + \mu r(x)y = 0,$$

respectively, on an interval $[a, b]$. If $p(a) = p(b) = 0$, show that

$$\int_a^b r(x)\, f(x)\, g(x)\, dx = 0,$$

provided $\lambda \neq \mu$. (The functions $f$ and $g$ are said to be *orthogonal with respect to the weight function r on [a, b]*.)

6. The equation

$$xy'' + (1 - x)y' + ny = 0,$$

where $n$ is a nonnegative integer, is known as Laguerre's equation of order $n$.

(a) Show that the differential equation above possesses the polynomial solution

$$L_n(x) = \sum_{k=0}^{n} \frac{n!(-1)^k x^k}{(k!)^2(n - k)!}$$

(The function $L_n$ is known as the Laguerre polynomial of degree $n$.)

(b) Show that

$$\int_0^\infty e^{-x} L_n(x)\, L_m(x)\, dx = 0$$

if $m \neq n$. Suggestion: observe that Laguerre's equation can be written as

$$(xe^{-x}y')' + ne^{-x}y = 0.$$

## 7.7   BESSEL FUNCTIONS

Our study of Bessel functions, the main topic of this section, requires a knowledge of another function, the gamma function. We define the *gamma function* $\Gamma$ by means of the formula[5]

[5] The improper integral diverges if $x \leq 0$.

$$\Gamma(x) = \int_0^\infty t^{x-1} e^{-t} \, dt, \quad x > 0. \tag{7.58}$$

The two properties

$$\Gamma(1) = 1, \tag{7.59}$$

$$\Gamma(x + 1) = x\Gamma(x), \quad x > 0 \tag{7.60}$$

are easily verified. From the definition (7.58), we see that

$$\Gamma(1) = \int_0^\infty e^{-t} \, dt = 1.$$

Verification of the other property involves an integration by parts. We have

$$\Gamma(x + 1) = \lim_{T \to \infty} \int_0^T t^x e^{-t} \, dt$$

$$= \lim_{T \to \infty} \left\{ [-t^x e^{-t}]_0^T + x \int_0^T t^{x-1} \, dt \right\}$$

$$= x\,\Gamma(x).$$

From the two properties (7.59) and (7.60) we see that

$$\Gamma(2) = 1 \cdot \Gamma(1) = 1,$$
$$\Gamma(3) = 2 \cdot \Gamma(2) = 2 \cdot 1,$$
$$\Gamma(4) = 3 \cdot \Gamma(3) = 3 \cdot 2 \cdot 1.$$

In general

$$\Gamma(n + 1) = n! \tag{7.61}$$

for every nonnegative integer $n$, as can be shown by mathematical induction. The relation

$$\Gamma(x) = \frac{\Gamma(x + 1)}{x}$$

is valid for $x > 0$. However, the right-hand side of this equation is defined for $x > -1$. We *define* $\Gamma(x)$ for $-1 < x < 0$ as

$$\Gamma(x) = \frac{\Gamma(x + 1)}{x}. \tag{7.62}$$

Replacing $x$ by $x + 1$ in this formula, we have

$$\Gamma(x + 1) = \frac{\Gamma(x + 2)}{x + 1}, \qquad -2 < x < -1. \tag{7.63}$$

Then from formulas (7.62) and (7.63) we have

$$\Gamma(x) = \frac{\Gamma(x + 2)}{x(x + 1)}$$

for $x > -1$ and $x = 0$. This formula serves to define $\Gamma(x)$ for $-2 < x < -1$. Continuing in this way, we have

$$\Gamma(x) = \frac{\Gamma(x + k)}{x(x + 1)(x + 2)\cdots(x + k - 1)} \tag{7.64}$$

for $x > -k$, $x \neq 0, -1, -2, \ldots, -k + 1$. A graph of the gamma function is shown in Fig. 7.1.

The differential equation

$$x^2 y'' + xy' + (x^2 - \alpha^2)y = 0, \tag{7.65}$$

where $\alpha$ is a nonnegative real constant, is known as *Bessel's equation of order* $\alpha$. It has a regular singular point at $x = 0$. We shall seek series solutions of the form

$$y(x) = x^s \sum_{n=0}^{\infty} A_n x^n. \tag{7.66}$$

Some computation shows that a function of the form (7.66) is a solution if

$$[s(s - 1) + s - \alpha^2]A_0 x^s + [(s + 1)s + s + 1 - \alpha^2]A_1 x^{s-1}$$

$$+ \sum_{n=2}^{\infty} \{[(n + s)(n + s - 1) + (n + s) - \alpha^2]A_n + A_{n-2}\}x^{n+s} = 0.$$

Thus $s$ must be a root of the equation

$$s^2 - \alpha^2 = 0 \tag{7.67}$$

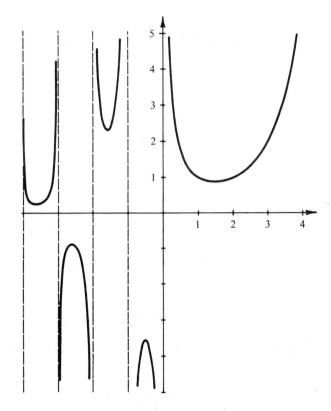

**Figure 7.1**

and the coefficients $A_i$ must satisfy the relations

$$[(s+1)^2 - \alpha^2]A_1 = 0, \qquad (7.68)$$

$$[(n+s)^2 - \alpha^2]A_n = -A_{n-2}, \quad n \geq 2. \qquad (7.69)$$

The exponents of Bessel's equation at $x = 0$, as determined from Eq. (7.67), are $s_1 = \alpha$ and $s_2 = -\alpha$.

If $s = \alpha$, the relations (7.68) and (7.69) require that $A_1 = 0$ and

$$n(n + 2\alpha)A_n = -A_{n-2}, \quad n \geq 2.$$

Thus we have $A_1 = A_3 = A_5 = \cdots = 0$ and

$$A_2 = -\frac{1}{2(2 + 2\alpha)} A_0,$$

$$A_4 = -\frac{1}{4(4+2\alpha)} A_2 = \frac{1}{2 \cdot 4(2+2\alpha)(4+2\alpha)} A_0,$$

$$\cdots\cdots\cdots\cdots\cdots\cdots\cdots\cdots\cdots\cdots\cdots\cdots\cdots\cdots\cdots\cdots$$

$$A_{2m} = \frac{(-1)^m}{2 \cdot 4 \cdots (2m)(2+2\alpha)(4+2\alpha) \cdots (2m+2\alpha)} A_0$$

$$= \frac{(-1)^m}{2^{2m}m!(1+\alpha)(2+\alpha) \cdots (m+\alpha)} A_0.$$

A solution is

$$y_1(x) = A_0\, x^\alpha \left[ 1 + \sum_{m=1}^{\infty} \frac{(-1)^m (x/2)^{2m}}{m!(1+\alpha)(2+\alpha) \cdots (m+\alpha)} \right],$$

where the power series converges for all $x$. Choosing

$$A_0 = \frac{1}{2^\alpha \Gamma(1+\alpha)},$$

we obtain a specific solution, known as the *Bessel function of the first kind* of order $\alpha$. We denote it by $J_\alpha$. Thus

$$J_\alpha(x) = \sum_{m=0}^{\infty} \frac{(-1)^m (x/2)^{2m+\alpha}}{m!\,\Gamma(m+\alpha+1)}. \tag{7.70}$$

From this formula we see that $J_0(0) = 1$ and $J_\alpha(0) = 0$ for $\alpha > 0$. Graphs of $J_0$ and $J_1$ are shown in Fig. 7.2.

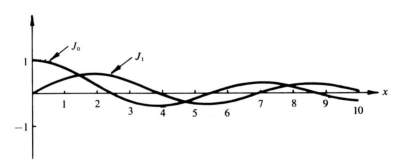

**Figure 7.2**

The difference $s_1 - s_2 = 2\alpha$ is an integer whenever $2\alpha$ is an integer. It turns out that Bessel's equation possesses a solution of the form (7.66) corresponding to $s_2 = -\alpha$ except when $\alpha$ is a nonnegative integer. A particular second solution is

$$J_{-\alpha}(x) = \sum_{m=0}^{\infty} \frac{(-1)^m (x/2)^{2m-\alpha}}{m!\,\Gamma(m - \alpha + 1)}. \tag{7.71}$$

Notice that $J_{-\alpha}(x)$ becomes infinite as $x$ approaches zero through positive values. When $\alpha$ is an integer, there is a solution corresponding to $s_2 = -\alpha$ that is of the form

$$y_2(x) = J_\alpha(x) \ln x + x^{-\alpha} \sum_{m=0}^{\infty} B_m x^m.$$

This solution also becomes infinite as $x$ tends to zero through positive values. Thus the only solutions that are bounded near $x = 0$ are those that are multiples of $J_\alpha$. Because of this fact, in many applications interest is centered on these solutions.

The Bessel functions of the first kind satisfy a number of recurrence relations. One of them is

$$\frac{d}{dx}[x^\alpha J_\alpha(x)] = x^\alpha J_{\alpha-1}(x). \tag{7.72}$$

Although this relation is not so easy to *discover* from the definition (7.70), it is easily verified. We have

$$\frac{d}{dx}[x^\alpha J_\alpha(x)] = \frac{d}{dx} \sum_{m=0}^{\infty} \frac{(-1)^m x^{2m+2\alpha}}{2^{2m+\alpha} m!\,\Gamma(m + \alpha + 1)}$$

$$= \sum_{m=0}^{\infty} \frac{(-1)^m x^{2m+2\alpha-1}}{2^{2m+\alpha-1} m!\,\Gamma(m + \alpha)}$$

$$= x^\alpha \sum_{m=0}^{\infty} \frac{(-1)^m (x/2)^{2m+\alpha-1}}{m!\,\Gamma(m + \alpha)}$$

$$= x^\alpha J_{\alpha-1}(x).$$

In similar fashion it can be shown that

$$\frac{d}{dx}[x^{-\alpha} J_\alpha(x)] = -x^{-\alpha} J_{\alpha+1}(x). \tag{7.73}$$

The relations (7.70) and (7.71) are equivalent to the relations

$$J'_\alpha(x) = J_{\alpha-1}(x) - \frac{\alpha}{x} J_\alpha(x) \tag{7.74}$$

$$J'_\alpha(x) = -J_{\alpha+1}(x) + \frac{\alpha}{x} J_\alpha(x), \tag{7.75}$$

respectively. Adding, we obtain the formula

$$J'_\alpha(x) = \tfrac{1}{2}[J_{\alpha-1}(x) - J_{\alpha+1}(x)]. \tag{7.76}$$

Subtraction yields the relation

$$J_{\alpha+1}(x) = \frac{2\alpha}{x} J_\alpha(x) - J_{\alpha-1}(x). \tag{7.77}$$

Because of this last relation, it is only necessary to tabulate the functions $J_\alpha$ only for $0 \le \alpha < 2$. In particular, every function $J_n$, with $n$ an integer, can be expressed in terms of $J_0$ and $J_1$. Extensive tables of these two functions have been compiled. See, for example, Jahnke and Emde (1945).

### Exercises for Section 7.7

1.  Given that $\Gamma(\tfrac{1}{2}) = \sqrt{\pi}$, find:
    (a) $\Gamma(\tfrac{3}{2})$    (b) $\Gamma(\tfrac{5}{2})$    (c) $\Gamma(-\tfrac{1}{2})$    (d) $\Gamma(-\tfrac{3}{2})$
2.  If $x$ is not zero or a negative integer, verify that

$$x(x + 1)(x + 2) \cdots (x + n)\Gamma(x) = \Gamma(x + n + 1)$$

    for every positive integer $n$.

3.  Show that $J_\alpha(0) = 0$ if $\alpha > 0$, and that $J_0(0) = 1$.

4.  Use the series definition to calculate the following quantities, correct to three decimal places.
    (a) $J_0(0.2)$    (b) $J_1(0.2)$    (c) $J_2(0.2)$

5.  Show that

$$\frac{d}{dx} [x^{-\alpha} J_\alpha(x)] = -x^{-\alpha} J_{\alpha+1}(x)$$

6. (a) Express $J_3(x)$ in terms of $J_0(x)$ and $J_1(x)$.

   (b) Express $J_2'(x)$ in terms of $J_0(x)$ and $J_1(x)$.

7. Show that

   (a) $\int x^{\alpha+1} J_\alpha(x)\, dx = x^{\alpha+1} J_{\alpha+1}(x) + c$,

   (b) $\int x^{1-\alpha} J_\alpha(x)\, dx = -x^{1-\alpha} J_{\alpha-1}(x) + c$

8. (a) Show that the change of dependent variable $u = x^{1/2} y$ in Bessel's equation (7.65) leads to the equation

$$ u'' + \left( 1 + \frac{1 - 4\alpha^2}{x^2} \right) u = 0 $$

   (b) Use the result of part (a) and the fact that $\Gamma(\tfrac{1}{2}) = \sqrt{\pi}$ to show that

$$ J_{1/2}(x) = \sqrt{\frac{2}{\pi x}} \sin x , $$

$$ J_{-1/2}(x) = \sqrt{\frac{2}{\pi x}} \cos x . $$

   (c) Use the result of part (b) to express $J_{3/2}(x)$ in terms of elementary functions.

9. Show that the variable changes

$$ t = ax^r, \qquad y = x^s u , $$

   in Bessel's equation

$$ t^2 \frac{d^2 u}{dt^2} + t \frac{du}{dt} + (t^2 - \alpha^2) u = 0 , $$

   lead to the equation

$$ x^2 \frac{d^2 y}{dx^2} + (1 - 2s)x \frac{dy}{dx} + [(s^2 - r^2\alpha^2) + a^2 r^2 x^{2r}] y = 0 . $$

10. Use the result of Exercise 9 to express the general solution of the given equation in terms of Bessel functions

   (a) $y'' + x^2 y = 0$    (b) $x^2 y'' + 5xy' + (9x^6 - 12)y = 0$

**11.** Let $\tilde{y}$ be any solution of Bessel's equation on $(0, \infty)$ and let

$$\tilde{u}(x) = x^{1/2}\,\tilde{y}(x).$$

(a) Show that $\tilde{u}$ is a solution of the nonhomogeneous equation

$$u'' + u = \frac{4\alpha^2 - 1}{x^2}\,\tilde{u}(x).$$

(b) Use the method of variation of parameters to show that

$$\tilde{u}(x) = c\,\sin(x - k) + (4\alpha^2 - 1)\int_{x_0}^{x} \frac{\sin(x - t)}{t^2}\,\tilde{u}(t)\,dt,$$

where $c$ and $k$ are constants and $x_0 > 0$.

(c) Use the inequality of Exercise 3, Section 5.11, to show that $|\tilde{u}(x)| \le M$, $M$ a constant, for $x \ge x_0$. Thus $|\tilde{y}(x)| \le Mx^{-1/2}$ for $x \ge x_0$. In particular, $\tilde{y}(x) \to 0$ as $x \to \infty$.

# VIII

Existence and
Uniqueness of Solutions

## 8.1 PRELIMINARIES

Let us consider an initial value problem for a first-order differential equation,

$$y' = f(x, y), \qquad y(x_0) = k, \tag{8.1}$$

where the function $f$ is defined in some region $D$ of the $xy$-plane that contains the point $(x_0, k)$. We now pose the general questions of whether the problem has a solution, and if it does, whether there can be more than one solution. In other words, we want to know if a solution *exists*, and if it does, whether the solution is *unique*.

In a specific case where it is possible to find a formula that describes the set of all solutions, it may be easy to answer these questions. Our concern is with a more general situation. Consider, for example, the problem

$$y' = x^2 + y^4, \qquad y(0) = 1. \tag{8.2}$$

Here we cannot find a simple formula for the general solution of the differential equation. Nevertheless, it can be shown that a solution of the initial value problem exists and that it is unique.

It will be necessary to impose some restrictions on the function $f$ in the problem (8.1). One is that $f$ be continuous. We shall also require that $f$

satisfy a condition known as a *Lipschitz condition*. A function $f(x,y)$ is said to satisfy a Lipschitz condition with respect to $y$ in a region $D$ of the $xy$-plane if there exists a positive number $K$ such that

$$|f(x, y_1) - f(x, y_2)| \le K|y_1 - y_2| \qquad (8.3)$$

whenever the points $(x, y_1)$ and $(x, y_2)$ both lie in $D$. The number $K$ is called a *Lipschitz constant* for $f$.

As an example, let us consider a function of the form

$$f(x, y) = p(x)\,y + q(x),$$

where $p$ and $q$ are defined on a closed interval $[a, b]$. If $p$ is continuous in $[a, b]$, then $f$ satisfies a Lipschitz condition in the region $D$ that consists of the points $(x, y)$ for which $a \le x \le b$. For $|p|$ has a maximum value $K$ on $[a, b]$ and hence

$$|f(x, y_1) - f(x, y_2)| = |p(x)(y_1 - y_2)| \le K|y_1 - y_2|$$

for $(x, y_1)$ and $(x, y_2)$ in $D$.

Next, suppose that $f(x, y)$ is any function that is continuous, along with its first partial derivative with respect to $y$, on a rectangle $R$ of the form

$$R = \{(x, y) : a \le x \le b, \quad c \le y \le d\}.$$

Then $f$ satisfies a Lipschitz condition in $R$. To see this, let $(x, y_1)$ and $(x, y_2)$ be in $R$. We may use the mean value theorem to write

$$f(x, y_1) - f(x, y_2) = f_y(x, y_3)(y_1 - y_2),$$

where $y_3$ is between $y_1$ and $y_2$. If $K$ is the maximum value of $|f_y|$ on $R$, then

$$|f(x, y_1) - f(x, y_2)| \le K|y_1 - y_2|.$$

In order to study the questions posed about the initial value problem, it will be convenient to reformulate that problem in a manner now to be described. If a function $u$ is a solution of the initial value problem (8.1) on an interval $\mathscr{I}$ that contains $x_0$, then

$$u'(x) = f[x, u(x)] \qquad (8.4)$$

for $x$ in $\mathscr{I}$, and

$$u(x_0) = k. \qquad (8.5)$$

If $f$ is continuous in a region that contains the points $(x, u(x))$ for $x$ in $\mathscr{I}$, then $u'$ is continuous on $\mathscr{I}$. Integrating both members of Eq. (8.4) from $x_0$ to $x$, we have

$$u(x) = y_0 + \int_{x_0}^{x} f[t, u(t)]\, dt \qquad (8.6)$$

for $x$ in $\mathscr{I}$. Thus a solution of the initial value problem (8.1) is also a solution of the integral equation (8.6). On the other hand, suppose that $u$ is a continuous function that satisfies Eq. (8.6). Setting $x = x_0$ in that equation, we see that $u(x_0) = y_0$. Upon differentiating both members of the equation, we see that $u'(x) = f[x, u(x)]$. Hence the initial value problem (8.1) is equivalent to the integral equation (8.6). In the next section we shall exhibit a method for the construction of a solution of the integral equation (8.6).

## Exercises for Section 8.1

1. Let $f(x, y) = xy^2$ for all $(x, y)$.
   (a) Show that $f$ satisfies a Lipschitz condition on any rectangle of the form $a \leq x \leq b$, $c \leq y \leq d$.
   (b) Show that $f$ satisfies a Lipschitz condition on any *bounded* region. (A region is bounded if it is contained in the interior of some circle.)
   (c) Show that $f$ does not satisfy a Lipschitz condition in the region that consists of the entire $xy$-plane.

2. If $f(x, y)$ satisfies a Lipschitz condition with respect to $y$ in a region $D$, show that $f$ is a continuous function of $y$ for each fixed $x$.

3. If $f(x, y)$ and $f_y(x, y)$ are continuous and $f_y(x, y)$ is bounded on a convex region $D$, show that $f$ satisfies a Lipschitz condition on $D$. (A region $D$ is convex if for every pair of points in $D$ the line segment joining the points is in $D$.) Where in your proof is the assumption of convexity needed?

4. Let $f(x, y)$ and $g(x, y)$ satisfy Lipschitz conditions with respect to $y$ in a region $D$. Show that
   (a) $f + g$ satisfies a Lipschitz condition.
   (b) $fg$ does not necessarily satisfy a Lipschitz condition.

5. If $f(x, y) = y^{2/3}$, show that $f$ does not satisfy a Lipschitz condition in any rectangle of the form $|x| \leq a$, $|y| \leq b$.

6. If $f(x, y) = (\sin y)/(1 + x^2)$ for all $(x, y)$, show that $f$ satisfies a Lipschitz condition.

7. Find an integral equation that is equivalent to the given initial value problem.

   (a)  $y' = xy + y^2$,   $y(1) = 2$        (b)  $y' = \sin xy$,   $y(0) = 1$

8. Find a solution of the given integral equation.

   (a)  $y(x) = 5 + \int_0^x [e^t + y(t)] \, dt$

   (b)  $y(x) = -1 + \int_0^x [y(t)]^2 \sin t \, dt$

## 8.2   SUCCESSIVE APPROXIMATIONS

As explained at the end of Section 8.1, we can establish the existence of a solution of the initial value problem

$$y' = f(x, y), \qquad y(x_0) = k, \tag{8.7}$$

by showing that the integral equation

$$y(x) = k + \int_{x_0}^x f[t, y(t)] \, dt \tag{8.8}$$

has a continuous solution.

**Theorem 8.1**   Let the function $f$ be continuous and satisfy a Lipschitz condition on the rectangle

$$R = \{(x, y) : |x - x_0| \le a, \quad |y - k| \le b\}.$$

If $M$ is a positive number such that $|f(x, y)| \le M$ for $(x, y)$ in $R$ (the existence of such a number $M$ is assured because $f$ is continuous and hence bounded on the closed rectangle) and if

$$\alpha = \min(a, b/M), \tag{8.9}$$

then the initial value problem (8.7) has a solution on the interval $[x_0 - \alpha, x_0 + \alpha]$. The solution of the problem is unique in the following sense. If $u$ and $v$ are both solutions on an interval $\mathscr{J}$ that contains $x_0$, then $u(x) = v(x)$ for all $x$ in $\mathscr{J}$.

PROOF   The proof of the existence of a solution involves the construction of a sequence of functions which can be shown to converge to a solution of the problem. The functions of the sequence are sometimes called *successive approximations,* and hence the title of this section. The name of Picard is associated with this method. We define the functions of our sequence by means of the formulas

$$y_0(x) = k, \tag{8.10}$$

$$y_n(x) = k + \int_{x_0}^{x} f[t, y_{n-1}(t)] \, dt \tag{8.11}$$

for $n \geq 1$ and for $|x - x_0| \leq \alpha$. Actually, for $y_n$ to be well defined, $y_{n-1}$ must be defined and continuous and the points $(x, y_{n-1}(x))$, $|x - x_0| \leq \alpha$, must lie in the rectangle $R$. We shall use mathematical induction to show that each function $y_n$ is defined, continuous, and satisfies $|y_n(x) - k| \leq b$ for $|x - x_0| \leq \alpha$.

Certainly the constant function $y_0$ is defined and continuous, and we have $|y_0(x) - k| = 0 < b$ for $|x - x_0| \leq \alpha$. Suppose that $y_k$ is defined and continuous and that $(x, y_k(x))$ is in $R$ for $|x - x_0| \leq \alpha$. Then $f[x, y_k(x)]$ is defined and continuous and $y_{k+1}$, where

$$y_{k+1}(x) = k + \int_{x_0}^{x} f[t, y_k(t)] \, dt,$$

is defined and continuous. Also

$$|y_{k+1}(x) - k| \leq \left| \int_{x_0}^{x} |f[t, y_k(t)]| \, dt \right| \leq \left| \int_{x_0}^{x} M \, dt \right|,$$

so that

$$|y_{k+1}(x) - k| \leq M |x - x_0| \leq M \alpha \leq b.$$

Thus $(x, y_{k+1}(x))$ is in $R$ for $|x - x_0| \leq \alpha$. Hence each function $y_n$ of the sequence is defined and continuous, and

$$|y_n(x) - k| \leq M \alpha \leq b$$

for $n = 0, 1, 2, \ldots$ .

The next order of business is to show that the sequence converges. In order to do this, it suffices to show that the infinite series

$$y_0 + \sum_{n=1}^{\infty} (y_n - y_{n-1}) \tag{8.12}$$

converges, because the $n$th partial sum of this series is $y_n$. We shall use a comparison test. From Eq. (8.11), with $n = 1$, we have

$$|y_1(x) - k| \leq \left| \int_{x_0}^x |f(t, k)|\, dt \right| \leq M |x - x_0|.$$

Next, since

$$y_2(x) = k + \int_{x_0}^x f[t, y_1(t)]\, dt,$$

$$y_1(x) = k + \int_{x_0}^x f[t, y_0(t)]\, dt,$$

we have (subtracting)

$$y_2(x) - y_1(x) = \int_{x_0}^x \{f[t, y_1(t)] - f[t, y_0(t)]\}\, dt.$$

If $K$ is a Lipschitz constant for $f$, then

$$|f[t, y_1(t)] - f(t, y_0)| \leq K |y_1(t) - y_0(t)| \leq KM |t - x_0|.$$

Consequently, we see that

$$|y_2(x) - y_1(x)| \leq KM \left| \int_{x_0}^x |t - x_0|\, dt \right|, \quad |x - x_0| \leq \alpha.$$

By considering the cases $x \geq x_0$ and $x < x_0$ separately, we find that

$$|y_2(x) - y_1(x)| \leq \frac{KM |x - x_0|^2}{2}.$$

It can be shown by mathematical induction (Exercise 4) that

$$|y_n(x) - y_{n-1}(x)| \leq \frac{MK^{n-1} |x - x_0|^n}{n!} \tag{8.13}$$

for $|x - x_0| \leq \alpha$ and $n = 1, 2, 3, \ldots$ . Thus

$$|y_n(x) - y_{n-1}(x)| \leq \frac{MK^{n-1} \alpha^n}{n!} \tag{8.14}$$

for $|x - x_0| \leq \alpha$ and $n \geq 1$. The series of constants

$$|k| + \sum_{n=1}^{\infty} \frac{MK^{n-1}\alpha^n}{n!}$$

converges (as can be shown by the ratio test). Hence the series (8.12) converges and the sequence $\{y_n\}$ converges. Let us denote the limit function by $y$. Since $|y_n(x) - k| \leq b$ for $|x - x_0| \leq \alpha$ we must have $|y(x) - k| \leq b$ for $|x - x_0| \leq \alpha$.

We must now show that the function $y$ is continuous and is a solution of the integral equation (8.9). The key to the investigation of these matters is the concept of *uniform convergence*. A sequence of functions $\{f_n\}$ is said to converge uniformly to a function $f$ on an interval $\mathcal{I}$ if for every positive number $\varepsilon$ there is an integer $N(\varepsilon)$ such that

$$|f_n(x) - f(x)| < \varepsilon \qquad (8.15)$$

whenever

$$n \geq N(\varepsilon) \qquad (8.16)$$

for all $x$ in $\mathcal{I}$. The integer $N$ may depend on $\varepsilon$ but not on the point of the interval. If the sequence of functions $\{f_n\}$ converges uniformly on $\mathcal{I}$, then for each $x$ in $\mathcal{I}$ the sequence of numbers $\{f_n(x)\}$ converges. For any given $\varepsilon$, there is one integer $N(\varepsilon)$ such that satisfaction of condition (8.16) insures that the criterion (8.15) holds for all $x$. An infinite series of functions,

$$\sum_{n=1}^{\infty} u_n$$

is said to converge uniformly on an interval if and only if its sequence of partial sums converges uniformly on the interval.

We now state, without proof, two standard theorems about uniform convergence. Proofs can be found in Taylor (1955).

**Theorem 8.2**  If a sequence or series of continuous functions converges uniformly on an interval, then the limit function is continuous on that interval.

**Theorem 8.3**  Suppose that each of the functions $u_n$, $n = 1, 2, 3, \ldots$, is defined on an interval $\mathcal{I}$. If there is a convergent series of nonnegative constants

$$\sum_{n=1}^{\infty} M_n$$

such that

$$|u_n(x)| \le M_n$$

for $n \ge 1$ and for all $x$ in $\mathscr{I}$, then the series

$$\sum_{n=1}^{\infty} u_n$$

converges uniformly on $\mathscr{I}$.

We now apply these theorems to our problem in differential equations. In view of the comparison (8.14), the series (8.12) converges uniformly on the interval $[x_0 - \alpha, x_0 + \alpha]$, by Theorem 8.3. Hence its sequence of partial sums $\{y_n\}$ converges uniformly on the same interval. Hence the limit function $y$ is continuous, by Theorem 8.2. Next, since

$$y_n(x) - k - \int_{x_0}^{x} f[t, y_{n-1}(t)] \, dt = 0$$

for every positive integer $n$, we may write

$$y(x) - k - \int_{x_0}^{x} f[t, y(t)] \, dt$$

$$= y(x) - y_n(x) + \int_{x_0}^{x} \{f[t, y_{n-1}(t)] - f[t, y(t)]\} \, dt.$$

Thus

$$\left| y(x) - k - \int_{x_0}^{x} f[t, y(t)] \, dt \right|$$

$$\le |y(x) - y_n(x)| + K \left| \int_{x_0}^{x} |y_{n-1}(t) - y(t)| \, dt \right|.$$

Given $\varepsilon > 0$, there is an integer $N(\varepsilon)$ such that

$$|y(x) - y_k(x)| < \frac{1}{2} \frac{\varepsilon}{1 + \alpha K}$$

whenever $n \ge N$ and $|x - x_0| \le \alpha$. Chosing $n = N + 1$, we see that

$$\left| y(x) - k - \int_{x_0}^{x} f[t, y(t)] \, dt \right| < \frac{1}{2} \frac{\varepsilon}{1 + \alpha K} + K \left| \int_{x_0}^{x} \frac{1}{2} \frac{\varepsilon}{1 + \alpha K} \, dt \right|$$

$$\leq \frac{1}{2} \frac{\varepsilon}{1 + \alpha K} + \frac{1}{2} \frac{\varepsilon K}{1 + \alpha K} |x - x_0|$$

$$\leq \varepsilon.$$

Since this is true for every positive number $\varepsilon$, we must have

$$y(x) - k - \int_{x_0}^{x} f[t, y(t)] \, dt = 0,$$

which we wished to show.

We now prove that the problem (8.7) can have but one solution. Suppose that $u$ and $v$ are both solutions on an interval $\mathscr{J}$ that contains $x_0$. Then

$$u(x) = k + \int_{x_0}^{x} f[t, u(t)] \, dt,$$

$$v(x) = k + \int_{x_0}^{x} f[t, v(t)] \, dt,$$

so that

$$u(x) - v(x) = \int_{x_0}^{x} \{f[t, u(t)] - f[t, v(t)]\} \, dt$$

for $x$ in $\mathscr{J}$. Using the fact that $f$ satisfies a Lipschitz condition, we have

$$|u(x) - v(x)| \leq K \left| \int_{x_0}^{x} |u(t) - v(t)| \, dt \right|$$

for $x$ in $\mathscr{J}$. An application of the lemma of Section 5.11 shows that $|u(x) - v(x)| = 0$ or $u(x) = v(x)$ for $x$ in $\mathscr{J}$. Thus $u$ and $v$ must be the same and there is only one solution.

The number $\alpha$ that is described in Theorem 8.1 may be small even when $a$ and $b$ are large. To illustrate this, we consider an example.

**Example 1**

$$y' = 200xy^2, \qquad y(0) = 1.$$

Here the functions $f(x, y) = 200xy^2$ and $f_y(x, y) = 400xy$ are continuous everywhere, and hence on any rectangle of the form $|x| \le a$, $|y - 1| \le b$. The solution of the initial value problem is given by the formula

$$y = \frac{1}{1 - 100x^2}.$$

This solution exists only on the interval $(-1/10, 1/10)$, so it is clear that $\alpha < 1/10$ no matter how $a$ and $b$ are chosen.

Usually the function $f$ and its partial derivatives will be continuous on a region $D$ that is not a rectangle. The region may consist of the entire plane, as in the preceding example. However, Theorem 8.1 can be applied by considering a rectangle that is contained in $D$. The theorem then assures the existence of a solution on some interval $[x_0 - \alpha, x_0 + \alpha]$, where $\alpha$ may be small. It may be possible to continue or extend the solution to the right of $x_0 + \alpha$, or to the left of $x_0 - \alpha$. For instance, suppose that $y$ is a solution on the interval $[x_0 - \alpha, x_0 + \alpha]$. If the point $P : (x_0 + \alpha, y(x_0 + \alpha))$ lies in the interior of the region $D$, then there is a rectangle with center at $P$ and contained in $D$. According to Theorem 8.1, there is a solution $\tilde{y}$, for which $\tilde{y}(x_0 + \alpha) = y(x_0 + \alpha)$, that exists on some interval $[x_0 + \alpha - \alpha_1, x_0 + \alpha + \alpha_1]$, where $\alpha_1 > 0$. By the uniqueness part of Theorem 8.1, $y$ and $\tilde{y}$ must coincide on the interval where both are defined. Thus the solution $y$ is continued to the right of $x_0 + \alpha$, up to $x_0 + \alpha + \alpha_1$. If the point $(x_0 + \alpha + \alpha_1, y(x_0 + \alpha + \alpha_1))$ is in $D$, the solution can be continued still farther to the right. It may happen that the solution can be continued to the right for all $x$ greater than $x_0$. If not, it can be shown[1] that the solution can be continued up to a point $x_1$ and that as $x$ approaches $x_1$ from the left either $y(x)$ becomes infinite or else the point $(x, y(x))$ approaches the boundary of the region $D$. Similar remarks apply to the continuation of the solution to the left of $x_0 - \alpha$.

If, in Theorem 8.1, we omit the hypothesis that $f$ satisfies a Lipschitz condition and assume only that $f$ is continuous on the rectangle $R$, it is still possible to prove that a solution exists. A different method of proof must be employed in this case. Also, the solution may not be unique, as the following example shows.

**Example 2**

$$y' = 3y^{2/3}, \qquad y(0) = 0.$$

Here it may readily be verified that $u$ and $v$, where $u(x) = 0$ and $v(x) = x^3$ for

[1] See, for example, Hurewicz (1958).

all $x$, are both solutions. Also, the function $w$, where $w(x) = 0$ for $x \le 0$ and $w(x) = x^3$ for $x > 0$, is another solution. If $f(x, y) = 3y^{2/3}$, then $f$ is continuous on every rectangle of the form $|x| \le a$, $|y| \le b$. However, $f$ cannot satisfy a Lipschitz condition on any such rectangle. (See, in this connection, Exercise 5, Section 8.1.)

## Exercises for Section 8.2

1. Use Theorem 8.1 to show that the given initial value problem has a unique solution. In parts (a), (b), and (c) find the solution and describe the interval on which it is defined.

   (a) $y' = \frac{2}{3}x^{1/3}y^{-1}$, $y(1) = -2$

   (b) $y' = y^2 e^x$, $y(0) = 1$

   (c) $y' = \dfrac{x^2 + 2y^2}{xy}$, $y(1) = 2$

   (d) $y' = x^2 + y^2$, $y(0) = 1$

   (e) $y' = \sin xy + 2y$, $y(0) = 0$

   (f) $y' = \dfrac{1}{x^2 - y^2}$, $y(1) = 0$

2. Find the functions $y_0$, $y_1$, and $y_2$ in the sequence of successive approximations (8.10) and (8.11) for the case of the initial value problem
$$y' = x + y^2 - 2, \quad y(0) = 1.$$

3. Work Problem 2 for the initial value problem
$$y' = \frac{x^2 - y^2}{x}, \quad y(1) = 0.$$

4. Use mathematical induction to establish the inequality (8.13).

5. Show that the given series of functions converges uniformly on the specified interval. Use Theorem 8.3.

   (a) $\displaystyle\sum_{n=1}^{\infty} \frac{\sin nx}{n^2}$, for all $x$

   (b) $\displaystyle\sum_{n=1}^{\infty} \frac{1}{n^2 + x^2}$, for all $x$

   (c) $\displaystyle\sum_{n=0}^{\infty} e^{-nx}$, $x \ge a$, $a > 0$.

   (d) $\displaystyle\sum_{n=0}^{\infty} \frac{x^n}{n!}$, $|x| \le a$, $a > 0$.

6. Find at least two solutions of the given initial value problem.

   (a) $y' = \frac{3}{2}y^{1/3}$, $y(0) = 0$

   (b) $y' = 3y^{2/3}(y^{1/3} + 1)$, $y(0) = 0$

7. If $y_0$ were taken to be any continuous function such that $|y_0(x) - k| \le b$ for $|x - x_0| \le a$, would the sequence of functions defined by Eq. (8.11) still converge to a solution of the initial value problem? Why?

## 8.3 FIRST-ORDER SYSTEMS

We consider an initial value problem for a first-order system,

$$x_i' = f(t, x_1, x_2, \ldots, x_n), \tag{8.17}$$

$$x_i(t_0) = k_i, \quad i = 1, 2, \ldots, n, \tag{8.18}$$

where the functions $f_i$ are defined on some region $D$ of $n + 1$ dimensional space that contains the point $(t_0, k_1, k_2, \ldots, k_n)$. Using vector notation, we may write our problem as

$$\mathbf{x}' = \mathbf{f}(t, \mathbf{x}), \qquad \mathbf{x}(t_0) = \mathbf{k}, \tag{8.19}$$

where

$$\mathbf{x} = \begin{bmatrix} x_1 \\ x_2 \\ \vdots \\ x_n \end{bmatrix}, \qquad \mathbf{f} = \begin{bmatrix} f_1 \\ f_2 \\ \vdots \\ f_n \end{bmatrix}, \qquad \mathbf{k} = \begin{bmatrix} k_1 \\ k_2 \\ \vdots \\ k_n \end{bmatrix}.$$

The vector function $\mathbf{f}$ is said to be continuous in the region $D$ if each of its components $f_i$ is continuous in $D$. We recall from Chapter 6 the definition of the norm of a vector. If $\mathbf{u} = (u_1, u_2, \ldots, u_n)$, then

$$|\mathbf{u}| = \max_{1 \le i \le n} |u_i|.$$

The vector function $\mathbf{f}$ is said to satisfy a Lipschitz condition in $D$ if there is a positive number $K$ such that

$$|\mathbf{f}(t, \mathbf{x}) - \mathbf{f}(t, \mathbf{y})| \le K |\mathbf{x} - \mathbf{y}|$$

whenever $(t, \mathbf{x})$ and $(t, \mathbf{y})$ are both in $D$. A real-valued function $g(t, \mathbf{x})$ is said to satisfy a Lipschitz condition in $D$ if there is a positive number $L$ such that

$$|g(t, \mathbf{x}) - g(t, \mathbf{y})| \le L |\mathbf{x} - \mathbf{y}|$$

whenever $(t, \mathbf{x})$ and $(t, \mathbf{y})$ are in $D$. We leave it to the reader to show that the vector function $\mathbf{f}$ satisfies a Lipschitz condition if and only if each of its components $f_i$ satisfies a Lipschitz condition. A practical criterion for ascertaining that a function satisfies a Lipschitz condition is as follows.

**Lemma** Let $g$ be a function of $n + 1$ real variables, with values $g(t, \mathbf{x})$ for $(t, \mathbf{x})$ in a " rectangle " $R$, where

$$R = \{(t, \mathbf{x}) : |t - t_0| \le a, |\mathbf{x} - \mathbf{k}| \le b\} .$$

If $g$ and its partial derivatives with respect to $x_1, x_2, \ldots, x_n$ are continuous in $R$ then $g$ satisfies a Lipschitz condition in $R$.
   The proof, for the case $n = 2$, is left as an exercise.
   The main result of this section may be stated as follows.

**Theorem 8.4** Let $\mathbf{f}$ be continuous and satisfy a Lipschitz condition on the rectangle

$$R = \{(t, \mathbf{x}) : |t - t_0| \le a, \quad |\mathbf{x} - \mathbf{k}| \le b\} .$$

If $M$ is a positive number such that $|\mathbf{f}(t, \mathbf{x})| \le M$ for $(t, \mathbf{x})$ in $R$ (such a number is guaranteed to exist because of the hypotheses on $\mathbf{f}$) and if

$$\alpha = \min(a, b/M) ,$$

then the initial value problem (8.19) has a solution on the interval

$$[t_0 - \alpha, t_0 + \alpha] .$$

If $\mathbf{u}$ and $\mathbf{v}$ are both solutions of the problem on an interval $\mathscr{J}$ that contains $t_0$, then $\mathbf{u}(t) = \mathbf{v}(t)$ for all $t$ in $\mathscr{J}$.

PROOF   We first observe that a vector function is a solution of the initial value problem (8.19) if and only if it is a solution of the integral equation

$$\mathbf{x}(t) = \mathbf{k} + \int_{t_0}^{t} \mathbf{f}[s, \mathbf{x}(s)] \, ds . \tag{8.20}$$

Next we define a sequence of successive approximations $\{\mathbf{x}_n\}$ by means of the relations

$$\mathbf{x}_0(t) = \mathbf{k}$$

$$\mathbf{x}_n(t) = \mathbf{k} + \int_{t_0}^{t} \mathbf{f}[s, \mathbf{x}_{n-1}(s)] \, ds, \quad n \ge 1 \tag{8.21}$$

for $|t - t_0| \leq \alpha$. The proof that this sequence converges to a solution of Eq. (8.20) is similar to the proof of Theorem 8.1. We leave the details as an exercise.

If the functions $f_i$ and their partial derivatives $\partial f_i/\partial x_j$ are continuous in a region $D$ that is not a rectangle, we may apply Theorem 8.4 by considering a rectangle that is contained in $D$. Then the existence of a solution on an interval $[t_0 - \alpha, t_0 + \alpha]$ is guaranteed. It may be possible to continue the solution to the right of $t_0 + \alpha$ (and to the left of $t_0 - \alpha$). It may be possible to continue the solution to the right for all $t \geq t_0$. If not it can be shown that the solution exists up to a point $t_1$ and that as $t$ approaches $t_1$ either $|\mathbf{x}(t)|$ becomes infinite or else the solution curve approaches the boundary of the region $D$.

As an example, let us consider the initial value problem

$$\frac{dx_1}{dt} = \cos(x_1 x_2 + t) - t^2,$$

$$\frac{dx_2}{dt} = \exp(x_1 t) + t^3 x_2^4,$$

$$x_1(0) = 2, \quad x_2(0) = -1.$$

Here

$$f_1(t, x_1, x_2) = \cos(x_1 x_2 + t) - t^2,$$

$$f_2(t, x_1, x_2) = \exp(x_1 t) + t^3 x_2^4,$$

and

$$\frac{\partial f_1}{\partial x_1} = -x_2 \sin(x_1 x_2 + t),$$

$$\frac{\partial f_1}{\partial x_2} = -x_1 \sin(x_1 x_2 + t),$$

$$\frac{\partial f_2}{\partial x_1} = t \exp(x_1 t),$$

$$\frac{\partial f_2}{\partial x_2} = 4t^3 x_2^3.$$

Each of these six functions is continuous for all $(t, x_1, x_2)$. On any rectangle of the form

$$|t| \leq a, \quad |x_1 - 2| \leq b, \quad |x_2 + 1| \leq b$$

Theorem 8.4 applies and the existence of a solution is assured on some interval $[-\alpha, \alpha]$. The positive number $\alpha$ may be small.

We now consider the special case of a *linear* system

$$\frac{dx_i}{dt} = \sum_{j=1}^{n} a_{ij} x_j + b_i,$$

$$x_i(t_0) = k_i, \quad i = 1, 2, \ldots, n,$$

$$(8.22)$$

where the functions $a_{ij}$ and $b_i$ are continuous on an interval $\mathscr{I}$ that contains $t_0$. In vector notation we have

$$\mathbf{x}' = A\mathbf{x} + b, \qquad \mathbf{x}(t_0) = \mathbf{k}, \qquad (8.23)$$

where $A$ is a matrix function and $\mathbf{b}$ a vector function. Writing

$$\mathbf{f}(t, \mathbf{x}) = A(t)\, \mathbf{x} + \mathbf{b}(t)$$

for $t$ in $\mathscr{I}$ and all $\mathbf{x}$, we have

$$|\mathbf{f}(t, \mathbf{x}) - \mathbf{f}(t, \mathbf{y})| = |A(t)(\mathbf{x} - \mathbf{y})| \le n|A(t)|\, |\mathbf{x} - \mathbf{y}|.$$

On any closed subinterval $\mathscr{J}$ that i.  itained in $\mathscr{I}$, there is a number $K$ such that $n|A(t)| \le K$. Hence $\mathbf{f}$ satisfies a Lipschitz condition. Application of the method of successive approximations, using the scheme (8.21), shows that a solution exists throughout the closed interval $\mathscr{J}$. (We leave the details as an exercise.) Since any point in $\mathscr{I}$ is contained in such a closed interval, the (unique) solution exists throughout $\mathscr{I}$. We summarize as follows.

**Theorem 8.5**  The initial value problem

$$\mathbf{x}' = A\mathbf{x} + \mathbf{b}, \qquad \mathbf{x}(t_0) = \mathbf{k},$$

where $A$ and $b$ are continuous on an interval $\mathscr{I}$ that contains $t_0$, possesses a solution on $\mathscr{I}$. If $\mathbf{u}$ and $\mathbf{v}$ are any two solutions that both exist an on interval that contains $t_0$, then $\mathbf{u}(t) = \mathbf{v}(t)$ for $t$ on that interval.

In the example

$$x_1' = 2t\, x_1 + e^{-t} x_2 - \sin t,$$

$$x_2' = (\cos t)x_1 + t^3 x_2 + (t^2 + 1)^{-1},$$

$$x_1(2) = 5, \quad x_2(2) = -4,$$

we have

$$A(t) = \begin{bmatrix} 2t & e^{-t} \\ \cos t & t^3 \end{bmatrix}, \qquad b(t) = \begin{bmatrix} -\sin t \\ (t^2 + 1)^{-1} \end{bmatrix}.$$

Since these matrix functions are continuous everywhere, the initial value problem possesses a solution that exists on the set of all real numbers.

Theorems 8.4 and 8.5 yield important results about single differential equations. An $n$th-order equation of the form

$$x^{(n)} = f[t, x, x', \ldots, x^{(n-1)}] \tag{8.24}$$

can be rewritten as a first-order system for the quantities

$$x_1 = x, \quad x_2 = x', \quad x_3 = x'', \ldots, \quad x_n = x^{(n-1)}.$$

For a function $x$ is a solution of Eq. (8.24) if and only if the ordered set of functions $(x_1, x_2, \ldots, x_2)$ is a solution of the system

$$
\begin{aligned}
x_1' &= x_2, \\
x_2' &= x_3, \\
&\cdots\cdots \\
x_{n-1}' &= x_n, \\
x_n' &= f(t, x_1, x_2, \ldots, x_n).
\end{aligned}
\tag{8.25}
$$

The derivation of the following theorem is left as an exercise.

**Theorem 8.6**  Let $f$ be a function of $n + 1$ variables that is continuous and satisfies a Lipschitz condition (with respect to its last $n$ arguments) on a rectangle

$$|t - t_0| \le a, \qquad |x_i - k_i| \le b, \qquad 1 \le i \le n.$$

Then there exists a positive number $\alpha$ such that on the interval $[t_0 - \alpha, t_0 + \alpha]$ there is a solution of Eq. (8.24) for which

$$x^{(i-1)}(t_0) = k_i, \quad 1 \le i \le n.$$

This solution is unique.

Theorem 8.5 yields a corresponding result for a single linear equation of the form

$$x^{(n)} + a_1 x^{(n-1)} + \cdots + a_{n-1} x' + a_n x = b. \tag{8.26}$$

**Theorem 8.7** Let the functions $a_i$ and $b$ be continuous on an interval $\mathscr{I}$ and let $t_0$ be any point of $\mathscr{I}$. Then there exists on $\mathscr{I}$ a solution of the differential equation (8.26) for which

$$x^{(i-1)}(t_0) = k_i, \qquad 1 \le i \le n.$$

This solution is unique.

The proof of this result, using Theorem 8.5, is left as an exercise.

## Exercises for Section 8.3

1. Let $\mathbf{f}(t, \mathbf{x})$ be a vector function. Show that $\mathbf{f}$ satisfies a Lipschitz condition in a region if and only if each of its components satisfies a Lipschitz condition in the region.

2. Give a proof of the lemma at the beginning of this section for the case $n = 2$. Suggestion: write

$$f_i(t, x_1, x_2) - f_i(t, y_1, y_2)$$
$$= f_i(t, x_1, x_2) - f_i(t, y_1, x_2) + f_i(t, y_1, x_2) - f_i(t, y_1, y_2)$$

for $i = 1, 2$, and use the mean value theorem.

3. Complete the proof of Theorem 8.4.

4. Given the initial value problem

$$\frac{dx_1}{dt} = -x_1^2, \qquad \frac{dx_2}{dt} = x_1 x_2, \qquad x_1(0) = 1, \quad x_2(0) = 5.$$

(a) Use Theorem 8.4 to show that the problem has a unique solution on some interval containing $t = 0$.

(b) Find the solution of the problem and indicate the interval on which it exists.

5. Work Problem 4 for the initial value problem

$$\frac{dx_1}{dt} = -2tx_1 x_2 + x_2^2, \qquad \frac{dx_2}{dt} = -2tx_2^2, \qquad x_1(1) = 0, \quad x_2(1) = 1.$$

6. Find the vector functions $\mathbf{x}_0$, $\mathbf{x}_1$, and $\mathbf{x}_2$ in the sequence of successive approximations (8.21) in the case of the initial value problem

$$x_1' = tx_1 - x_2^2, \quad x_2' = x_1 x_2 - 2, \quad x_1(0) = -2, \quad x_2(0) = 1.$$

7. Work Problem 6 for the initial value problem

$$x_1' = x_1^2 - x_2, \quad x_2' = x_1^2 + x_2^2, \quad x_1(1) = 0, \quad x_2(1) = 1.$$

8. Prove Theorem 8.5.

9. Use Theorem 8.5 to justify the claim that the given initial value problem possesses a (unique) solution on the interval $(0, \infty)$.

$$tx_1' = (\cos t + 1)x_1 = t^2 x_2 - 7, \quad x_1(1) = 2$$
$$tx_2' = (1 - t)x_1 + (\sin t)x_2 + e^t, \quad x_2(1) = 0.$$

10. Prove Theorem 8.6, using Theorem 8.4.

11. Prove Theorem 8.7, using Theorem 8.5.

12. Use Theorem 8.6 to show that the given initial value problem has a (unique) solution. Then find the solution and indicate the interval on which it exists.

(a) $2x \dfrac{d^2 x}{dt^2} = \left(\dfrac{dx}{dt}\right)^2 + 1, \quad x(0) = 1, \quad x'(0) = -1$

(b) $x \dfrac{d^2 x}{dt^2} = -\left(\dfrac{dx}{dt}\right)^2, \quad x(0) = 1, \quad x'(0) = -1$

13. Use Theorem 8.7 to show that the given initial value problem possesses a (unique) solution on the interval $(-1, \infty)$.

$$x'' - \frac{2}{t+1} x' + t^2 x = t^3 + 1, \qquad x(0) = 1, \quad x'(0) = 7.$$

# REFERENCES

CODDINGTON, E. A. (1961). "An Introduction to Ordinary Differential Equations." Prentice-Hall, Englewood Cliffs, New Jersey.

FINKBEINER, D. T. II (1966). "Introduction to Matrices and Linear Transformations," 2nd ed. Freeman, San Francisco.

HOFFMAN, K., and KUNZE, R. (1961). "Linear Algebra." Prentice-Hall, Englewood Cliffs, New Jersey.

HOHN, F. E. (1964). "Elementary Matrix Algebra," 2nd ed. MacMillan, New York.

HUREWICZ, W. (1958). "Lectures on Ordinary Differential Equations." M.I.T. Press, Cambridge, Massachusetts.

JAHNKE, E., and EMDE, F. (1945). "Tables of Functions," 4th ed. Dover, New York.

KAPLAN, W. (1958). "Ordinary Differential Equations." Addison-Wesley, Reading, Massachusetts.

KREIDER, D. L., KULLER, R. G., and OSTBERG, D. R. (1968). "Elementary Differential Equations." Addison-Wesley, Reading, Massachusetts.

KREIDER, D. L., KULLER, R. G., OSTBERG, D. R., and PERKINS, F. W. (1966). "An Introduction to Linear Analysis." Addison-Wesley, Reading, Massachusetts.

MURDOCK, D. C. (1957). "Linear Algebra for Undergraduates." Wiley, New York.

PIERCE, B. O. (1929). "A Short Table of Integrals." Ginn, New York.

RABENSTEIN, A. L. (1966). "Introduction to Ordinary Differential Equations." Academic Press, New York.

SMILEY, M. F. (1965). "Algebra of Matrices." Allyn and Bacon, Boston, Massachusetts.

TAYLOR, A. E. (1955). "Advanced Calculus." Blaisdell, New York.

# Answers to Selected Exercises

## CHAPTER 1

### Section 1.1

1. (a) First order, linear      (c) First order, nonlinear
   (e) Second order, nonlinear    (g) Third order, linear

2. (a) $y = x^2 - 3x + c$      (c) $y = \ln\left|\dfrac{x-4}{x}\right| + c$
   (e) $y = -\ln|\cos x| + c_1 x + c_2$    (g) $y = x^4 - x^3 + c_1 x^2 + c_2 x + c_3$

3. (a) $y = -5$      (c) $y = 2x^2 - 3x - 17$
   (e) $y = -x + 3$      (g) $y = -\cos x + 1$

5. (a) $y = ce^{-3x}$      (c) $y = ce^{x/3}$

8. $f'(1) = 5, \quad f''(1) = 22, \quad f'''(1) = 140$

### Section 1.2

1. $y = \pm(4x^2 + c)^{1/2}, \quad y = -(4x^2 + 5)^{1/2}$

3. $y = 1/(c - x^2)$ and $y = 0, \quad y = 1/(5 - x^2)$

5. $y = \dfrac{x + c}{1 - cx}, \quad y = \dfrac{7x + 1}{7 - x}$

7. $y = 1 \pm (e^x + c)^{1/2}, \quad y = 1 - (e^x + 8)^{1/2}$

9. $y = (ce^{\sin x} - 1)^{1/3}$

11. $y = \sin^{-1}(x + c) + 2n\pi$    and    $y = -\sin^{-1}(x + c) + (2n + 1)\pi$

13. $y = \dfrac{x}{\ln|x| + c}$ and $y = 0$      15. $y = \pm x(cx^2 + 1)^{1/2}$

**17.** $y = x \ln(cx^2 + 1)$

**19.** $\dfrac{1}{2}\left(cx^2 - \dfrac{1}{c}\right)$

**22.** (a)  $y = 2\tan(2x + c) - 4x + 1$

(c)  $y = 3x \pm (12x + c)^{1/2}$

**24.** (a)  $x - 1 = (y + 2)(\ln|y + 2| + c)$

## Section 1.3

**3.** $x^2 y^3 - 2xy^2 = c$

**5.** Not exact

**7.** $y = x^2 \pm (2x^4 + c)^{1/2}$

**9.** $y = \pm[x^2 \pm (x^4 + c)^{1/2}]^{1/2}$

**11.** $y = \ln[x \pm (2x^2 + c)^{1/2}]$

**13.** $y = [\sin x \pm (\sin^2 x + c)^{1/2}]^{-1}$

**17.** $\rho(x, y) = y^{-3},\quad y = [x \pm (4x^2 + c)^{1/2}]^{-1}$

**19.** $\rho(x, y) = xy^{-2},\quad y = x^{-2}[c \pm (c^2 + x^5)^{1/2}]$

**21.** $\rho(x, y) = x^{-2},\quad y/x + x^3 - y^4 = c$

**23.** $\rho(x, y) = y^{-2},\quad x/y + y^2 + x^3 = c$

## Section 1.4

**1.** $y = cx^{-2} + x^2,\quad y = 3x^{-2} + x^2$

**3.** $y = x^2 e^{-x}(3x + c)$

**5.** $y = e^{x^2}\left(c + \displaystyle\int e^{-x^2}\,dx\right),\quad y = e^{x^2}\left(be^{-a^2} + \displaystyle\int_a^x e^{-t^2}\,dt\right)$

**7.** $y = \ln x + c/\ln x$

**9.** $y = 1 + ce^{-x^2}$

**11.** $y = x(x + 1) + cx/(x + 1)$

**17.** $y = e^{3x}(cx^{-2} + x)^{-3},\quad y = 0$

**19.** $y = \pm x(c - 2\ln|x|)^{1/2}$

**21.** $y = \pm\{(x + 1)[c + (x - 1)^2]\}^{-1/2},\quad y = 0$

**23.** $y = \tan(1 + cx^{-2})$

## Section 1.5

**1.** $y = e^{-x} + k$

**3.** $x^2 + y^2 = k$

**5.** $y^4 = kx$

**7.** $y^3 + 3x^2 y = k$

**9.** $x^3 + y^3 = k$

**11.** $x^2 + 2xy - y^2 = k$

**13.** $y = x - 2\tan^{-1} x + k$

**15.** $y = \sqrt{3}x - 2\ln|2x + \sqrt{3}| + k$

**19.** $r = k\cos\theta$

**21.** $r^2 = k\sin 2\theta$

## Section 1.6

3. 4 gm

5. 20 years

7. $V = \pi(4 - t/12)^3)\, \text{ft}^3$

9. 4.77 min

11. $200 \exp\left[-\left(\dfrac{1}{50} + k\right)t\right]$

## Section 1.7

1. After $10(\ln 3)/\ln(3/2) = 27.1$ min

3. $u = 60 + \dfrac{1}{k} - t + \left(140 - \dfrac{1}{k}\right)e^{-kt}, \quad k = \dfrac{1}{10}\ln(8/7)$

5. (a) $x(t) = y(t) = a/(1 + akt)$      (b) $t = 1/(ka)$

7. $x = a\left[1 + \dfrac{t}{T}(2^{n-1} - 1)\right]^{-1/(n-1)}, \quad n > 1, \qquad x = a2^{-t/T}, \quad n = 1$

## Section 1.8

3. $x = \pm\dfrac{2}{3c_1}(c_1 t - 1)^{3/2} + c_2$

5. $x = -t - \dfrac{2}{c_1}\ln|c_1 t - 1| + c_2, \quad x = \pm t + c$

7. $x = \dfrac{t^2}{2} \pm \left\{\dfrac{t}{2}(t^2 + c_1)^{1/2} + \dfrac{c_1}{2}\ln|t + (t^2 + c_1)^{1/2}|\right\} + c_2$

9. $x = \dfrac{c_1}{8}t^4 - \dfrac{1}{2c_1}\ln|t| + c_2$

11. $x = \pm[c_1(t + c_2)^2 - c_1^{-1}]^{1/2}, \quad x = \pm(\pm 2t + c)^{1/2}$

13. $x = \pm(2t + c_2)^{1/2} + c_1, \quad x = c$

15. $x = \ln[(c_2 e^{c_1 t} - 1)/c_1], \quad x = \ln(t + c), \quad x = c$

17. $x = [t + c_2 \pm ((t + c_2)^2 + 4c_1)^{1/2}]/(2c_1), \quad x = -(t + c)^{-1}, \quad x = c$

19. $x = (c_2 e^{c_1 t} + 2)/c_1, \quad x = -2t + c, \quad x = c$

## Section 1.9

3. (a) $m\ddot{x} + c\dot{x}|\dot{x}| = -mg$

     (b) $t_1 = \left(\dfrac{m}{cg}\right)^{1/2}\tan^{-1}\left[\left(\dfrac{c}{mg}\right)^{1/2}v_0\right], \quad h = \dfrac{m}{2c}\ln\dfrac{cv_0^2 + mg}{mg}$

(c) $\quad v_1 = -v_0 \left( \dfrac{mg}{cv_0^2 + mg} \right)^{1/2}$

$$T = \left( \dfrac{m}{cg} \right)^{1/2} \tanh^{-1} \dfrac{v_0}{(v^2 + mg/c)^{1/2}}$$

$$= \left( \dfrac{m}{cg} \right)^{1/2} \cosh^{-1} \left( \dfrac{cv_0^2 + mg}{mg} \right)^{1/2}$$

5. (a) $m\ddot{x} = -c\dot{x}^n + mg$, where $x$ is the directed distance downward.

(b) $\quad v = \left( \dfrac{mg}{c} \right)^{1/n}$

7. (a) $\quad h = \dfrac{1}{2} \left( \dfrac{m}{cg} \right)^{1/2} \tan^{-1} \left[ \left( \dfrac{c}{mg} \right)^{1/2} v_0^2 \right]$

(b) $\quad h = \dfrac{1}{2} \left( \dfrac{m}{cg} \right)^{1/2} \tanh^{-1} \left[ \left( \dfrac{c}{mg} \right)^{1/2} v_0^2 \right]$

## CHAPTER 2

### Section 2.1

1. $x_1 = -1, \quad x_2 = 2$          3. No solution

5. $x_1 = c, \quad x_2 = 3c - 4$       7. $x_1 = 5, \quad x_2 = 3, \quad x_3 = -2$

9. $x_1 = 5c + 19, \quad x_2 = c, \quad x_3 = 7c + 22$

11. $x_1 = x_2 = x_3 = 0$

13. $x_1 = 2c - 4, \quad x_2 = c, \quad x_3 = -3c + 7$

15. $x_1 = c, \quad x_2 = 0, \quad x_3 = c, \quad x_4 = 0$

17. The only solution of the original system is $(2, 5)$. However each pair $(c, 7 - c)$ is a solution of the other system. Thus $(4, 3)$ is a solution of this new system, but is not a solution of the first system.

### Section 2.2

1. (a) $\begin{bmatrix} 2 & -1 \\ -1 & 3 \end{bmatrix}$

(c) $\begin{bmatrix} 0 & 1 & 2 \\ 1 & 1 & 1 \\ 2 & 0 & -1 \end{bmatrix}$

2. (a) $\begin{aligned} 3x_1 - 5x_2 &= 0 \\ -x_1 + 2x_2 &= 0 \end{aligned}$

(c) $\begin{aligned} x_1 + x_2 &= 0 \\ 5x_2 - x_3 &= 0 \\ 4x_2 + 2x_3 &= 0 \end{aligned}$

3. (a) $\begin{bmatrix} 6 & -15 \\ 3 & 0 \end{bmatrix}, \begin{bmatrix} -2 & 5 \\ -1 & 0 \end{bmatrix}, \begin{bmatrix} -4 & 10 \\ -2 & 0 \end{bmatrix}, \begin{bmatrix} 0 & 0 \\ 0 & 0 \end{bmatrix}$

5. $C + D = \begin{bmatrix} 4 & 6 \\ 8 & 10 \end{bmatrix}$

8. (a) $\begin{bmatrix} 2 \\ 5 \\ 0 \end{bmatrix}, \begin{bmatrix} 1 \\ 0 \\ 1 \end{bmatrix}, \begin{bmatrix} -3 \\ 0 \\ 4 \end{bmatrix}, \begin{bmatrix} 0 \\ 2 \\ 0 \end{bmatrix}$

9. (a) $[2 \quad 1 \quad -3 \quad 0]$,     $[5 \quad 0 \quad 0 \quad 2]$,     $[0 \quad 1 \quad 4 \quad 0]$

10. $A = \begin{bmatrix} 2 & 0 & 1 \\ -1 & 0 & 4 \\ 3 & 0 & 0 \\ 0 & 2 & 0 \end{bmatrix}$

12. (a) $-8, \sqrt{5}, 3\sqrt{5}$            (c) $-9, \sqrt{14}, \sqrt{26}$

13. (a) $1 + 4i, 1 - 4i, \sqrt{6}, \sqrt{5}$

## Section 2.3

1. $AB = \begin{bmatrix} 0 & 0 \\ -5 & 5 \end{bmatrix}$,          $BA = \begin{bmatrix} 3 & -1 \\ -6 & 2 \end{bmatrix}$

3. $AB = \begin{bmatrix} 7 & -1 & 0 \\ -5 & -7 & 5 \\ -4 & 1 & -1 \end{bmatrix}$,     $BA = \begin{bmatrix} -3 & 11 & -5 \\ 4 & 2 & 1 \\ -3 & -6 & 0 \end{bmatrix}$

5. $AB = \begin{bmatrix} 7 & 4 & 9 \\ 3 & 1 & 11 \end{bmatrix}$,       $BA$ is not defined

7. $AB$ is not defined,       $BA = \begin{bmatrix} 0 & 3 \\ 6 & -3 \\ 0 & 0 \end{bmatrix}$

9. Neither $AB$ nor $BA$ is defined.

11. (a) $A = \begin{bmatrix} 3 & 1 \\ 4 & -2 \end{bmatrix}$,     $\mathbf{x} = \begin{bmatrix} x_1 \\ x_2 \end{bmatrix}$,     $\mathbf{b} = \begin{bmatrix} 7 \\ -3 \end{bmatrix}$

     (c) $A = \begin{bmatrix} 2 & -1 & 1 \\ -1 & 1 & 5 \\ 2 & 1 & 0 \end{bmatrix}$,     $\mathbf{x} = \begin{bmatrix} x_1 \\ x_2 \\ x_3 \end{bmatrix}$,     $\mathbf{b} = \begin{bmatrix} 4 \\ -2 \\ 3 \end{bmatrix}$

12. (a) $2x_1 - 3x_2 = 2$          (c) $2x_1 - x_2 + 3x_3 = 1$
         $x_1 + 4x_2 = -5$                 $3x_2 - 2x_3 = -1$
                                         $x_1 \quad\quad + 4x_3 = 0$

13. $\displaystyle\sum_{i=1}^{4} \sum_{j=1}^{6} a_{ij}$    or    $\displaystyle\sum_{j=1}^{6} \sum_{i=1}^{4} a_{ij}$

## Section 2.4

**1.** (a), (c), (d)                     **3.** (d)

**5.** None (not a square matrix).

**7.** (a) $\begin{bmatrix} 2 & -2 & 4 \\ 0 & -3 & 2 \\ 0 & 0 & 0 \end{bmatrix}$          **8.** (a) $\begin{bmatrix} -6 & 0 & -3 \\ 0 & 0 & 9 \\ 8 & 0 & 6 \end{bmatrix}$

**12.**
(a) $\begin{bmatrix} 2 & 4 \\ -1 & 5 \end{bmatrix}$          (c) $\begin{bmatrix} 6 & 1 & 2 \\ 2 & 0 & 1 \\ 4 & 3 & 1 \end{bmatrix}$

## Section 2.5

**1.** There are $4! = 24$ permutations

**2.** (a) 2  (c) 0  (e) 5  (g) 3     **3.** (a) $-1$  (c) $-1$  (e) $-1$

**5.** (a) 22  (c) $-17$     **6.** (a) $-2$  (c) 29     **11.** $3x^2 + 2$

## Section 2.6

**5.** (a) $-10$                    **6.** (a) 178

**7.** (a) $-83$                    **8.** (a) $-324$

**9.** 252                          **11.** det $A$ is 0 or 1

## Section 2.7

**1.** (a) $A_{11} = 0,$    $A_{12} = 1,$    $A_{21} = -3,$  $A_{22} = 2$

  (c) $A_{11} = 2,$    $A_{12} = -7,$  $A_{13} = 1,$

   $A_{21} = 4,$    $A_{22} = -7,$  $A_{23} = -5$

   $A_{31} = 2,$    $A_{32} = 7,$   $A_{33} = 1$

**2.** (a) 20

**3.** $(-1)^N d_1 d_2 \cdots d_n,$    $N = \dfrac{n(n-1)}{2}$

## Section 2.8

**1.** $m = n$ and $A$ nonsingular       **3.** $x_1 = -43/6,$  $x_2 = 22/3$

**5.** Solutions $x_1 = c - 2,$  $x_2 = 3c$

**7.** $x_1 = -34/37, \quad x_2 = -25/37, \quad x_3 = 44/37$

**9.** Solutions $x_1 = -c - 3/2, \quad x_2 = 2c, \quad x_3 = 5c - 3/2$

**11.** This way the rounding errors are multiplied by a numerically smaller quantity.

## Section 2.9

**1.** If $AB = I$ then $\det A \cdot \det B = 1$ so $\det A \neq 0$.

**3.** $\begin{bmatrix} 4/3 & 1/3 \\ -5/3 & -2/3 \end{bmatrix}$

**5.** $\begin{bmatrix} 1/2 & -1/2 \\ -1/2 & 3/2 \end{bmatrix}$

**7.** $\begin{bmatrix} 11/3 & -4/3 & -2 \\ 2/3 & -1/3 & 0 \\ -4/3 & 2/3 & 1 \end{bmatrix}$

**9.** $\begin{bmatrix} 2/5 & 1/5 & 0 \\ -6/5 & -3/5 & 1 \\ -1 & -1 & 1 \end{bmatrix}$

**11.** $\begin{bmatrix} 2 & -6 & 1 & -8 \\ 0 & 1 & 0 & 1 \\ 1 & -3 & 1 & -3 \\ 0 & 1 & 0 & 2 \end{bmatrix}$

**15.** Any matrix of the form
$$\begin{bmatrix} 0 & x \\ 1/x & 0 \end{bmatrix}, \quad x \neq 0$$
is its own inverse.

# CHAPTER 3

## Section 3.1

**3.** Yes    **5.** No    **7.** No    **9.** Yes    **11.** Yes    **13.** Yes

**15.** Yes

## Section 3.2

**12.** No.    **14.** No. Let $U$ be the subspace of $R^2$ that consists of all vectors with first component zero.

## Section 3.3

**1.** (a) Dependent    (c) Dependent    (e) Independent
    (g) Dependent    (i) Dependent

**2.** (a) Independent    (c) Dependent    (e) Dependent

**5.** (b) No. Consider the subset of $R^2$ with elements $(1, 0), (0, 1), (1, 1)$.
    (d) No.

**11.** Yes

## Section 3.4

1. False

2. (a) Independent    (c) Independent if $m$ is odd, dependent if $m$ is even.
   (e) Dependent

4. $W(x) = -48e^{5x}$

## Section 3.5

2. (a) No (b) No    4. (a) No

5. Dimension 4. One basis consists of

$$\begin{bmatrix} 1 & 0 \\ 0 & 0 \end{bmatrix}, \qquad \begin{bmatrix} 0 & 1 \\ 0 & 0 \end{bmatrix}, \qquad \begin{bmatrix} 0 & 0 \\ 1 & 0 \end{bmatrix}, \qquad \begin{bmatrix} 0 & 0 \\ 0 & 1 \end{bmatrix}.$$

7. Dimension 2. Examples of bases are $\{1, x\}$, $\{x + 1, x - 1\}$

9. (a) $a = (v_1 - v_2)/3$,     $b = (v_1 + 2v_2)/3$

## Section 3.6

1. (a) $(-4, 3)$    (c) $(-9, 5)$

2. $(13, 8)$

3. (b) and (c) belong to the kernel.

## Section 3.7

1. (a) 0    (c) $x^2 \cos x - (2 + x) \sin x$

3. (a) and (c) belong to the kernel.

5. (a) $L_1 L_2 = D^2 + (2 + x)D + 2x$
       $L_2 L_1 = D^2 + (2 + x)D + (2x + 1)$
   (c) $L_1 L_2 = D^3 - (x + 1)D^2 + xD - 1$
       $L_2 L_1 = D^3 - (x + 1)D^2 + (x + 1)D - 1$

## Section 3.8

1. Every element is mapped into the zero element.

5. (a) 2    (b) 4

7. (a) 2    (b) 1    (c) A basis is $\{\mathbf{x}_1\}$, where $\mathbf{x}_1 = (1, -1, 1)$.

9. (a) 1    (b) 3    (c) A basis is $\{\mathbf{x}_1, \mathbf{x}_2, \mathbf{x}_3\}$, where $\mathbf{x}_1 = (0, 1, 0, 0)$,
   $\mathbf{x}_2 = (-1, 0, 1, 0)$, $\mathbf{x}_3 = (1, 0, 0, 1)$.

11. (a) 1    (b) 2    A basis is $\{\mathbf{x}_1, \mathbf{x}_2\}$, where $\mathbf{x}_1 = (0, 1, 0)$, $\mathbf{x}_2 = (1, 0, 2)$.

13. (a) 1    (b) 1    (c) A basis is $\{\mathbf{x}_1\}$, where $\mathbf{x}_1 = (3, 1)$.

## Section 3.9

1. (a) $(17, -5)$   (c) $(-3, 2)$    2. (a) $(1, -3, 2)$    3. $(-2, 4, 0)$

6. (a) $\begin{bmatrix} -11 & 10 \\ 6 & -6 \end{bmatrix}$    (b) $a_1' = -11a_1 + 10a_2$
$a_2' = 6a_1 - 6a_2$

10. $\begin{bmatrix} 0 & 1 & -1 \\ 1 & -2 & 3 \\ -1 & 2 & -2 \end{bmatrix}$

## Section 3.10

1. (a) $\dfrac{1}{5}\begin{bmatrix} -5 & 6 \\ 5 & 2 \end{bmatrix}$    (c) $\dfrac{1}{5}\begin{bmatrix} -6 & 1 \\ 22 & 3 \end{bmatrix}$

2. Two

3. (a) $\begin{bmatrix} 2 & -4 \\ -1 & 2 \\ 0 & 0 \end{bmatrix}$, rank one    (c) $\begin{bmatrix} 0 & 0 \\ 0 & 0 \\ 0 & 0 \end{bmatrix}$, rank zero

5. (a) $cA$   (b) $A + B$

## Section 3.11

1. (a) and (d) are orthogonal
7. $x_1' = \frac{1}{2}(-\sqrt{3}x_1 - x_2), \quad x_2' = \frac{1}{2}(-x_1 + \sqrt{3}x_2)$

# CHAPTER 4

## Section 4.1

1. $\lambda_1 = -2, (a, -a), \lambda_2 = 3, (b, 4b),$
$K = \begin{bmatrix} 1 & 1 \\ -1 & 4 \end{bmatrix}, \quad D = \operatorname{diag}(-2, 3)$

3. $\lambda_1 = i, (2a, (1 + i)a), \quad \lambda_2 = -i, (2b, (1 - i)b)$
$K = \begin{bmatrix} 2 & 2 \\ 1 + i & 1 - i \end{bmatrix}, \quad D = \operatorname{diag}(i, -i)$

5. $\lambda_1 = \lambda_2 = -1, (a, a)$.
Not similar to a diagonal matrix.

7. $\lambda_1 = 1, (a, 0, 0), \quad \lambda_2 = 2, (b, 0, -b), \quad \lambda_3 = -2, (-c, 4c, 5c)$

9. $\lambda_1 = \lambda_2 = \lambda_3 = 2, (a, b, c)$

## Section 4.2

1. $\lambda_1 = 1, \quad m_1 = 1, \quad n_1 = 1$
   $\lambda_2 = -1, \quad m_1 = 2, \quad n_1 = 1$
   Not similar to a diagonal matrix

3. $\lambda_1 = 0, \quad m_1 = n_1 = 1$
   $\lambda_2 = 2, \quad m_2 = n_2 = 2$
   Similar to a diagonal matrix

5. $\lambda_1 = 1, \quad m_1 = 3, \quad n_1 = 1$
   Not similar to a diagonal matrix

## Section 4.3

1. $K = \dfrac{1}{\sqrt{5}} \begin{bmatrix} 2 & -1 \\ 1 & 2 \end{bmatrix}, \qquad\qquad D = \operatorname{diag}(10, -5)$

3. $K = \dfrac{1}{7} \begin{bmatrix} 2 & 6 & 3 \\ -6 & 3 & -2 \\ 3 & 2 & -6 \end{bmatrix}, \qquad D = \operatorname{diag}(49, 0, 0)$

5. (a) $\dfrac{1}{3}(1, 2, -2), \quad \dfrac{1}{3\sqrt{2}}(4, -1, 1)$

## Section 4.4

1. (a) $\begin{bmatrix} -3 & -2 \\ -2 & 5 \end{bmatrix}$  (c) $\begin{bmatrix} -1 & 3 & 1 \\ 3 & 0 & -2 \\ 1 & -2 & 5 \end{bmatrix}$

2. (a) $2x_1^2 - 3x_2^2 - 2x_1x_2$
   (c) $2x_1^2 + x_2^2 - x_3^2 + 4x_2x_3 + 2x_3x_1 - 8x_1x_2$

3. $9x_1'^2 + 4x_2'^2 = 36, \quad$ ellipse

5. $(x_1' - 2x_2' + \sqrt{2})(x_1' + 2x_2' + \sqrt{2}) = 0, \quad$ intersecting lines

7. $(x_1' + 2)^2 + (x_2' - 1)^2 + 4x_3'^2 = 4, \quad$ ellipsoid

9. $3x_1'^2 + 3x_2'^2 - x_3'^2 + 4 = 0, \quad$ hyperboloid of two sheets

11. $x_1'^2 - x_2'^2 = 4x_3', \quad$ hyperbolic paraboloid

## Section 4.5

1. (a) $\begin{bmatrix} 9 & 1 \\ -2 & 5 \end{bmatrix}$

**2.** (a) $\begin{bmatrix} 1 & 4 \\ 4 & -11 \end{bmatrix}$   (c) $\begin{bmatrix} 128 & 256 \\ 64 & 128 \end{bmatrix}$

**3.** (a) $\dfrac{1}{3}\begin{bmatrix} 2 & 1 \\ 1 & -1 \end{bmatrix}$   (c) Singular

## Section 4.6

**7.** $\begin{bmatrix} 1 & 1 & 2 \\ 0 & 1 & 4 \\ 0 & 0 & 1 \end{bmatrix}$

**9.** $e^A = \dfrac{1}{2}\begin{bmatrix} 3e - e^{-1} & -e + e^{-1} \\ 3e - 3e^{-1} & -e + 3e^{-1} \end{bmatrix}$

# CHAPTER 5

## Section 5.1

**1.** $y = 0$

**3.** (a) $y = 2 \sin x$   (c) $y = 0$

**5.** It is the zero function

## Section 5.2

**1.** (a) $(D^2 - 3D + 2)y = 0$   (c) $(D^3 - 3D^2 - D + 1)y = 0$

**2.** (a) $(D - 1)(D + 2)y = 0$   (c) $(D - 1)^2(D + 2)y = 0$

**3.** (a) $(D^2 - D - 2)y = 0$   (c) $(D^3 - 4D^2 + 4D)y = 0$

**5.** $y = c_1 e^{2x} + c_2 e^{3x}$

**7.** $y = c_1 e^x + c_2 e^{-x} + c_3 e^{-5x}$

**9.** Not possible

## Section 5.3

**3.** (a) $\cos 3x + i \sin 3x$   (c) $e^{2x}(\cos 3x - i \sin 3x)$

(e) $\dfrac{1}{2}(e^{2ix} + e^{-2ix})$   (g) $\dfrac{1}{2i}(e^{ix} - e^{-ix})$

**7.** $e^{-x} \cos 2x, \quad e^{-x} \sin 2x$

**8.** (a) $\cos 3x, \quad \sin 3x$   (c) $e^{2x} \cos 2x, \quad e^{2x} \sin 2x$

**9.** $2 \cos 2x - 4 \sin 2x, \quad 4 \cos 2x + 2 \sin 2x$

**11.** $\cos(2 \ln x), \quad \sin(2 \ln x)$

## Section 5.4

1. $y = c_1 e^{-2x} + c_2 e^{3x}$
3. $y = c_1 + c_2 e^{-2x}$
5. $y = c_1 + c_2 e^x + c_3 e^{-4x}$
7. $y = (c_1 + c_2 x)e^{-x}$
9. $y = (c_1 + c_2 x + c_3 x^2)e^{2x}$
11. $y = c_1 + c_2 x + c_3 e^{-x}$
13. $y = c_1 \cos 3x + c_2 \sin 3x$
15. $y = e^{3x}(c_1 \cos 2x + c_2 \sin 2x)$
17. $y = (c_1 + c_2 x)\cos x + (c_3 + c_4 x)\sin x$
19. $y = (c_1 + c_2 x)e^{2x} + c_3 \cos \sqrt{2}x + c_4 \sin \sqrt{2}x$
21. $y = -3e^x + 2e^{3x}$
23. $y = \cos 2x - 2 \sin 2x$
25. $y = 3 - e^{-x}$
29. (a) $y'' + 4y' + 4y = 0$    (c) $y'' + 4y = 0$
    (e) $(D^2 + 9)^2 y = 0$

## Section 5.5

1. $y = c_1 x^2 + c_2 x^{-1}$
3. $y = c_1 + c_2 x^{1/3}$
5. $y = (c_1 + c_2 \ln x)x^{1/2}$
7. $y = c_1 + c_2 \ln x + c_3 x$
9. $y = c_1 \cos(2 \ln x) + c_2 \sin(2 \ln x)$
11. $y = c_1 x + c_2 \cos(\ln x) + c_3 \sin(\ln x)$
13. $y = 4x^{-1} - 3x^{-2}$
15. $y = \cos(2 \ln x) + 2 \sin(2 \ln x)$
17. $y = cx$
21. (a) $y = [c_1 + c_2 \ln(x - 3)](x - 3)^{-1}$
22. (a) All zeros must have positive real parts.

## Section 5.6

1. $y = c_1 e^x + c_2 e^{-x} + \sin 2x$
3. (a) $y = c_1 e^{2x} + 2e^{5x}$    (c) $y = c_1 \cos x + c_2 \sin x + 2e^x$
7. (a) $y = c_1 e^x + c_2 e^{3x} + 2e^{2x} + e^{-x}$

## Section 5.7

1. $y = c_1 e^x + c_2 e^{-3x} + e^{2x}$,    $y = 3e^x + e^{-3x} + e^{2x}$
3. $y = c_1 e^{-x} + c_2 e^{-2x} + (6x - 5)e^x$
5. $y = c_1 e^{-x} + c_2 e^{-2x} - \cos 2x + 3 \sin 2x$,    $y = -\cos 2x + 3 \sin 2x$

7. $y = c_1 e^{3x} + c_2 e^{-2x} - \frac{1}{3}$

9. $y = (c_1 + c_2 x)e^x + c_3 e^{-x} + \frac{2}{5}(\cos 2x - 2 \sin 2x)$

11. $y = c_1 e^{-x} + c_2 e^{-2x} - 5xe^{-2x}$

13. $y = c_1 e^{-x} + c_2 e^{2x} + (x^2 + 2x/3)e^{-x}$

15. $y = c_1 e^{-x} + c_2 e^{-3x} + (x^3 - 3x^2/2 + 3x/2)e^{-x}$

17. $y = c_1 + c_2 e^{-x} + x^3 - 3x^2 + 6x$

19. $y = e^{-x}(c_1 \cos 2x + c_2 \sin 2x) + xe^{-x} \sin 2x$

22. (a)  $y = c_1 e^x + c_2 e^{2x} + e^{-x}$

    (c)  $y = c_1 e^{-x} + c_2 e^{2x} - 3 \cos x - \sin x$

    (e)  $y = (c_1 + c_2 x)e^x + c_3 e^{-x} + 3e^{2x}$

24. (a)  $y = c_1 e^{2x} + c_2 e^{-x} + 2xe^{2x}$

    (c)  $y = c_1 \cos x + c_2 \sin x + 2x \sin x$

25. $y = c_1 x^{-2} + c_2 x^3 + x^4$          27. $y = c_1 x + c_2 x^3 - 2$

29. $y = c_1 x + c_2 x^{-2} + 2x \ln x$

## Section 5.8

1. $y = c_1 e^x + c_2 e^{-x} - 1 + e^x \ln(1 + e^{-x}) - e^{-x} \ln(1 + e^x)$

3. $y = (c_1 + c_2 x)e^{-x} + x^2 e^{-x}(2 \ln |x| - 3)$

5. $y = c_1 \cos x + c_2 \sin x - x \cos x + (\sin x)\ln |\sin x|$

7. $y = e^{-x}(c_1 \cos x + c_2 \sin x - 1 - \cos x + \sin x \ln |\sec x + \tan x|)$

9. $y = (c_1 + c_2 x + c_3 x^2)e^x - 2xe^x \ln |x|$

13. $y = c_1 x + c_2 xe^{1/x}$

15. $y = c_1 e^x + c_2 x^{1/2} e^x + xe^x$

## Section 5.9

1. (a)  $x = c_1 e^{-\alpha t} + c_2 e^{-\beta t}$      (b)  $x = \dfrac{x_0}{\beta - \alpha}(\beta e^{-\alpha t} - \alpha e^{-\beta t})$

    (c)  $x = \dfrac{v_0}{\beta - \alpha}(e^{-\alpha t} - e^{-\beta t})$

## Section 5.10

1. (a)  $I = \dfrac{E_0}{R}(1 - e^{-Rt/L})$

    (b)  $E_0(1 - e^{-Rt/L})$ across the resistance, $E_0 e^{-Rt/L}$ across the inductance.

**3.**  $I = \dfrac{A\omega_1}{D} \sin(\omega_1 t + \alpha + \beta)$,  $\beta = \sin^{-1} \dfrac{c^{-1} - \omega_1^2 L}{D}$,

$D = [(c^{-1} - \omega_1^2 L)^2 + (\omega_1 R)^2]^{1/2}$

**5.**  $I = -Q_0 \omega \sin \omega t$,  $\omega = (LC)^{-1/2}$

**7.**  (a)  $E = \dfrac{AR^2 C\omega}{1 + (RC\omega)^2} e^{-t/(RC)} + \dfrac{AR}{1 + (RC\omega)^2} (\sin \omega t - RC\omega \cos \omega t)$

  (b)  $\dfrac{A}{1 + (RC\omega)^2} (\sin \omega t - RC\omega \cos \omega t)$

  (c)  $\dfrac{ARC\omega}{1 + (RC\omega)^2} (\cos \omega t + RC\omega \sin \omega t)$

## CHAPTER 6

### Section 6.1

**3.**  Solve the first equation for $x_1$, then the second for $x_2$, and then the third for $x_3$.

**5.**  (a)  $x_1' = -k_1 x_1$,  $x_2' = -k_2 x_1' - k_3 x_2$

  (b)  $x_1(t) = ae^{-k_1 t}$,  $x_2(t) = \left( b - \dfrac{ak_1 k_2}{k_3 - k_1} \right) e^{-k_3 t} + \dfrac{ak_1 k_2}{k_3 - k_1} e^{-k_1 t}$

**7.**  $m_1 \ddot{x}_1 = km_1 m_2/(x_1 - x_2)^2$,  $m_2 \ddot{x}_2 = -km_1 m_2/(x_1 - x_2)^2$

### Section 6.2

**1.**  $x_1' = 4x_1 + \cos t + e^t$,      $x_1(t_0) = k_1$,  $x_2(t_0) = k_2$
   $x_2' = 3x_1 + e^t$

**3.**  $u_1' = u_3 + e^t$                    $u_1 = x_1$,  $x_1(t_0) = k_1$
   $u_2' = u_3$                        $u_2 = x_2$,  $x_2(t_0) = k_2$
   $u_3' = u_1 + u_2 + u_3 + \sin t$      $u_3 = x_2'$,  $x_2'(t_0) = k_3$

**5.**  $u_1' = u_2$,                    $u_1 = x_1$,  $x_1(t_0) = k_1$
   $u_2' = u_3$,                      $u_2 = x_1'$,  $x_1'(t_0) = k_2$
   $u_3' = u_2 - u_1 u_5 + \sin t$,      $u_3 = x_1''$,  $x_1''(t_0) = k_3$
   $u_4' = u_5$,                     $u_4 = x_2$,  $x_2(t_0) = k_4$
   $u_5' = u_3 - u_4 u_2 - \cos t$,      $u_5 = x_2'$,  $x_2'(t_0) = k_5$

**7.**  $u_1' = u_2$,    $u_2' = tu_2 - u_1^2 + \sin t$

**9.**  $u_1' = u_2$,    $u_2' = u_3$,   $u_3' = u_3 - u_1 + e^t$

## Section 6.3

3. $x_1 = -2c_1e^{-t} - 3c_2e^{-2t} - 3e^{-3t}, \quad x_1 = -6e^{-2t} - 3e^{-3t}$
   $x_2 = c_1e^{-t} + c_2e^{-2t} + 2e^{-3t}, \quad x_2 = 2e^{-2t} + 2e^{-3t}$

5. $x_1 = c_1e^{-t} + 3e^{-2t}$
   $x_2 = -2c_1te^{-t} + c_2e^{-t} + 12e^{-2t}$

7. $x_1 = 5c_1\cos t + 5c_2\sin t + e^{-t}$
   $x_2 = (c_1 + 2c_2)\cos t + (-2c_1 + c_2)\sin t + c_3e^{-2t} + 2e^{-t}$

9. $x_1 = c_1e^{-t} + 4c_2e^{-2t} + c_3\cos t + c_4\sin t, \quad x_1 = e^{-t} + \sin t$
   $x_2 = c_1e^{-t} + 5c_2e^{-2t}, \quad x_2 = e^{-t}$

11. $x_1 = 5c_1e^{-t} - 2c_2e^{-2t} + c_3e^t, \quad x_1 = -2e^{-2t} - e^t$
    $x_2 = 2c_1e^{-t}, \qquad\qquad\qquad x_2 = 0$
    $x_3 = -2c_1e^{-t} + 3c_2e^{-2t}, \qquad x_3 = 3e^{-2t}$

13. $x_1 = 2c_1 + c_2e^{-t} + c_3e^t + t^2$
    $x_2 = c_1 + c_2e^{-t} - t$
    $x_3 = -2c_1 - 2c_2e^{-t} - c_3e^t$

15. $x_1 = 7c_1e^t + c_2\cos 2t + c_3\sin 2t$
    $x_2 = 2c_1e^t + c_2\cos 2t + c_3\sin 2t$
    $x_3 = 3c_1e^t + (c_2 + c_3)\cos 2t + (-c_2 + c_3)\sin 2t$

17. $x_1 = c_1e^{-2t} + c_2e^{-t} + \cos t$
    $x_2 = c_1e^{-2t} + c_3e^{-t}$
    $x_3 = c_1e^{-2t} + \sin t$

## Section 6.4

1. (a) $x_1' = 2x_1 - x_2 + e^t$      (c) $x_1' = 2x_1 - x_2 + e^{2t}$
       $x_2' = 3x_1 + 3e^{-2t}$            $x_2' = x_2 + x_3$
                                       $x_3' = 3x_1 + 2x_2 - e^t$

2. (a) $A(t) = \begin{bmatrix} -2 & 1 \\ -1 & -1 \end{bmatrix}, \qquad b(t) = \begin{bmatrix} \cos 2t \\ -2\sin 2t \end{bmatrix}$

   (c) $A(t) = \begin{bmatrix} 2 & 1 & -1 \\ 1 & -1 & 0 \\ 0 & 1 & 2 \end{bmatrix}, \qquad b(t) = \begin{bmatrix} 2e^{-t} \\ -e^{-t} \\ 0 \end{bmatrix}$

## Section 6.5

1. (a) Independent        (c) Dependent
2. (a) Dependent         (c) Dependent

**3.** (a)  Independent                    (c)  Independent

**5.** (a)  Yes                            (b)  No

**7.** (a)  The rank of $U(t)$ must be less than $m$ for each $t$ in $\mathscr{I}$.

   (b)  The vector functions must be linearly independent.

**8.** (c)  $x(t) = c_1 \mathbf{u}_1(t) + c_2 \mathbf{u}_2(t)$, where $c_1$ and $c_2$ are arbitrary constants.

   (d)  $x(t) = -2\begin{bmatrix} e^{2t} \\ e^{2t} \end{bmatrix} + \begin{bmatrix} 3e^{3t} \\ 2e^{3t} \end{bmatrix}$

## Section 6.6

**1.** $x(t) = c_1 e^{2t}\begin{bmatrix} 1 \\ 1 \end{bmatrix} + c_2 e^{-3t}\begin{bmatrix} 4 \\ -1 \end{bmatrix},$      $x(t) = 2e^{2t}\begin{bmatrix} 1 \\ 1 \end{bmatrix} - e^{-3t}\begin{bmatrix} 4 \\ -1 \end{bmatrix}$

**3.** $x(t) = c_1\left(\begin{bmatrix} 1 \\ -1 \end{bmatrix}\cos 2t - \begin{bmatrix} 1 \\ 0 \end{bmatrix}\sin 2t\right) + c_2\left(\begin{bmatrix} 1 \\ 0 \end{bmatrix}\cos 2t + \begin{bmatrix} 1 \\ -1 \end{bmatrix}\sin 2t\right)$

  $x(t) = \begin{bmatrix} 1 \\ 3 \end{bmatrix}\cos 2t + \begin{bmatrix} 7 \\ -4 \end{bmatrix}\sin 2t$

**5.** $x(t) = c_1\begin{bmatrix} 1 \\ 0 \end{bmatrix}e^{-2t} + c_2\begin{bmatrix} 0 \\ 1 \end{bmatrix}e^{-2t}$

**7.** $x(t) = c_1\begin{bmatrix} 1 \\ 0 \\ 2 \end{bmatrix} + c_2\begin{bmatrix} 1 \\ 1 \\ -1 \end{bmatrix}e^{-t} + c_3\begin{bmatrix} 1 \\ 0 \\ -1 \end{bmatrix}e^{-3t}$

**9.** $x(t) = c_1\begin{bmatrix} 1 \\ -1 \\ -1 \end{bmatrix} + c_2\begin{bmatrix} 0 \\ 1 \\ 0 \end{bmatrix}e^{-2t} + c_3\begin{bmatrix} 3 \\ 0 \\ -1 \end{bmatrix}e^{-2t}$

**11.** $x(t) = c_1\begin{bmatrix} 0 \\ 1 \\ -1 \end{bmatrix}e^{2t} + c_2 e^{2t}\left(\begin{bmatrix} 5 \\ -2 \\ 5 \end{bmatrix}\cos t + \begin{bmatrix} 0 \\ 1 \\ 0 \end{bmatrix}\sin t\right)$

  $+ c_3 e^{2t}\left(\begin{bmatrix} 0 \\ -1 \\ 0 \end{bmatrix}\cos t + \begin{bmatrix} 5 \\ -2 \\ 5 \end{bmatrix}\sin t\right)$

## Section 6.7

**3.** (a)  $\begin{bmatrix} e^{4t} & 0 \\ 0 & e^{-t} \end{bmatrix}$          **8.** (a)  $x(t) = \begin{bmatrix} 3e^t \\ 5e^{2t} \end{bmatrix}$

**9.** (a)  $x(t) = \begin{bmatrix} 3\exp(t-t_0) \\ 5\exp 2(t-t_0) \end{bmatrix}$

## Section 6.8

**1.**  $e^{tA} = \frac{1}{2}(A + I)e^t - \frac{1}{2}(A - I)e^{-t} = \frac{1}{2}e^t\begin{bmatrix} 1 & 1 \\ 1 & 1 \end{bmatrix} - \frac{1}{2}e^{-t}\begin{bmatrix} -1 & 1 \\ 1 & -1 \end{bmatrix}$

$\mathbf{x}(t) = e^t\begin{bmatrix} 2 \\ 2 \end{bmatrix} + e^{-t}\begin{bmatrix} 1 \\ -1 \end{bmatrix}$

**3.**  $e^{tA} = [I + t(A - I)]e^t = e^t\begin{bmatrix} 1 & 0 \\ 0 & 1 \end{bmatrix} + te^t\begin{bmatrix} -1 & -1 \\ 1 & 1 \end{bmatrix}$

$\mathbf{x}(t) = e^t\begin{bmatrix} 2 \\ -3 \end{bmatrix} + te^t\begin{bmatrix} 1 \\ -1 \end{bmatrix}$

**5.**  $e^{tA} = (\cos 3t)I + \frac{1}{3}(\sin 3t)A = (\cos 3t)\begin{bmatrix} 1 & 0 \\ 0 & 1 \end{bmatrix} + \frac{1}{3}(\sin 3t)\begin{bmatrix} 1 & -5 \\ 2 & -1 \end{bmatrix}$

$\mathbf{x}(t) = (\cos 3t)\begin{bmatrix} 3 \\ 0 \end{bmatrix} + (\sin 3t)\begin{bmatrix} 1 \\ 2 \end{bmatrix}$

**7.**  $e^{tA} = -\frac{1}{2}e^t(A^2 - A - 2I) + \frac{1}{6}e^{-t}(A^2 - 3A + 2I) + \frac{1}{3}e^{2t}(A^2 - I)$

$= -\frac{1}{2}e^t\begin{bmatrix} -2 & 0 & 2 \\ 2 & 0 & -2 \\ 0 & 0 & 0 \end{bmatrix} + \frac{1}{6}e^{-t}\begin{bmatrix} -2 & -2 & 2 \\ 6 & 6 & -6 \\ -2 & -2 & 2 \end{bmatrix}$

$+ \frac{1}{3}e^{2t}\begin{bmatrix} 1 & 1 & 2 \\ 0 & 0 & 0 \\ 1 & 1 & 2 \end{bmatrix}$

$\mathbf{x}(t) = e^t\begin{bmatrix} 6 \\ -6 \\ 0 \end{bmatrix} + e^{-t}\begin{bmatrix} -2 \\ 6 \\ -2 \end{bmatrix} + e^{2t}\begin{bmatrix} -1 \\ 0 \\ -1 \end{bmatrix}$

**9.**  $e^{tA} = e^t[I + t(A - I) + \frac{1}{2}t^2(A - I)^2]$

$= e^t\begin{bmatrix} 1 & 0 & 0 \\ 0 & 1 & 0 \\ 0 & 0 & 1 \end{bmatrix} + te^t\begin{bmatrix} 2 & -2 & 1 \\ 2 & -2 & 1 \\ -4 & 4 & 0 \end{bmatrix} + t^2e^t\begin{bmatrix} -2 & 2 & 0 \\ -2 & 2 & 0 \\ 0 & 0 & 0 \end{bmatrix}$

$\mathbf{x}(t) = e^t\begin{bmatrix} 1 \\ 1 \\ 0 \end{bmatrix}$

**11.**  $e^{tA} = \frac{1}{4}(A - 2I)^2 + \frac{1}{4}e^{2t}A(4I - A)[I + t(A - 2I)]$

$= \frac{1}{4}\begin{bmatrix} 6 & 2 & -2 \\ 0 & 0 & 0 \\ 6 & 2 & -2 \end{bmatrix} + \frac{1}{4}e^{2t}\begin{bmatrix} -2 & -2 & 2 \\ 0 & 4 & 0 \\ -6 & -2 & 6 \end{bmatrix}$

$\mathbf{x}(t) = e^{2t}\begin{bmatrix} 0 \\ 2 \\ 2 \end{bmatrix}$

**14.**  $y = -2e^x + 5xe^x$

## Section 6.9

1. $x(t) = c_1 \begin{bmatrix} 1 \\ 1 \end{bmatrix} e^t + c_2 \begin{bmatrix} 2 \\ 1 \end{bmatrix} e^{2t} + \begin{bmatrix} 1 \\ 1 \end{bmatrix} e^{-t}$

3. $x(t) = c_1 \left( \begin{bmatrix} 2 \\ 0 \end{bmatrix} \cos 2t - \begin{bmatrix} 0 \\ -1 \end{bmatrix} \sin 2t \right) + c_2 \left( \begin{bmatrix} 0 \\ -1 \end{bmatrix} \cos 2t + \begin{bmatrix} 2 \\ 0 \end{bmatrix} \sin 2t \right)$

$\quad + \begin{bmatrix} 3 \\ 1 \end{bmatrix} e^{-2t}$

5. $x(t) = c_1 \begin{bmatrix} 1 \\ 3 \end{bmatrix} e^t + c_2 \begin{bmatrix} 1 \\ 1 \end{bmatrix} e^{-t} + \begin{bmatrix} -1 \\ -1 \end{bmatrix} \cos t + \begin{bmatrix} 1 \\ 1 \end{bmatrix} \sin t$

7. $x(t) = c_1 \begin{bmatrix} 1 \\ 0 \\ 1 \end{bmatrix} + c_2 \begin{bmatrix} 2 \\ 1 \\ 2 \end{bmatrix} e^t + c_3 \begin{bmatrix} 2 \\ -1 \\ 4 \end{bmatrix} e^{-t} + \begin{bmatrix} -1 \\ -1 \\ 0 \end{bmatrix} e^{2t}$

11. $x(t) = c_1 \begin{bmatrix} 1 \\ -1 \end{bmatrix} e^{-2t} + c_2 \begin{bmatrix} 1 \\ -6 \end{bmatrix} e^{3t} + \begin{bmatrix} -1 \\ 0 \end{bmatrix} e^{-2t} + \begin{bmatrix} 2 \\ -2 \end{bmatrix} t e^{-2t}$

13. $x(t) = c_1 \left( \begin{bmatrix} 1 \\ 0 \end{bmatrix} \cos t + \begin{bmatrix} 0 \\ 1 \end{bmatrix} \sin t \right) + c_2 \left( \begin{bmatrix} 0 \\ -1 \end{bmatrix} \cos t + \begin{bmatrix} 1 \\ 0 \end{bmatrix} \sin t \right)$

$\quad + \begin{bmatrix} -1 \\ 0 \end{bmatrix} \cos t + \begin{bmatrix} -2 \\ 0 \end{bmatrix} \sin t + \begin{bmatrix} 2 \\ -1 \end{bmatrix} t \cos t + \begin{bmatrix} 1 \\ 2 \end{bmatrix} t \sin t$

15. $x_p(t) = \frac{1}{2}(A + I) \begin{bmatrix} 2te^t \\ 4te^t \end{bmatrix} - \frac{1}{2}(A - I) \begin{bmatrix} e^t - e^{-t} \\ 2(e^t - e^{-t}) \end{bmatrix}$

17. $x_p(t) = (A + I) \begin{bmatrix} t \\ t^2 \end{bmatrix} - A \begin{bmatrix} 1 - e^{-t} \\ 2(t - 1 + e^{-t}) \end{bmatrix}$

19. $x_p(t) = \frac{1}{2}(A + I) \begin{bmatrix} -\cos t + \sin t + e^t \\ -\sin t - \cos t + e^t \end{bmatrix} - \frac{1}{2}(A - I) \begin{bmatrix} \cos t + \sin t - e^{-t} \\ \sin t - \cos t + e^{-t} \end{bmatrix}$

21. $x_p(t) = \begin{bmatrix} 2te^t \\ 3t^2 e^t \\ 0 \end{bmatrix} + (A - I) \begin{bmatrix} t^2 e^t \\ t^3 e^t \\ 0 \end{bmatrix}$

## Section 6.10

1. (b) In the equation $P(r) = ar^4 + br^2 + c = 0$, $b^2 - 4ac = (m_2 k_1 - m_1 k_2)^2 + m_2 k_2 (m_2 k_2 + 2m_1 k_2 + 2m_2 k_1)$, which is positive. Hence $-b \pm (b^2 - 4ac)^{1/2}$ is negative.

3. $m_1 \ddot{x}_1 = -k_1 x_1 + c(\dot{x}_2 - \dot{x}_1)$, $\qquad m_2 \ddot{x}_2 = -k_2 x_2 - c(\dot{x}_2 - \dot{x}_1)$

5. (a) $v_0 (2h/g)^{1/2}$ $\qquad\qquad$ (b) $y = -\frac{1}{2} gx^2/v_0^2 + h$

7. (a) $(v_0^2/g) \sin 2\alpha$, $\qquad\qquad$ (b) $y = -\dfrac{g}{2(v_0 \cos \alpha)^2} x^2 + (\tan \alpha)x$

9. $m\ddot{x} = -kx(x^2 + y^2)^{-3/2}$, $\qquad m\ddot{y} = -ky(x^2 + y^2)^{-3/2}$

## Section 6.11

**1.** $LI_1'' + R_1 I_1' + \dfrac{1}{C}(I_1 - I_2) = 0,\qquad R_2 I_2' + \dfrac{1}{C}(I_2 - I_1) = E'(t)$

$I_1(0) = I_1'(0) = 0,\qquad I_2(0) = E(0)/R_2$

**3.** $L(I_1' - I_2') + R_1 I_1 = 0,\quad L(I_2' - I_1') + R_2 I_2 = E,+\quad I_1(0) = E/(R_1 + R_2),$

$I_1(t) = 2e^{-t/6},\qquad I_2(t) = 6 - 4e^{-t/6}$

**5.** $L_1 I_1' + (R_1 + R_2)I_1 - R_2 I_2 = 0,$

$R_2 I_2' - R_2 I_1' + \left(\dfrac{1}{C_1} + \dfrac{1}{C_2}\right)I_2 - \dfrac{1}{C_2} I_3 = 0$

$R_3 I_3' + \dfrac{1}{C_2}(I_3 - I_2) = E'(t),\qquad I_1(0) = I_2(0) = 0,\qquad I_3(0) = E(0)/R_3$

**7.** $\left(\dfrac{1}{R_1} + \dfrac{1}{R_2}\right)E_1' - \dfrac{1}{R_2} E_2' + \dfrac{1}{L} E_1 = 0,\qquad \left(\dfrac{1}{R_2} + \dfrac{1}{R_3}\right)E_2 - \dfrac{1}{R_2} E_1 = I(t)$

$E_1(0) = \dfrac{R_1 R_3 I(0)}{R_1 + R_2 + R_3}$

## CHAPTER 7

## Section 7.1

**1.** $(0, 2)$        **3.** $(1, 3)$

**5.** All $x$        **7.** $(-1, 1)$

**9.** $f(x) + g(x) = \displaystyle\sum_{n=0}^{\infty} (n + 2)(x - 2)^n,\quad |x - 2| < 1$

$f(x)g(x) = \dfrac{1}{2} \displaystyle\sum_{n=0}^{\infty} (n + 1)(n + 2)(x - 2)^n,\quad |x - 2| < 1$

**11.** $f(x) + g(x) = \displaystyle\sum_{n=0}^{\infty} \dfrac{n^2 + 2n + 2}{n + 1} x^n,\quad |x| < 1$

$f(x)g(x) = \displaystyle\sum_{n=0}^{\infty} \left(\sum_{k=0}^{n} \dfrac{n - k + 1}{k + 1}\right)x^n,\quad |x| < 1$

**13.** $\displaystyle\sum_{n=0}^{\infty} (n + 2)(n + 1)x^n$        **15.** $\displaystyle\sum_{n=1}^{\infty} (n^2 - 2n + 3)x^n$

**17.** $f'(x) = \displaystyle\sum_{n=1}^{\infty} \dfrac{(-1)^n n x^{n-1}}{2^n(n + 1)},\qquad f''(x) = \displaystyle\sum_{n=2}^{\infty} \dfrac{(-1)^n n(n - 1)x^{n-2}}{2^n(n + 1)},\quad |x| < 2$

## Section 7.2

**1.**  $3 + 13(x - 2) + 9(x - 2)^2 + 2(x - 2)^3$

**3.**  $\displaystyle\sum_{n=1}^{\infty} \frac{(-1)^{n+1}}{n}(x - 1)^n$

**5.**  $\displaystyle 1 + \tfrac{1}{2}x + \sum_{n=2}^{\infty} (-1)^{n+1}\frac{1 \cdot 3 \cdot 5 \cdots (2n - 3)}{2^n n!}\,x^n$

**7.**  $\displaystyle\sum_{n=0}^{\infty} (-1)^n \left(\frac{x}{4}\right)^n, \quad |x| < 4$

**9.**  $\displaystyle\frac{2}{3}\sum_{n=0}^{\infty} [1 - (-\tfrac{1}{2})^{n+1}]x^n, \quad |x| < 1$

**11.**  $\displaystyle\sum_{n=0}^{\infty} (n + 1)x^n, \quad |x| < 1$

**13.**  $\displaystyle\sum_{n=0}^{\infty} \frac{x^{2n+1}}{2n + 1}, \quad |x| < 1$

**15.**  $\displaystyle x\frac{d}{dx}(xe^x) = (x^2 + x)e^x$

**17.**  $\displaystyle\frac{d}{dx}\left[x\int_0^x \frac{1}{1 - t}\,dt\right] = \frac{x}{1 - x} - \ln(1 - x)$

## Section 7.3

**1.**  (a)  $x = -2, 0, 1$      (c)  None

**3.**  $\displaystyle y = A_0\left[1 + \sum_{m=1}^{\infty} \frac{x^{2m}}{2^m 1 \cdot 3 \cdot 5 \cdots (2m - 1)}\right]$

$\displaystyle + A_1 \sum_{m=1}^{\infty} \frac{x^{2m-1}}{2^{2m-2}(m - 1)!}, \quad \text{for all } x$

**5.**  $\displaystyle y = A_0\left[1 + \sum_{m=1}^{\infty} \frac{(-1)^m 2^m m!\, x^{2m}}{1 \cdot 3 \cdot 5 \cdots (2m - 1)}\right]$

$\displaystyle + A_1 \sum_{m=1}^{\infty} (-1)^{m+1}\frac{1 \cdot 3 \cdot 5 \cdots (2m - 1)}{2^{m-1}(m - 1)!}\,x^{2m-1}, \qquad |x| < \sqrt{2}$

**7.**  $\displaystyle y = A_0 \sum_{m=0}^{\infty} \frac{x^{3m}}{3^m m!} + A_1 \sum_{m=0}^{\infty} \frac{x^{3m+1}}{1 \cdot 4 \cdot 7 \cdots (3m + 1)}, \quad \text{for all } x$

**9.**  $y = A_0(1 + \tfrac{1}{2}x^2 - \tfrac{1}{6}x^3 + \tfrac{1}{8}x^4 + \cdots) + A_1(x + \tfrac{1}{6}x^3 - \tfrac{1}{12}x^4 + \cdots), \; |x| < 1$

**11.**  $\displaystyle y = A_0 \sum_{m=0}^{\infty} \frac{(-1)^m (x - 1)^{2m}}{2^m m!} + A_1 \sum_{m=1}^{\infty} \frac{(-1)^{m+1}(x - 1)^{2m-1}}{1 \cdot 3 \cdot 5 \cdots (2m - 1)}, \quad \text{for all } x$

13. $y = A_0\left[1 - 3(x - 2)^2 + (x - 2)^4 + 3 \sum\limits_{m=3}^{\infty} \dfrac{(x - 2)^{2m}}{(2m - 3)(2m - 1)}\right]$
$+ A_1[(x - 2) - (x - 2)^3], \quad |x - 2| < 1$

## Section 7.4

1. (a) $0, -1$    (c) $-1/2$

3. $y = c_1 x\left[1 + \sum\limits_{n=1}^{\infty} \dfrac{x^n}{n!\,5 \cdot 7 \cdot 9 \cdots (2n + 3)}\right]$
$+ c_2 x^{-1/2}\left[1 - x - \sum\limits_{n=2}^{\infty} \dfrac{x^n}{n!\,1 \cdot 3 \cdot 5 \cdots (2n - 3)}\right]$

5. $y = c_1 x^{1/3}\left[1 + \sum\limits_{n=1}^{\infty} \dfrac{(-1)^n x^n}{n!\,4 \cdot 7 \cdot 10 \cdots (3n + 1)}\right]$
$+ c_2\left[1 + \sum\limits_{n=1}^{\infty} \dfrac{(-1)^n x^n}{n!\,2 \cdot 5 \cdot 8 \cdots (3n - 1)}\right]$

7. $y = c_1 x^{1/3} \sum\limits_{m=0}^{\infty} 4\,\dfrac{(-1)^m x^{2m}}{2^m m!\,(6m + 4)} + c_2 x^{-1}$

9. $y = c_1 x^{-1} \sum\limits_{n=0}^{\infty} \dfrac{(n + 1)!\,x^n}{1 \cdot 3 \cdot 5 \cdots (2n + 1)} + c_2 x^{-3/2} \sum\limits_{n=0}^{\infty} \dfrac{1 \cdot 3 \cdot 5 \cdots (2n + 1)}{4^n n!} x^n$

11. $y = c_1(x - 1)^{1/3}[1 - \tfrac{1}{5}(x - 1)]$
$+ c_2(x - 1)^{-1/3}\left[1 + \sum\limits_{n=1}^{\infty} \dfrac{(-5)(-2)1 \cdot 4 \cdots (3n - 8)}{3^n n!\,1 \cdot 4 \cdot 7 \cdots (3n - 2)} (x - 1)^n\right]$

14. $y = A_0 \sum\limits_{m=0}^{\infty} \dfrac{x^{-2m}}{2^m m!} + A_1 \sum\limits_{m=1}^{\infty} \dfrac{x^{-2m+1}}{1 \cdot 3 \cdot 5 \cdots (2m - 1)}$

## Section 7.5

1. $y_1(x) = x \sum\limits_{n=0}^{\infty} \dfrac{x^n}{(n!)^2}, \qquad y_2(x) = y_1(x)\ln x + x \sum\limits_{n=1}^{\infty} B_n x^n$

3. $y_1(x) = 1 - x^2 + \tfrac{1}{8}x^4, \qquad y_2(x) = y_1(x)\ln x + \sum\limits_{n=1}^{\infty} B_n x^n$

5. $y_1(x) = x^{-2}\left[1 + \sum\limits_{n=1}^{6} \dfrac{(-6)(-5) \cdots (n - 7)}{n!} x^n\right]$,
$y_2(x) = y_1(x)\ln x + x^{-2} \sum\limits_{n=1}^{\infty} B_n x^n$

7. $y_1(x) = x\left[1 + \dfrac{1}{2}\sum\limits_{n=1}^{\infty} \dfrac{n + 2}{n!} x^n\right], \qquad y_2(x) = y_1(x)\ln x + \sum\limits_{n=0}^{\infty} B_n x^n$

9. $y_1(x) = x(1 - 2x + x^2)$,     $y_2(x) = y_1(x)\ln x + x^{-1} \sum_{n=0}^{\infty} B_n x^n$

11. $y_1(x) = x^{-1}(1 - 3x^2 + 3x^4 - x^6)$,

$$y_2(x) = x^{-2}\left[1 + \sum_{m=1}^{\infty} \frac{(-7)(-5)(-3) \cdots (2m-9)}{1 \cdot 3 \cdot 5 \cdots (2m-1)} x^{2m}\right]$$

13. $y_1(x) = 1 + 2 \sum_{n=1}^{\infty} \frac{x^n}{(n+2)!}$,     $y_2(x) = x^{-2}(1 + x)$

15. $y_1(x) = x \sum_{n=0}^{\infty} x^n = \frac{x}{1-x}$,     $y_2(x) = x^{-3}(1 + x + x^2 + x^3)$

## Section 7.7

1. (a) $\sqrt{\pi/2}$                     (c) $-2\sqrt{\pi}$
4. (a) 0.990                              (c) 0.100

6. $J_3(x) = \frac{8 - x^2}{x^2} J_1(x) - \frac{4}{x} J_0(x)$,

8. (c) $J_{3/2}(x) = \left(\frac{2}{\pi x}\right)^{1/2} \left(\frac{\sin x}{x} - \cos x\right)$

10. (a) $y = x^{1/2}[c_1 J_{1/4}(x^2/2) + c_2 J_{-1/4}(x^2/2)]$

## CHAPTER 8

## Section 8.1

7. (a) $y(x) = 2 + \int_1^x \{ty(t) + [y(t)]^2\}\, dt$

8. (a) $y = (5 + x)e^x$

## Section 8.2

1. (a) $y = -(x^{4/3} + 3)^{1/2}$,   for all $x$
   (c) $y = x(5x^2 - 1)^{1/2}$,   $x > 1/\sqrt{5}$
2. $y_0(x) = 1$,   $y_1(x) = 1 - x + \frac{1}{2}x^2$,
   $y_2(x) = 1 - x - \frac{1}{2}x^2 + \frac{2}{3}x^3 - \frac{1}{4}x^4 + x\frac{1}{20}{}^5$
6. (a) $y_1(x) = 0$ for all $x_1$;   $y_2(x) = 0$ for $x \le 0$ and $y_2(x) = x^{3/2}$ for $x > 0$
7. Yes

## Section 8.3

4.  $x_1(t) = (t + 1)^{-1}, \quad x_2(t) = 5(t + 1), \quad t > -1$

6.  $\mathbf{x}_0(t) = \begin{bmatrix} -2 \\ 1 \end{bmatrix}, \qquad \mathbf{x}_1(t) = \begin{bmatrix} -2 - t - t^2 \\ 1 - 4t \end{bmatrix},$

$\mathbf{x}_2(t) = \begin{bmatrix} -2 - t + 3t^2 - \frac{17}{3}t^3 - \frac{1}{4}t^4 \\ 1 - 4t + \frac{7}{2}t^2 + t^3 + t^4 \end{bmatrix}$

12. (a) $x(t) = \frac{1}{2}(t^2 - 2t + 2)$,   for all $t$

# Index

Numbers in parentheses indicate exercises.